M. Nikravesh, L. A. Zadeh, J. Kacprzyk (Eds.)

Soft Computing for Information Processing and Analysis

Studies in Fuzziness and Soft Computing, Volume 164

Editor-in-chief
Prof. Janusz Kacprzyk
Systems Research Institute
Polish Academy of Sciences
ul. Newelska 6
01-447 Warsaw
Poland
E-mail: kacprzyk@ibspan.waw.pl

Further volumes of this series can be found on our homepage: springeronline.com

Vol. 148. C. Martin-Vide, V. Mitrana, G. Păun (Eds.)
Formal Languages and Applications, 2004
ISBN 3-540-20907-7

Vol. 149. J.J. Buckley
Fuzzy Statistics, 2004
ISBN 3-540-21084-9

Vol. 150. L. Bull (Ed.)
Applications of Learning Classifier Systems, 2004
ISBN 3-540-21109-8

Vol. 151. T. Kowalczyk, E. Pleszczyńska, F. Ruland (Eds.)
Grade Models and Methods for Data Analysis, 2004
ISBN 3-540-21120-9

Vol. 152. J. Rajapakse, L. Wang (Eds.)
Neural Information Processing: Research and Development, 2004
ISBN 3-540-21123-3

Vol. 153. J. Fulcher, L.C. Jain (Eds.)
Applied Intelligent Systems, 2004
ISBN 3-540-21153-5

Vol. 154. B. Liu
Uncertainty Theory, 2004
ISBN 3-540-21333-3

Vol. 155. G. Resconi, J.L. Jain
Intelligent Agents, 2004
ISBN 3-540-22003-8

Vol. 156. R. Tadeusiewicz, M.R. Ogiela
Medical Image Understanding Technology, 2004
ISBN 3-540-21985-4

Vol. 157. R.A. Aliev, F. Fazlollahi, R.R. Aliev
Soft Computing and its Applications in Business and Economics, 2004
ISBN 3-540-22138-7

Vol. 158. K.K. Dompere
Cost-Benefit Analysis and the Theory of Fuzzy Decisions – Identification and Measurement Theory, 2004
ISBN 3-540-22154-9

Vol. 159. E. Damiani, L.C. Jain, M. Madravia
Soft Computing in Software Engineering, 2004
ISBN 3-540-22030-5

Vol. 160. K.K. Dompere
Cost-Benefit Analysis and the Theory of Fuzzy Decisions – Fuzzy Value Theory, 2004
ISBN 3-540-22161-1

Vol. 161. N. Nedjah, L. de Macedo Mourelle (Eds.)
Evolvable Machines, 2005
ISBN 3-540-22905-1

Vol. 162. N. Ichalkaranje, R. Khosla, L.C. Jain
Design of Intelligent Multi-Agent Systems, 2005
ISBN 3-540-22913-2

Vol. 163. A. Ghosh, L.C. Jain (Eds.)
Evolutionary Computation in Data Mining, 2005
ISBN 3-540-22370-3

Masoud Nikravesh
Lotfi A. Zadeh
Janusz Kacprzyk (Eds.)

Soft Computing for Information Processing and Analysis

Springer

Prof. Masoud Nikravesh
University of California
Dept. Electrical Engineering and Computer Science – EECS
94720 Berkeley, CA
USA
E-mail: nikravesh@cs.berkeley.edu

Prof. Janusz Kacprzyk
Systems Research Institute
Polish Academy of Sciences
ul. Newelska 6
01-447 Warsaw
Poland
E-mail: kacprzyk@ibspan.waw.pl

Prof. Lotfi A. Zadeh
University of California
Div. Computer Science
Lab. Electronics Research
Soda Hall 387
94720-1776 Berkeley, CA
USA
E-mail: zadeh@cs.berkeley.edu

ISSN 1434-9922
ISBN 3-540-22930-2 Springer Berlin Heidelberg New York

Library of Congress Control Number: 2004111073

Springer is a part of Springer Science+Business Media
springeronline.com

Printed in Germany

Typesetting: data delivered by editors
Cover design: E. Kirchner, Springer-Verlag, Heidelberg
Printed on acid free paper 62/3020/M - 5 4 3 2 1 0

Preface

World Wide Web search engines have become the most heavily-used online services, with millions of searches performed each day. Their popularity is due, in part, to their ease of use. The central tasks for the most of the search engines can be summarize as 1) query or user information request- do what I mean and not what I say!, 2) model for the Internet, Web representation-web page collection, documents, text, images, music, etc, and 3) ranking or matching function-degree of relevance, recall, precision, similarity, etc.

Design of any new intelligent search engine should be at least based on two main motivations:

i· The web environment is, for the most part, unstructured and imprecise. To deal with information in the web environment what is needed is a logic that supports modes of reasoning which are approximate rather than exact. While searches may retrieve thousands of hits, finding decision-relevant and query-relevant information in an imprecise environment is a challenging problem, which has to be addressed.

ii· Another, and less obvious, is deduction in an unstructured and imprecise environment given the huge stream of complex information.

Search engines, with Google at the top, have many remarkable capabilities. But what is not among them is the deduction capability—the capability to synthesize an answer to a query by drawing on bodies of information which are resident in various parts of the knowledge base. It is this capability that differentiates a question-answering system, Q/A system for short, from a search engine.

Construction of Q/A systems has a long history in AI. Interest in Q/A systems peaked in the seventies and eighties, and began to decline when it became obvious that the available tools were not adequate for construction of systems having significant question-answering capabilities. However, Q/A systems in the form of domain-restricted expert systems have proved to be of value, and are growing in versatility, visibility and importance.

Search engines as we know them today owe their existence and capabilities to the advent of the Web. A typical search engine is not designed to come up with answers to queries exemplified by "How many Ph.D. degrees in computer science were granted by Princeton University in 1996?" or "What is the name and affiliation of the leading eye surgeon in Boston?" or "What is the age of the oldest son of the President of Finland?" or "What is the fastest way of getting from Paris to London?"

Upgrading a search engine to a Q/A system is a complex, effort-intensive, open-ended problem. Semantic Web and related systems for upgrading quality of search may be viewed as steps in this direction. But what may be argued, as is done in the following, is that existing tools, based as they are on bivalent logic and probability theory, have intrinsic limitations. The principal obstacle is the nature of world knowledge.

The centrality of world knowledge in human cognition, and especially in reasoning and decision-making, has long been recognized in AI. The Cyc system of Douglas Lenat is a repository of world knowledge. The problem is that much of world knowledge consists of perceptions. Reflecting the bounded ability of sensory organs, and ultimately the brain, to resolve detail and store information, perceptions are intrinsically imprecise. More specifically, perceptions are f-granular in the sense that (a) the boundaries of perceived classes are fuzzy; and (b) the perceived values of attributes are granular, with a granule being a clump of values drawn together by indistinguishability, similarity, proximity or functionality. What is not widely recognized is that f-granularity of perceptions put them well beyond the reach of computational bivalent-logic-based theories. For example, the meaning of a simple perception described as "Most Swedes are tall," does not admit representation in predicate logic and/or probability theory.

Dealing with world knowledge needs new tools. A new tool which is suggested for this purpose is the fuzzy-logic-based method of computing with words and perceptions (CWP), with the understanding that perceptions are described in a natural language. A concept which plays a key role in CWP is that of Precisiated Natural Language (PNL). It is this language that is the centerpiece of our approach to reasoning and decision-making with world knowledge.

A concept which plays an essential role in PNL is that of precisiability. More specifically, a proposition, p, in a natural language, NL, is PL precisiable, or simply precisiable, if it is translatable into a mathematically well-defined language termed precisiation language, PL. Examples of precisiation languages are: the languages of propositional logic; predicate logic; modal logic; etc.; and Prolog; LISP; SQL; etc. These languages are based on bivalent logic. In the case of PNL, the precisiation language is a fuzzy-logic-based language referred to as the Generalized Constraint Language (GCL). By construction, GCL is maximally expressive.

A basic assumption underlying GCL is that, in general, the meaning of a proposition, p, in NL may be represented as a generalized constraint of the form X isr R, where X is the constrained variable; R is the constraining relation, and r is a discrete-valued variable, termed modal variable, whose values define the modality of the constraint, that is, the way in which R constrains X. The principal modalities are; possibilistic (r=blank); probabilistic (r=p); veristic (r=v); usuality (r=u); fuzzy random set (r=rs); fuzzy graph (r=fg); and Pawlak set (r=ps). In general, X, R and r are implicit in p. Thus, precisiation of p, that is, translation of p into GCL, involves explicitation of X, R and r. GCL is generated by (a) combining generalized constraints; and (b) generalized constraint propagation, which is governed by the rules of inference in fuzzy logic. The translation of p expressed as a generalized constraint is referred to as the GC-form of p, GC(p). GC(p) may be viewed as a generalization of the concept of logical form. An abstraction of the GC-form is referred to as a protoform (prototypical form) of p, and is denoted as PF(p). For example, the protoform of p: "Most Swedes are tall" is Q A's are B's, where A and B are labels of fuzzy sets, and Q is a fuzzy quantifier. Two propositions p and q are said to be PF-equivalent if they have identical protoforms. For example, "Most Swedes are tall," and "Not many professors are rich," are PF-equivalent. In effect, a protoform of p is its deep semantic structure. The protoform language, PFL, consists of protoforms of elements of GCL.

With the concepts of GC-form and protoform in place, PNL may be defined as a subset of NL which is equipped with two dictionaries: (a) from NL to GCL; and (b) from GCL to PFL. In addition, PNL is equipped with a multiagent modular deduction database, DDB, which contains rules of deduction in PFL. A simple example of a rule of deduction in PFL which is identical to the compositional rule of inference in fuzzy logic, is: if X is A and (X, Y) is B then Y is A∘B, where A∘B is the composition of A and B, defined by $\mu_B(v) = \sup_u(\mu_A(u) \wedge \mu_B(u,v))$, where μ_A and μ_B are the membership functions of A and B, respectively, and $\wedge$ is min or, more generally, a T-norm. The rules of deduction in DDB are organized into modules and submodules, with each module and submodule associated with an agent who controls execution of rules of deduction and passing results of execution.

In our approach, PNL is employed in the main to represent information in the world knowledge database (WKD). For example, the items:

If X/Person works in Y/City then it is likely that X lives in or near Y
If X/Person lives in Y/City then it is likely that X works in or near Y

are translated into GCL as:

Distance (Location (Residence (X/Person), Location (Work (X/Person) isu near,

where isu, read as ezoo, is th e usuality constraint. The corresponding protoform is:

F (A(B(X/C), A(E(X/C)) isu G.

A concept which plays a key role in organization of world knowledge is that of an epistemic (knowledge-directed) lexicon (EL). Basically, an epistemic lexicon is a network of nodes and weighted links, with node i representing an object in the world knowledge database, and a weighted link from node i to node j representing the strength of association between i and j. The name of an object is a word or a composite word, e.g., car, passenger car or Ph.D. degree. An object is described by a relation or relations whose fields are attributes of the object. The values of an attribute may be granulated and associated with granulated probability and possibility distributions. For example, the values of a granular attribute may be labeled small, medium and large, and their probabilities may be described as low, high and low, respectively. Relations which are associated with an object serve as PNL-based descriptions of the world knowledge about the object. For example, a relation associated with an object labeled Ph.D. degree may contain attributes labeled Eligibility, Length.of.study, Granting.institution, etc. The knowledge associated with an object may be context-dependent. What should be stressed is that the concept of an epistemic lexicon is intended to be employed in representation of world knowledge — which is largely perception- based—rather than Web knowledge, which is not.

As a very simple illustration of the use of an epistemic lexicon, consider the query "How many horses received the Ph.D. degree from Princeton University in 1996." No existing search engine would come up with the correct answer, "Zero, since a horse cannot be a recipient of a Ph.D. degree." To generate the correct answer, the attribute Eligibility in the Ph.D. entry in EL should contain the condition "Human, usually over twenty years of age."

In conclusion, the main thrust of the fuzzy-logic-based approach to question-answering which is outlined here, is that to achieve significant question-answering capability it is necessary to develop methods of dealing with the reality that much of world knowledge—and especially knowledge about underlying probabilities is perception-based. Dealing with perception-based information is more complex and more effort-intensive than dealing with measurement-based information. In this instance, as in many others, complexity is the price that has to be paid to achieve superior performance.

In this context, Berkeley Initiative in Soft Computing (BISC), University of California, Berkeley formed a Technical Committee to organize a Meeting entitled "Fuzzy Logic and the Internet: Enhancing the Power of the Internet" to understand the significance of the fields accomplishments, new developments and future directions. In addition, the Technical Committee selected and invited over 100 scientists (and industry experts as technical committee members) from the related disciplines to participate in the Meeting "State of the Art Assessment and New Directions for Research" which took place at the University of California, Berkeley, in August 14-18, 2001 and Dec 14-19, 2003. We would like to thank the authors of the papers and gratefully acknowledge their effort.

The chapters of the book are evolved from presentations made by the participants at the Meeting. The papers include reports from the front of soft computing in the Internet industry and address the problems of the fields by considering a very important topic of search engine, fuzzy query, decision analysis and support system, e-business and e-commerce.

The book provides a collection of twenty-one(21) articles including web intelligence, search engines and navigations, perception based information processing, fuzzy ontology and thesauri, user modeling and personal information provision, Agents, feature selection, association rules, cognitive maps, analogical reasoning, information network, , semantic web/net, web-assistant and agents, knowledge representation, content-based information retrieval, information organization, and causality.

We would like to take this opportunity to thank all the contributors and reviewers of the articles. We also wish to acknowledge our colleagues who have contributed to the area directly or indirectly related to the content of this book. Finally, we gratefully acknowledge the BTexact technologies -- specially, Dr. Ben Azvine and Dr. Nader Azarmi-- for the financial and technical support, which made the Meeting and book possible

Masoud Nikravesh, Lotfi A Zadeh and Janusz Kacprzyk
*Berkeley Initiative in Soft Computing (**BISC**)*
Department of Electrical Engineering and Computer Sciences
University of California, Berkeley
CA 94720-1776;
Zadeh@cs.berkeley.edu
Telephone (Zadeh): 510-642-4959; Fax: 510-642-1712
Nikravesh@cs.berkeley.edu
Telephone (Nikravesh): 510-643-4522; Fax: 510-642-5775

June 2004
Berkeley, California
USA

Web Intelligence, World Knowledge and Fuzzy Logic

Lotfi A. Zadeh
BISC Program, Computer Sciences Division, EECS Department
University of California, Berkeley, CA 94720, USA
Email: zadeh@cs.berkeley.edu
BISC Program URL: http://www-bisc.cs.berkeley.edu/
http://zadeh.cs.berkeley.edu/
Tel.(office): (510) 642-4959
Fax (office): (510) 642-1712

Abstract: Existing search engines—with Google at the top—have many remarkable capabilities; but what is not among them is deduction capability—the capability to synthesize an answer to a query from bodies of information which reside in various parts of the knowledge base. In recent years, impressive progress has been made in enhancing performance of search engines through the use of methods based on bivalent logic and bivalent-logic-based probability theory. But can such methods be used to add nontrivial deduction capability to search engines, that is, to upgrade search engines to question-answering systems? A view which is articulated in this note is that the answer is "No." The problem is rooted in the nature of world knowledge, the kind of knowledge that humans acquire through experience and education.

It is widely recognized that world knowledge plays an essential role in assessment of relevance, summarization, search and deduction. But a basic issue which is not addressed is that much of world knowledge is perception-based, e.g., "it is hard to find parking in Paris," "most professors are not rich," and "it is unlikely to rain in midsummer in San Francisco." The problem is that (a) perception-based information is intrinsically fuzzy; and (b) bivalent logic is intrinsically unsuited to deal with fuzziness and partial truth.

To come to grips with the fuzziness of world knowledge, new tools are needed. The principal new tool—a tool which is briefly described in their note—is Precisiated Natural Language (PNL). PNL is based on fuzzy logic and has the capability to deal with partiality of certainty, partiality of possibility and partiality of truth. These are the capabilities that are needed to be able to draw on world knowledge for assessment of relevance, and for summarization, search and deduction.

1. Introduction

In moving further into the age of machine intelligence and automated reasoning, we have reached a point where we can speak, without exaggeration, of systems which have a high machine IQ (MIQ) (Zadeh, [17]). The Web and especially search engines—with Google at the top—fall into this category. In the context of the Web, MIQ becomes Web IQ, or WIQ, for short.
Existing search engines have many remarkable capabilities. However, what is not among them is the deduction capability—the capability to answer a query by a synthesis of information which resides in various parts of the knowledge base. A question-answering system is by definition a system which has this capability. One of the principal goals of Web intelligence is that of upgrading search engines to question-answering systems. Achievement of this goal requires a quantum jump in the WIQ of existing search engines [1].

Can this be done with existing tools such as the Semantic Web [3], Cyc [8], OWL [13] and other ontology-centered systems [12, 14]—tools which are based on bivalent logic and bivalent-logic-based probability theory? It is beyond question that, in recent years, very impressive progress has been made through the use of such tools. But can we achieve a quantum jump in WIQ? A view which is advanced in the following is that bivalent-logic- based methods have intrinsically limited capability to address complex problems which arise in deduction from information which is pervasively ill-structured, uncertain and imprecise.
The major problem is world knowledge—the kind of knowledge that humans acquire through experience and education [5]. Simple examples of fragments of world knowledge are: Usually it is hard to find parking near the campus in early morning and late afternoon; Berkeley is a friendly city; affordable housing is nonexistent in Palo Alto; almost all professors have a Ph.D. degree; most adult Swedes are tall; and Switzerland has no ports.
Much of the information which relates to world knowledge—and especially to underlying probabilities—is perception-based (Fig. 1). Reflecting the bounded ability of sensory organs, and ultimately the brain, to resolve detail and store information, perceptions are intrinsically imprecise. More specifically, perceptions are f-granular in the sense that (a) the boundaries of perceived classes are unsharp; and (b) the values of perceived attributes are granular, with a granule being a clump of values drawn together by indistinguishability, similarity, proximity or functionality [2].

Imprecision of perception-based information is a major obstacle to dealing with world knowledge through the use of methods based on bivalent logic and bivalent-logic-based probability theory—both of which are intolerant of imprecision and partial truth. What is the basis for this contention? A very simple example offers an explanation.

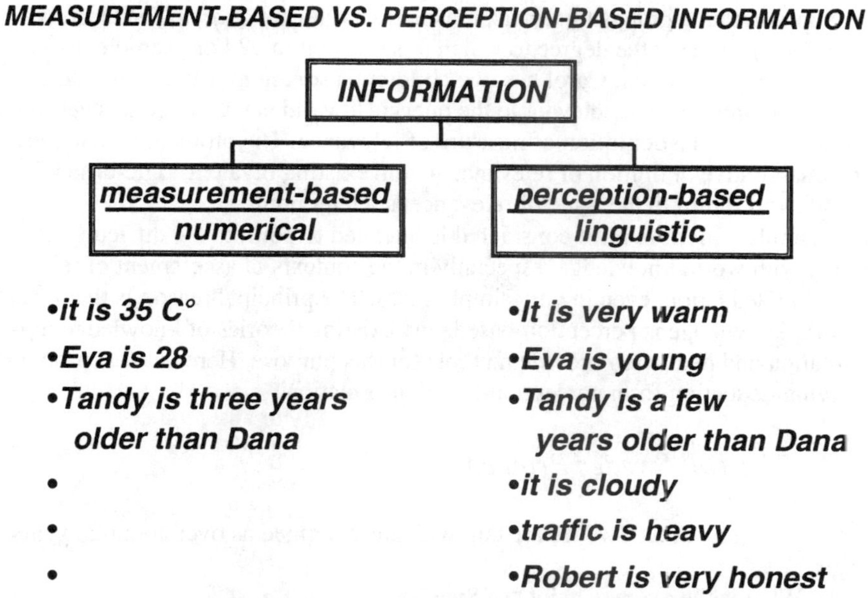

Figure 1. Measurement-based and perception-based information

Suppose that I have to come up with a rough estimate of Pat's age. The information which I have is: (a) Pat is about ten years older than Carol; and (b) Carol has two children: a son, in mid-twenties; and a daughter, in mid-thirties.

How would I come up with an answer to the query *q*: How old is Pat? First, using my world knowledge, I would estimate Carol's age, given (b); then I would add "about ten years," to the estimate of Carol's age to arrive at an estimate of Pat's age. I would be able to describe my estimate in a natural language, but I would not be able to express it as a number, interval or a probability distribution.

How can I estimate Carol's age given (b)? Humans have an innate ability to process perception-based information—an ability that bivalent-logic-based methods do not have; nor would such methods allow me to add "about ten years" to my estimate of Carol's age.

There is another basic problem—the problem of relevance. Suppose that instead of being given (a) and (b), I am given a collection of data which includes (a) and (b), and it is my problem to search for and identify the data that are relevant to the query. I came across (a). Is it relevant to *q*? By itself, it is not. Then I came across (b). Is it relevant to *q*? By itself, it is not. Thus, what I have to recognize is that, in isolation, (a) and (b) are irrelevant, that is, uninformative, but, in combination, they are relevant i.e., are informative. It is not difficult to recognize this in the simple example under consideration, but when we have a large database, the problem of identifying the data which in combination are relevant to the query, is very complex.

Suppose that a proposition, *p*, is, in isolation, relevant to *q*, or, for short, is *i*-relevant to *q*. What is the degree to which *p* is relevant to *q*? For example, to what degree is the proposition: Carol has two children: a son, in mid-twenties, and a daughter, in mid-thirties, relevant to the query: How old is Carol? To answer this question, we need a definition of measure of relevance. The problem is that there is no quantitative definition of relevance within existing bivalent-logic-based theories. We will return to the issue of relevance at a later point.

The example which we have considered is intended to point to the difficulty of dealing with world knowledge, especially in the contexts of assessment of relevance and deduction, even in very simple cases. The principal reason is that much of world knowledge is perception-based, and existing theories of knowledge representation and deduction provide no tools for this purpose. Here is a test problem involving deduction form perception-based information.

The Tall Swedes Problem

Perception: Most adult Swedes are tall, with adult defined as over about 20 years in age.

Query: What is the average height of Swedes?

A fuzzy-logic-based solution of the problem will be given at a later point.

A concept which provides a basis for computation and reasoning with perception-based information is that of Precisiated Natural Language (PNL) [15]. The capability of PNL to deal with perception-based information suggests that it may play a significant role in dealing with world knowledge. A quick illustration is the Carol example. In this example, an estimate of Carol's age, arrived at through the use of PNL would be expressed as a bimodal distribution of the form

$$\text{Age(Carol)} \quad \text{is} \quad ((P_1,V_1) + \ldots + (P_n,V_n)) \quad , i=1, \ldots, n$$

where the V_i are granular values of Age, e.g., less than about 20, between about 20 and about 30, etc.; P_i is a granular probability of the event (Age(Carol) is V_i), i=1, …, n; and + should be read as "and." A brief description of the basics of PNL is presented in the following.

2. Precisiated Natural Language (PNL)

PNL deals with perceptions indirectly, through their description in a natural language, Zadeh [15]. In other words, in PNL a perception is equated to its description in a natural language. PNL is based on fuzzy logic—a logic in which everything is, or is allowed to be, a matter of degree. It should be noted that a natural language is, in effect, a system for describing perceptions.

The point of departure in PNL is the assumption that the meaning of a proposition, *p*, in a natural language, NL, may be represented as a generalized constraint of the form (Fig. 2)

$$X \text{ is} r \ R,$$

• ***standard constraint:*** **X ∈ C**
• ***generalized constraint: X isr R***

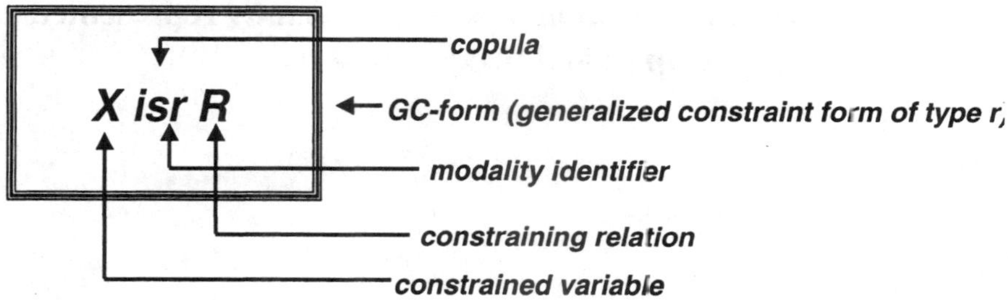

• ***X= (X₁ , ..., Xₙ)***
• ***X may have a structure: X=Location (Residence(Carol))***
• ***X may be a function of another variable: X=f(Y)***
• ***X may be conditioned: (X/Y)***
• **r := / ≤ .../ ⊂/ ⊃/ blank / v / p / u / rs / fg / ps / ...**

Figure 2. Generalized Constraint

where *X* is the constrained variable, *R* is a constraining relation which, in general, is not crisp (bivalent); and *r* is an indexing variable whose values define the modality of the constraint. The principal modalities are: possibilistic (*r*=blank); veristic(*r*=*v*); probabilistic(*r*=*p*); random set(*r*=*rs*); fuzzy graph (*r*=*fg*); usuality (*r*=*u*); and Pawlak set (*r*=*ps*). The set of all generalized constraints together with their combinations, qualifications and rules of constraint propagation, constitutes the Generalized Constraint Language (GCL). By construction, GCL is maximally expressive. In general, *X*, *R* and *r* are implicit in *p*. Thus, in PNL the meaning of *p* is precisiated through explicitation of the generalized constraint which is implicit in *p*, that is, through translation into GCL (Fig. 3). Translation of *p* into GCL is exemplified by the following.

(a) Monika is young ⟶ Age(Monika) is young,

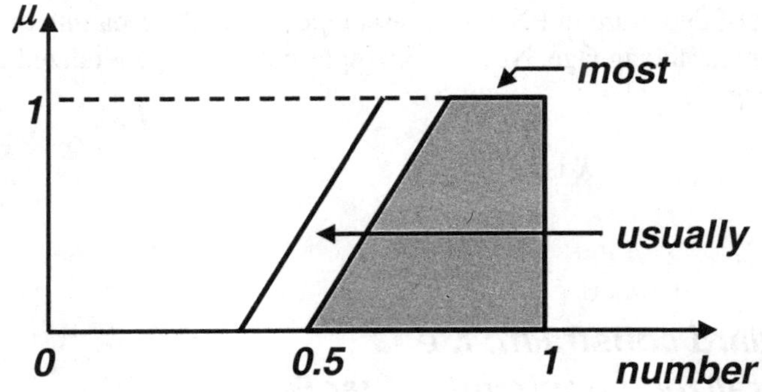

Figure 3. Calibration of *most* and *usually* represented as trapezoidal fuzzy numbers.

where young is a fuzzy relation which is characterized by its membership function μ_{young}, with $\mu_{young}(u)$ representing the degree to which a numerical value of age, u, fits the description of age as "young."

(b) Carol lives in a small city near San Francisco ⟶ Location (Residence(Carol)) is SMALL [City; μ] ∩ NEAR [City; μ],

where SMALL [City; μ] is the fuzzy set of small cities; NEAR [City; μ] is the fuzzy set of cities near San Francisco; and ∩ denotes fuzzy set intersection (conjunction).

(c) Most Swedes are tall ⟶ ΣCount(tall.Swedes/Swedes) is most.

In this example, the constrained variable is the relative Σcount of tall Swedes among Swedes, and the constraining relation is the fuzzy quantifier "most," with "most" represented as a fuzzy number (Fig. 3). The relative Σcount, ΣCount(A/B), is defined as follows. If A and B are fuzzy sets in a universe of discourse $U=\{u_i, \ldots, u_n\}$, with the grades of membership of u_i in A and B being μ_i and v_i, respectively, then by definition the relative ΣCount of A in B is expressed as

$$\Sigma\text{Count}(A/B) = \frac{\Sigma_i \mu_i \wedge \nu_i}{\sum_i \nu_i},$$

where the conjunction, ^, is taken to be min. More generally, conjunction may be taken to be a *t*-norm [11].

What is the rationale for introducing the concept of a generalized constraint? Conventional constraints are crisp (bivalent) and have the form $X \varepsilon C$, where X is the constrained variable and C is a crisp set. On the other hand, perceptions are f-granular, as was noted earlier. Thus, there is a mismatch between f-granularity of perceptions and crisp constraints. What this implies is that the meaning of a perception does not lend itself to representation as a collection of crisp constraints; what is needed for this purpose are generalized constraints or, more generally, an element of the Generalized Constraint Language (GCL) [15].

Representation of the meaning of a perception as an element of GCL is a first step toward computation with perceptions—computation which is needed for deduction from data which are resident in world knowledge. In PNL, a concept which plays a key role in deduction is that of a protoform—an abbreviation of "prototypical form," [15].

If p is a proposition in a natural language, NL, its protoform, PF(p), is an abstraction of p which places in evidence the deep semantic structure of p. For example, the protoform of "Monika is young," is "$A(B)$ is C," where A is abstraction of "Age," B is abstraction of "Monika" and C is abstraction of "young." Similarly,

Most Swedes are tall $\longrightarrow$ Count(A/B) is Q,

where A is abstraction of "tall Swedes," B is abstraction of "Swedes" and Q is abstraction of "most." Two propositions, p and q, are protoform-equivalent, or PF-equivalent, for short, if they have identical protoforms. For example, p: Most Swedes are tall, and q: Few professors are rich, are PF-equivalent.

The concept of PF-equivalence suggests an important mode of organization of world knowledge, namely protoform-based organization. In this mode of organization, items of knowledge are grouped into PF-equivalent classes. For example, one such class may be the set of all propositions whose protoform is $A(B)$ is C, e.g., Monika is young. The partially instantiated class Price(B) is low, would be the set of all objects whose price is low. As will be seen in the following, protoform-based organization of world knowledge plays a key role in deduction form perception-based information.

Basically, abstraction is a means of generalization. Abstraction is a familiar and widely used concept. In the case of PNL, abstraction plays an especially important role because PNL abandons bivalence. Thus, in PNL, the concept of a protoform is not limited to propositions whose meaning can be represented within the conceptual structure of bivalence logic.

In relation to a natural language, NL, the elements of GCL may be viewed as precisiations of elements of NL. Abstractions of elements of GCL gives rise to

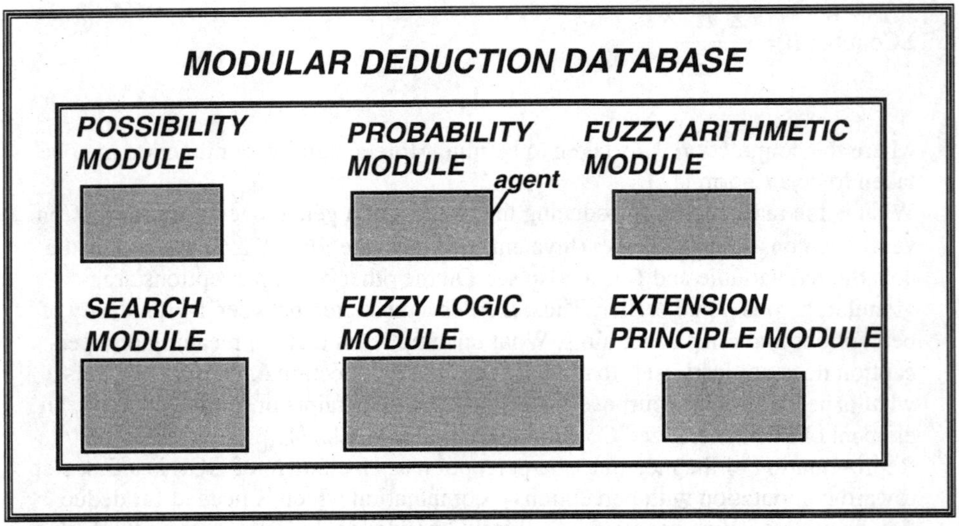

Figure 7. Basic structure of PNL: modular deduction database

where $A \circ B$ is the composition of A and B, defined in the computational part, in which μ_A, μ_B and $\mu_{A \circ B}$ are the membership functions of A, B and $A \circ B$, respectively. Similarly, a rule drawn from probability is

Prob (X is A) is B
Prob (X is C) is D

symbolic part

$$\mu_D(v) = max_q(\mu_B(\int_U \mu_A(u)g(u)du))$$

subject to: $v = \int_U \mu_C(u)g(u)du$

$$\int_U g(u)du = 1$$

computational part

where D is defined in the computational part and g is the probability density function of X.

what is referred to as Protoform Language, PFL, (Fig. 4). A consequence of the concept of PF-equivalence is that the cardinality of PFL is orders of magnitude smaller than that of GCL, or, equivalently, the set of precisiable propositions in NL. The small cardinality of PFL plays an essential role in deduction.

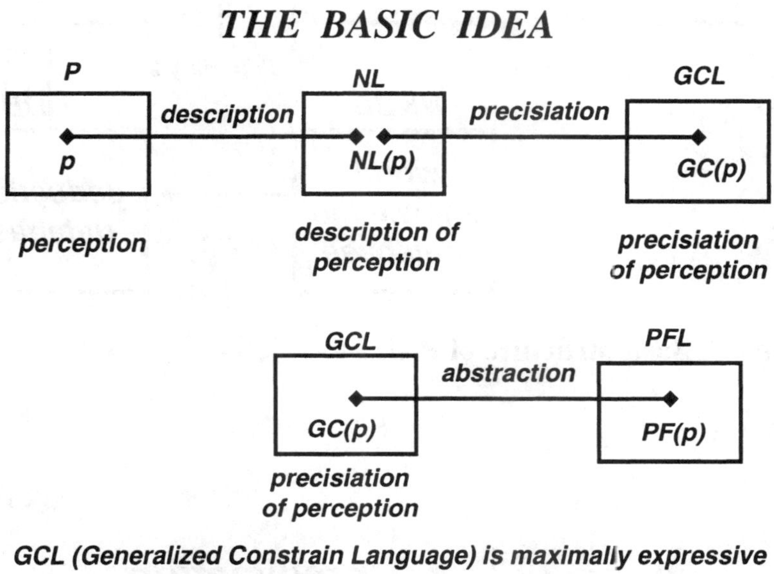

Figure 4. Precisiation and abstraction

The principal components of the structure of PNL (Fig. 5) are: (1) a dictionary from NL to GCL; (2) a dictionary from GCL to PFL (Fig. 6); (3) a multiagent, modular deduction database, DDB; and (4) a world knowledge database, WKDB. The constituents of DDB are modules, with a module consisting of a group of protoformal rules of deduction, expressed in PFL (Fig. 7), which are drawn from a particular domain, e.g., probability, possibility, usuality, fuzzy arithmetic, fuzzy logic, search, etc. For example, a rule drawn from fuzzy logic is the compositional rule of inference [11], expressed as

$$\frac{\begin{array}{c} X \text{ is } A \\ (X, Y) \text{ is } B \end{array}}{Y \text{ is } A \circ B}$$

symbolic part

$$\mu_{A \circ B}(v) = max_u(\mu_A(u) \wedge \mu_B(u,v))$$

computational part

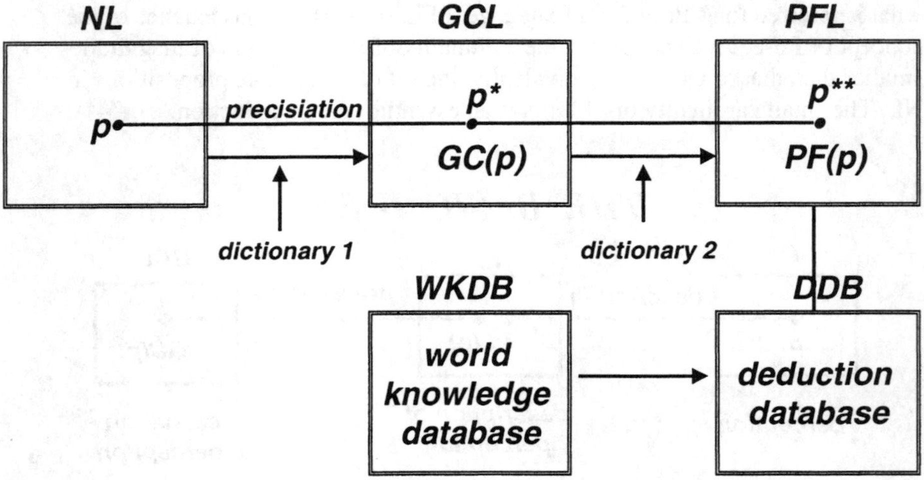

Figure 5. Basic structure of PNL

1:

proposition in NL	precisiation
p	*p* (GC-form)*
most Swedes are tall	*Σ Count (tall.Swedes/Swedes) is most*

2:

precisiation	protoform
p (GC-form)*	*PF(p*)*
Σ Count (tall.Swedes/Swedes) is most	*Q A's are B's*

Figure 6. Structure of PNL: dictionaries

The rules of deduction in DDB are, basically, the rules which govern propagation of generalized constraints. Each module is associated with an agent whose function is that of controlling execution of rules and performing embedded computations. The top-level agent controls the passing of results of computation from a module to other modules. The structure of protoformal, i.e., protoform-based, deduction is shown in Fig. 5. A simple example of protoformal deduction is shown in Fig. 8.

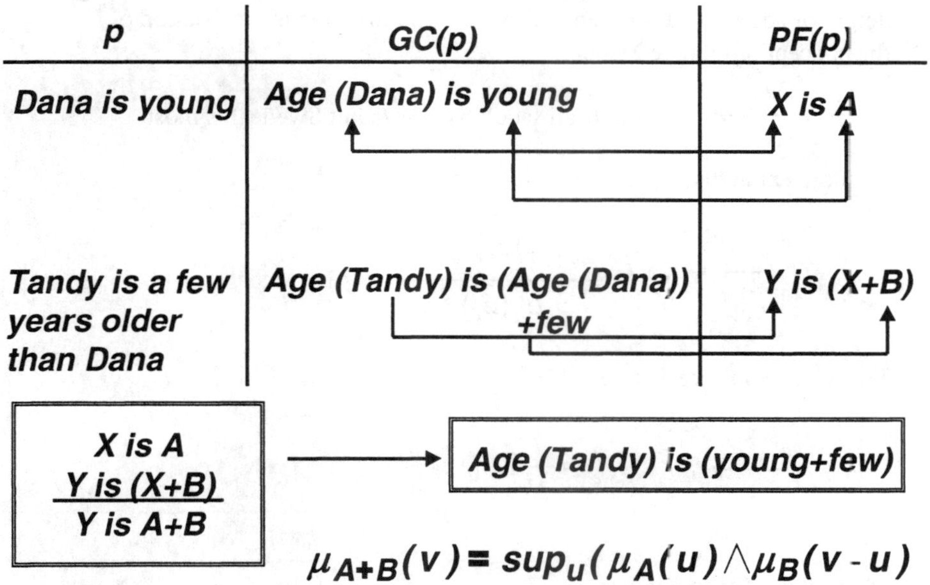

Figure 8. Example of protoformal reasoning

The principal deduction rule in PNL is the extension principle [19]. Its symbolic part is expressed as

$$\frac{f(X) \text{ is } A}{g(X) \text{ is } B}$$

in which the antecedent is a constraint on X through a function of X, $f(X)$; and the consequent is the induced constraint on a function of X, $g(X)$.

The computational part of the rule is expressed as

$$\mu_B(v) = \sup_u(\mu_A(f(u))$$

subject to

$$v = g(u)$$

To illustrate the use of the extension principle, we will consider the *Tall Swedes Problem*:
p: Most adult Swedes are tall
q: What is the average height of adult Swedes?

Let $P_1(u_1, \ldots, u_N)$ be a population of Swedes, with the height of u_i being h_i, i=1, …, N, and $\mu_a(u_i)$ representing the degree to which u_i is an adult. The average height of adult Swedes is denoted as h_{ave}. The first step is precisiation of p, using the concept of relative ΣCount:

$$p \longrightarrow \Sigma\text{Count}(\text{tall} \wedge \text{adult.Swedes}/\text{adult.Swedes}) \text{ is most}$$

or, more explicitly,

$$p \longrightarrow \frac{\sum_i \mu_{tall}(h_i) \wedge \mu_a(u_i)}{\Sigma_i \mu_a(u_i)} \quad \text{is} \quad \text{most.}$$

The next step is precisiation of q:

$$q \longrightarrow h_{ave} = \frac{1}{N}\Sigma_i h_i$$

h_{ave} is ?B,

where B is a fuzzy number
Using the extension principle, computation of h_{ave} as a fuzzy number is reduced to the solution of the variational problem.

$$\mu_B(v) = sup_h(\mu_{most}(\frac{\Sigma_i \mu_{tall}(h_i) \wedge \mu_a(u_i)}{\Sigma_i \mu_a(u_i)})) \qquad , h = (h_1, \ldots, h_N)$$

subject to

$$v = \frac{1}{N}(\Sigma_i h_i).$$

Note that we assume that a problem is solved once it is reduced to the solution of a well-defined computational problem.

3. PNL as a definition language

One of the important functions of PNL is that of serving as a high level definition language. More concretely, suppose that I have a concept, C, which I wish to define. For example, the concept of distance between two real-valued functions, f and g, defined on the real line.

The standard approach is to use a metric such as L_1 or L_2. But a standard metric may not capture my perception of the distance between f and g. The PNL-based approach would be to describe my perception of distance in a natural language and then precisiate the description.

This basic idea opens the door to (a) definition of concepts for which no satisfactory definitions exist, e.g., the concepts of causality, relevance and rationality, among many others, and (b) redefinition of concepts whose existing definitions do not provide a good fit to reality. Among such concepts are the concepts of similarity, stability, independence, stationarity and risk.

How can we concretize the meaning of "good fit?" In what follows, we do this through the concept of cointension.

More specifically, let U be a universe of discourse and let C be a concept which I wish to define, with C relating to elements of U. For example, U is a set of buildings and C is the concept of tall building. Let $p(C)$ and $d(C)$ be, respectively, my perception and my definition of C. Let $I(p(C))$ and $I(d(C))$ be the intensions of $p(C)$ and $d(C)$, respectively, with "intension" used in its logical sense, [6, 7] that is, as a criterion or procedure which identifies those elements of U which fit $p(C)$ or $d(C)$. For example, in the case of tall buildings, the criterion may involve the height of a building.

Informally, a definition, $d(C)$, is a good fit or, more precisely, is cointensive, if its intension coincides with the intension of $p(C)$. A measure of goodness of fit is the degree to which the intension of $d(C)$ coincides with that of $p(C)$. In this sense, cointension is a fuzzy concept. As a high level definition language, PNL makes it possible to formulate definitions whose degree of cointensiveness is higher than that of definitions formulated through the use of languages based on bivalent logic.

A substantive exposition of PNL as a definition language is beyond the scope of this note. In what follows, we shall consider as an illustration a relatively simple version of the concept of relevance.

4. Relevance

We shall examine the concept of relevance in the context of a relational model such as shown in Fig. 9. For concreteness, the attributes $A_1, \ldots, A_n$ may be interpreted as symptoms and D as diagnosis. For convenience, rows which involve the same value of D are grouped together. The entries are assumed to be labels of fuzzy sets. For example, A_5 may be blood pressure and a_{53} may be "high."

RELEVANCE, REDUNDANCE AND DELETABILITY

DECISION TABLE

Name	A_1	A_j	A_n	D
$Name_1$	a_{11}	a_{1j}	a_{in}	d_1
.	.	.	.	.
$Name_k$	a_{k1}	a_{kj}	a_{kn}	d_1
$Name_{k+1}$	$a_{k+1, 1}$	$a_{k+1, j}$	$a_{k+1, n}$	d_2
.	.	.	.	.
$Name_l$	a_{l1}	a_{lj}	a_{ln}	d_l
.	.	.	.	.
$Name_n$	a_{m1}	a_{mj}	a_{mn}	d_r

A_j: j th symptom

a_{ij}: value of j th symptom of Name

D: diagnosis

Figure 9. A relational model of decision

An entry represented as * means that the entry in question is conditionally redundant in the sense that its value has no influence on the value of *D* (Fig. 10).

REDUNDANCE ⟶ DELETABILITY

Name	A_1	A_j	A_n	D
.	.	.	.	.
$Name_r$	a_{r1}	*	a_{rn}	d_2
.	.	.	.	.

A_j is conditionally redundant for $Name_r$, A, is a_{r1}, A_n is a_{rn} If D is d_s for all possible values of A_j in *

A_j is redundant if it is conditionally redundant for all values of Name

- ***compactification algorithm (Zadeh, 1976); Quine-McCluskey algorithm***

Figure 10. Conditional redundance and redundance

An attribute, A_j, is redundant, and hence deletable, if it is conditionally redundant for all values of Name. An algorithm, termed compactification, which identifies all deletable attributes is described in Zadeh [18]. Compactification algorithm is a generalization of the Quine-McCluskey algorithm for minimization of switching circuits. The Redact algorithm in the theory of rough sets [10] is closely related to the compactification algorithm.

Name	A_1	A_j	A_n	D
Name $_r$	.	a_{ij}	.	d_1 . d_1
Name $_{i+s}$	.	a_{ij}	.	d_2 . d_2

(A_j is a_{ij}) is irrelevant (uninformative)

Figure 11. Irrelevance

The concept of relevance (informativeness) is weaker and more complex than that of redundance (deletability). As was noted earlier, it is necessary to differentiate between relevance in isolation (*i*-relevance) and relevance as a group. In the following, relevance should be interpreted as *i*-relevance.

D is ?d   if A_j is a_{rj}

constraint on A_j induces a constraint on D
example: (blood pressure is high) constrains D
(A_j is a_{rj})   is uniformative if D is unconstrained

A_j is irrelevant if it A_j is uniformative for all a_{rj}

irrelevance $\not\longrightarrow$ deletability

Figure 12. Relevance and irrelevance

A value of A_1, say a_{rj}, is irrelevant (uninformative) if the proposition A_j is a_{rj} does not constrain D (Fig. 11). For example, the knowledge that blood pressure is high may convey no information about the diagnosis (Fig. 12).

An attribute, A_j, is irrelevant (uninformative) if, for all a_{rj}, the proposition A_j is a_{rj} does not constrain D. What is important to note is that irrelevance does not imply deletability, as redundance does. The reason is that A_j may be i-irrelevant but not irrelevant in combination with other attributes. An example is shown in Fig. 13.

As defined above, relevance and redundance are bivalent concepts, with no degrees of relevance or redundance allowed. But if the definitions in question are interpreted as idealizations, then a measure of the departure from the ideal could be used as a measure of the degree of relevance or redundance. Such a measure could be defined through the use of PNL. In this way, PNL may provide a basis for defining relevance and redundance as matters of degree, as they are in realistic settings. However, what should be stressed is that our analysis is limited to relational models. Formalizing the concepts of relevance and redundance in the context of the Web is a far more complex problem—a problem for which no cointensive solution is in sight.

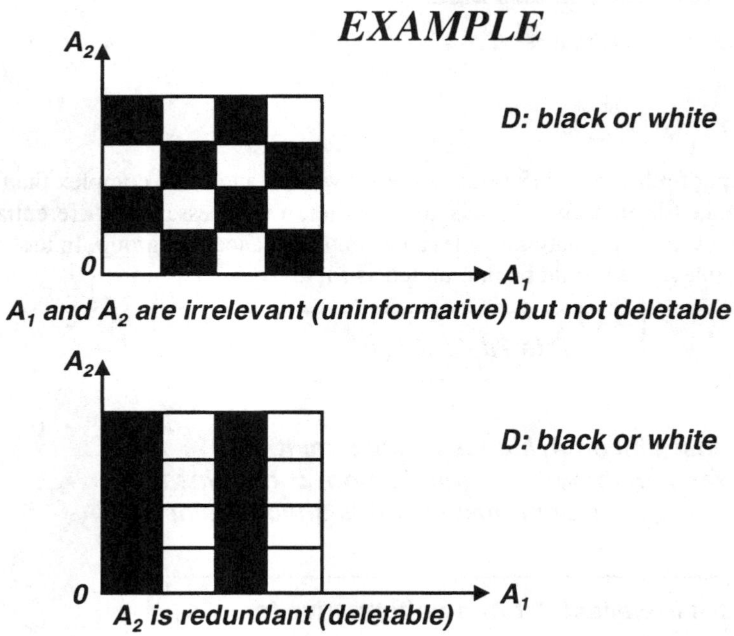

Figure 13. Redundance and irrelevance

5. Concluding remark

Much of the information which resides in the Web—and especially in the domain of world knowledge—is imprecise, uncertain and partially true. Existing bivalent-logic-based methods of knowledge representation and deduction are of limited effectiveness in dealing with information which is imprecise or partially true. To deal with such information, bivalence must be abandoned and new tools, such as PNL, should be employed. What is quite obvious is that, given the deeply entrenched tradition of basing scientific theories on bivalent logic, a call for abandonment of bivalence is not likely to meet a warm response. Abandonment of bivalence will eventually become a reality but it will be a gradual process.

References

1. Arjona, J.; Corchuelo, R.; Pena, J. and Ruiz, D. 2003. Coping with Web Knowledge. Advances in Web Intelligence. Springer-Verlag Berlin Heidelberg, 165-178.
2. Bargiela, A. and Pedrycz, W. 2003. Granular Computing—An Introduction. Kluwer Academic Publishers: Boston, Dordrecht, London.
3. Berners-Lee, T.; Hendler, J. and Lassila, O. 2001. The Semantic Web. *Scientific American.*
4. Chen, P.P. 1983. Entity-Relationship Approach to Information Modeling and Analysis. North Holland.
5. Craven, M.; DiPasquo, D.; Freitag, D.; McCallum, A.; Mitchell, T.; Nigam, K. and Slattery, S. 2000. Learning to Construct Knowledge Bases from the World Wide Web. *Artificial Intelligence* 118 (1-2): 69-113.
6. Cresswell, M. J. 1973. *Logic and Languages*. London, U.K.: Methuen.
7. Gamat, T. F. 1996. *Language, Logic and Linguistics*. University of Chicago Press.
8. Lenat, D. B. 1995.cyc: A Large-Scale Investment in Knowledge Infrastructure. *Communications of the ACM* 38(11): 32-38.
9. Novak, V. and Perfilieva, I., eds. 2000. Discovering the World with Fuzzy Logic. Studies in Fuzziness and Soft Computing. Heidelberg New York: Physica-Verlag.
10. Pawlak, Z. 1991. *Rough Sets: Theoretical Aspects of Reasoning about Data*, Kluwer Academic, Dordrecht.
11. Pedrycz, W., and Gomide, F. 1998. *Introduction to Fuzzy Sets*. Cambridge, Mass.: MIT Press.
12. Smith, B. and Welty, C. 2002. What is Ontology? Ontology: Towards a new synthesis. *Proceedings of the Second International Conference on Formal Ontology in Information Systems.*
13. Smith, M. K.; Welty, C.; McGuinness, D., eds. 2003. OWL Web Ontology Language Guide. W3C Working Draft 31.
14. Sowa, J. F., in Albertazzi, L. 1999. Ontological Categories. *Shapes of Forms: From Gestalt Psychology and Phenomenology to Ontology and Mathematics*, Dordrecht: Kluwer Academic Publishers, 307-340.
15. Zadeh, L.A. 2001. A New Direction in AI—Toward a Computational Theory of Perceptions. *AI Magazine* 22(1): 73-84.
16. Zadeh, L.A. 1999. From Computing with Numbers to Computing with Words—From Manipulation of Measurements to Manipulation of Perceptions. *IEEE Transactions on Circuits and Systems* 45(1): 105-119.
17. Zadeh, L.A. 1994. Fuzzy Logic, Neural Networks, and Soft Computing, *Communications of the ACM—AI*, Vol. 37, pp. 77-84.
18. Zadeh, L.A. 1976. A fuzzy-algorithmic approach to the definition of complex or imprecise concepts, *Int. Jour. Man-Machine Studies 8*, 249-291.

19.Zadeh, L.A. 1975. The concept of a linguistic variable and its application to approximate reasoning, Part I: *Inf. Sci.8*, 199-249, 1975; Part II: *Inf. Sci. 8*, 301-357, 1975; Part III: *Inf. Sci. 9*, 43-80.

Towards More Powerful Information Technology via Computing with Words and Perceptions: Precisiated Natural Language, Protoforms and Linguistic Data Summaries

Janusz Kacprzyk[1,2] and Sławomir Zadrożny[1]

[1] Systems Research Institute, Polish Academy of Sciences,
ul. Newelska 6, 01–447 Warsaw, Poland
[2] Warsaw School of Information Technology (WSISiZ)
ul. Newelska 6, 01–447 Warsaw, Poland
{kacprzyk, zadrozny}@ibspan.waw.pl

Summary. We show how Zadeh's idea of computing with words and perceptions, based on his concept of a precisiated natural language (PNL), can lead to a new direction in the use of natural language in data mining, linguistic data(base) summaries. We emphasize the relevance of Zadeh's another idea, that of a protoform, and show that various types of linguistic data summaries may be viewed as items in a hierarchy of protoforms of summaries. We briefly present an implementation for a sales database of a computer retailer as a convincing example that these tools and techniques are implementable and functional. These summaries involve both data from an internal database of the company and data downloaded from external databases via the Internet.

1 Introduction

Computing with words and perceptions, introduced by Zadeh in the mid-1990s, and best and most comprehensively presented in Zadeh and Kacprzyk's books [?], is a new "technology" in the representation, processing and solving of various real life problems when a human being is a crucial element. Such problems are omnipresent. The basic philosophy and rationale of computing with words and perceptions is that since for a human being natural language is the only fully natural way of communication and articulation, and also the only tool to express perceptions that are so characteristic for human beings, then maybe it could be expedient to try to "directly" use (elements of) natural language in the formulation, processing and solution of problems considered to maintain a higher human consistence, hence a higher implementability.

A prerequisite for computing with words is to have some way to formally represent elements of natural language used. Zadeh proposed to use here the

PNL (precisiated natural language). Basically, in PNL statements about values, relations, etc. between variables are represented by constraints. In the conventional case, a statement is, e.g., that the value of variable x belongs to a set X. In PNL, statements - written "x isr R" - may be different, and correspond to numeric values, intervals, possibility disctributions, verity distributions, probability distributions, usuality qualified statements, rough sets representations, fuzzy relations, etc. For our purposes, usuality qualified representation will be of special relevance. Basically, it says "x is usually R" that is meant as "in most cases, x is R". PNL may play various roles among which crucial are: description of perceptions, definition of sophisticated concepts, a language for perception based reasoning, etc. More details and insights may be found in Zadeh's articles earlier in this volume

Recently, Zadeh introduced the concept of a protoform. For our purposes, one should notice that most perceptions are summaries. For instance, a perception like "most Swedes are tall" is a summary. It can be represented in Zadeh's notation as "most As are Bs". This can be employed for reasoning under various assumptions. For instance, if we know that "x is A", we can deduce that, e.g. "it is likely that x is B", we can ask about an average height of a Swede, etc. One can go a step further, and define a protoform as an abstracted summary. In our case, this would be "QAs are Bs". Notice that we now have a more general, deinstantiated form of our point of departure (most Swedes are tall), and also of "most As are Bs". Needless to say that most human reasoning is protoform based, and the availability of such a more general representation is very valuable, and provides tools that can be used in many cases.

Here, we show that the concept of a precisiated natural language, and in particular of a protoform, viewed from the perspective of the computing with words and perceptions, can be of use in data mining, and more generally in attempts at a more effective and efficient use of vast information resources. We show the idea of a linguistic data summarization as a type of data mining that is very characteristic for human needs and comprehension abilities. Generally, data summarization is still an unsolved problem though many techniques are available. However, they make little use of human perceptions and natural language as, e.g., summarizing statistics, exemplified by the average, median, minimum, maximum, α-percentile, etc.

In this chapter we discuss an approach based on the concept of a *linguistic data(base) summary* that has been originally proposed by Yager [36, 37] and further developed mainly by Kacprzyk and Yager [16], and Kacprzyk, Yager and Zadrożny [17]. The essence of such linguistic data summaries is that a set of data, e.g., concerning employees, with (numeric) data on their age, sex, salaries, seniority, etc., can be summarized linguistically with respect to a selected attribute or attributes, say age and salaries, by linguistically quantified propositions, e.g., "almost all employees are well qualified", "most young employees are well paid", etc. which are simple, extremely human consistent and intuitive, and do summarize in a concise yet very informative form what we may be interested in.

We present the essence of such summaries, mainly from the perspective of Zadehs computing with words and perception paradigm (cf. Zadeh and Kacprzyk [41]) that can provide a general theoretical framework which is implementable, as shown in works mentioned above. In particular, we indicate the use of Zadehs concept of a protoform of a fuzzy linguistic summary (cf. Zadeh [40], Kacprzyk and Zadrożny [20]) that can provide an easy generalization, portability and scalability.

We present a number of approaches to mining of linguistic summaries. First, those based on Kacprzyk and Zadrożnys [21, 26] idea of an interactive approach to linguistic summaries in which the determination of a class of summaries of interest is done via Kacprzyk and Zadrożny's [19, 22] FQUERY for Access, a fuzzy querying add-on to Microsoft Access©. We show that by relating a range of types of linguistic summaries to fuzzy queries, with various known and sought elements, we can arrive at a hierarchy of protoforms of linguistic data summaries.

As an example we will show an implementation of the data summarization system proposed for the derivation of linguistic data summaries in a sales database of a computer retailer

Our general discussion and implementation should convince the reader that the use of a broadly perceived paradigm of computing with words and perceptions, equipped with a newly introduced concept of a protoform, may be a proper tool for being able to more intelligently manage with huge amounts of data we face in the present world.

2 Linguistic Data Summaries via Fuzzy Logic with Linguistic Quantifiers

The linguistic summary is meant as a sentence [in a (quasi)natural language] that subsumes the very essence (from a certain point of view) of a set of data. Here this set is assumed to be numeric, large and not comprehensible in its original form by the human being. In Yagers approach (cf. Yager [36], Kacprzyk and Yager [16], and Kacprzyk, Yager and Zadrożny [17]) we have:

- $Y = \{y_1, \ldots, y_n\}$ is a set of objects (records) in a database, e.g., the set of workers;
- $A = \{A_1, \ldots, A_m\}$ is a set of attributes characterizing objects from Y, e.g., salary, age, etc. in a database of workers, and $A_j(y_i)$ denotes a value of attribute A_j for object y_i.

A linguistic summary of data set D consists of:

- a summarizer S, i.e. an attribute together with a linguistic value (fuzzy predicate) defined on the domain of attribute A_j (e.g. low salary for attribute salary);

- a quantity in agreement Q, i.e. a linguistic quantifier (e.g. most);
- truth (validity) T of the summary, i.e. a number from the interval $[0,1]$ assessing the truth (validity) of the summary (e.g. 0.7); usually, only summaries with a high value of T are interesting;
- optionally, a qualifier R, i.e. another attribute together with a linguistic value (fuzzy predicate) defined on the domain of attribute A_k determining a (fuzzy subset) of Y (e.g. young for attribute age).

Thus, the linguistic summary may be exemplified by

$$T(\text{most of employees earn low salary}) = 0.7 \tag{1}$$

A richer form of the summary may include a qualifier as in, e.g.,

$$T(\text{most of young employees earn low salary}) = 0.7 \tag{2}$$

Thehe core of a linguistic summary is a *linguistically quantified proposition* in the sense of Zadeh [39], the one corresponding to (1)written as

$$Qy\text{'s are } S \tag{3}$$

and the one corresponding to (2) written as

$$QRy\text{'s are } S \tag{4}$$

The T, i.e., the truth value of (3) or (4), m may be calculated by using either original Zadehs calculus of linguistically quantified statements (cf. [39]), or other interpretations of linguistic quantifiers (cf. Liu and Kerre [31]), including Yagers OWA operators [38] and Dubois et al. OWmin operators [5], or via generalized quantifier, cf. Hájek and Holeňa[12] or Glöckner [11].

Using Zadeh's [39] fuzzy logic based calculus of linguistically quantified propositions, a (proportional, nondecreasing) linguistic quantifier Q is assumed to be a fuzzy set in the interval $[0,1]$ as, e.g.

$$\mu_Q(x) = \begin{cases} 1 & \text{for } x \geq 0.8 \\ 2x - 0.6 & \text{for } 0.3 < x < 0.8 \\ 0 & \text{for } x \leq 0.3 \end{cases} \tag{5}$$

and then

$$\text{truth}(Qy\text{'s are } S) = \mu_Q[\frac{1}{n}\sum_{i=1}^{n}\mu_S(y_i)] \tag{6}$$

or

$$\text{truth}(QRy\text{'s are } S) = \mu_Q[\frac{\sum_{i=1}^{n}(\mu_R(y_i) \wedge \mu_S(y_i))}{\sum_{i=1}^{n}\mu_R(y_i)}] \tag{7}$$

Both the fuzzy predicates S and R are of a simplified, atomic form referring to one attribute, and they can be extended to cover more sophisticated summaries involving some confluence of various attribute values as, e.g, "young and well paid". Clearly, when we try to linguistically summarize data, the most interesting are non-trivial, human-consistent summarizers (concepts) as, e.g.: productive workers, difficult orders, etc. to be definedand by a complicated combination of attributes.

Recently, Zadeh [40] introduced a relevant concept of a *protoform* which is defined as a more or less abstract prototype (template) of a linguistically quantified proposition. The most abstract protoforms correspond to (3) and (4), while (1) and (2) are examples of fully instantiated protoforms. Thus, evidently, protoforms form a hierarchy, where higher/lower levels correspond to more/less abstract protoforms. Going down this hierarchy one has to instantiate particular components of (3) and (4), i.e., quantifier Q and fuzzy predicates S and R. The instantiation of the former one boils down to the selection of a quantifier. The instantiation of fuzzy predicates requires the choice of attributes together with linguistic values (atomic predicates) and a structure they form when combined using logical connectives. This leads to a theoretically infinite number of potential protoforms. However, for the purposes of mining of linguistic summaries, there are obviously some limits on a reasonable size of a set of summaries that should be taken into account. These results from a limited capability of the user in the interprenion of summaries as well as from the computational point of view.

The concept of a protoform may provide a guiding paradigm for the design of a user interface supporting the mining of linguistic summaries. It may be assumed that the user specifies a protoform of linguistic summaries sought. Basically, the more abstract protoform the less should be assumed about summaries sought, i.e., the wider range of summaries is expected by the user. There are two limit cases, where:

- a totally abstract protoform is specified, i.e., (4),
- all elements of a protoform are totally specified as given linguistic terms,

and in the former case the system has to construct all possible summaries (with all possible linguistic components and their combinations) for the context of a given database (table) and present to the user those verifying the validity to a degree higher than some threshold. In the second case, the whole summary is specified by the user and the system has only to verify its validity. Thus, the former case is usually more interesting from the point of view of the user but at the same time more complex from the computational point of view. There is a number of intermediate cases that may be more practical. In Table 1 basic types of protoforms/linguistic summaries are shown, corresponding to protoforms of a more and more abstract form.

Basically, each of fuzzy predicates S and R may be defined by listing its atomic fuzzy predicates (i.e., pairs of "attribute/linguistic value") and structure, i.e., how these atomic predicates are combined. In Table 1 S (or R) corresponds to the full description of both the atomic fuzzy predicates (referred to as linguistic values, for short) as well as the structure. For example: "Q *young* employees earn a *high salary*" is a protoform of Type 2, while

Table 1. Classification of protoforms/linguistic summaries

Type	Protoform	Given	Sought
0	QRy's are S	All	validity T
1	Qy's are S	S	Q
2	QRy's are S	S and R	Q
3	Qy's are S	Q and structure of S	linguistic values in S
4	QRy's are S	Q, R and structure of S	linguistic values in S
5	QRy's are S	Nothing	S, R and Q

"Most employees earn a *"?" salary*" is a protoform of Type 3. In the first case the system has to select a linguistic quantifier (usually from a predefined dictionary) that when put in place of Q makes the resulting linguistically quantified proposition valid to the highest degree, and in the second case, the linguistic quantifier as well as the *structure* of summarizer S are given and the system has to choose a linguistic value to replace the question mark ("?") yielding a linguistically quantified proposition as valid as possible.

Thus, the use of protoforms makes it possible to devise a uniform procedure to handle a wide class of linguistic data summaries so that the system can be easily adaptable to a variety of situations, users interests and preferences, scales of the project, etc.

Usually, most interesting are linguistic summaries required by a summary of Type 5. They may be interpreted as fuzzy IF-THEN rules, and many interpretations are proposed (cf., e.g., Dubois and Prade [7]) there are considered many possible interpretations for fuzzy rules), and some of them were directly discussed in the context of linguistic summaries (cf. Section 3.3).

There are many views on the idea of a linguistic summary, for instance a fuzzy functional dependency, a gradual rule, even a typical value. Though they do reflect the essence of a human perception of what a linguistci summary should be, they are beyond the scope of this paper which focuses on a different approach.

3 Mining of Linguistic Data Summaries

In the process of mining of linguistic summaries, at the one extreme, the system may be responsible for both the construction and verification of summaries (which corresponds to Type 5 protoforms/summaries given in Table 1). At the other extreme, the user proposes a summary and the system only verifies its validity (which corresponds to Type 0 protoforms/summaries in Table 1). The former approach seems to be more attractive and in the spirit of data mining meant as the discovery of interesting, unknown regularities in data. On the other hand, the latter approach, obviously secures a better interpretability of the results. Thus, we will discuss now the possibility to employ a flexible querying interface for the purposes of linguistic summarization of data, and indicate the implementability of a more automatic approach.

3.1 A fuzzy querying add-on for formulating linguistic summaries

In Kacprzyk and Zadrożnys [21, 26] approach, the interactivity, i.e. a user assistance, in the mining of linguistic summaries is a key point, and is in the definition of summarizers (indication of attributes and their combinations). This proceeds via a user interface of a fuzzy querying add-on. In Kacprzyk and Zadrożny [19, 22, 27], a conventional database management system is used with a fuzzy querying tool, FQUERY for Access. An important component of this tool is a dictionary of linguistic terms to be used in queries. They include fuzzy linguistic values and relations as well as fuzzy linguistic quantifiers. There is a set of built-in linguistic terms, but the user is free to add his or her own. Thus, such a dictionary evolves in a natural way over time as the user is interacting with the system. For example, an SQL query searching for *troublesome orders* may take the following WHERE clause:

WHERE *Most* of the conditions are met out of
PRICE*ORDERED-AMOUNT IS *Low*
DISCOUNT IS *High*
ORDERED-AMOUNT IS *Much Greater Than* ON-STOCK

Obviously, the condition of such a fuzzy query directly correspond to summarizer S in a linguistic summary. Moreover, the elements of a dictionary are perfect building blocks of such a summary. Thus, the derivation of a linguistic summary of type (3) may proceed in an interactive (user-assisted) way as follows:

- the user formulates a set of linguistic summaries of interest (relevance) using the fuzzy querying add-on,
- the system retrieves records from the database and calculates the validity of each summary adopted, and
- a most appropriate linguistic summary is chosen.

Referring to Table 1, we can observe that Type 0 as well as Type 1 linguistic summaries may be easily produced by a simple extension of FQUERY for Access. Basically, the user has to construct a query, a candidate summary, and it is to be determined which fraction of rows matches that query (and which linguistic quantifier best denotes this fraction, in case of Type 1). For Type 3 summaries, a query/summarizer S consists of only one simple condition built of the attribute whose typical (exceptional) value is sought. For example, using: $Q =$ "most" and $S =$ "age=?" we look for a typical value of "age". From the computational point of view Type 5 summaries represent the most general form considered: fuzzy rules describing dependencies between specific values of particular attributes.

The summaries of Type 1 and 3 have been implemented as an extension to Kacprzyk and Zadrożnys [23, 24, 25] FQUERY for Access.

3.2 Linguistic summaries and fuzzy association rules

The discovery of general, Type 5 rules is difficult, and some simplifications about the structure of fuzzy predicates and/or quantifier are needed, for instance to obtain association rules which have been initially defined for binary valued attributes as (cf. Agraval and Srikant [1]):

$$A_1 \wedge A_2 \wedge \ldots \wedge A_n \longrightarrow A_{n+1} \tag{8}$$

and note that much earlier origins of that concept are mentioned in the work by Hájek and Holeňa [12]).

Such an association rule states that if in a database row all the attributes from $\{A_1, A_2, \ldots, A_n\}$ take on value 1, then also the attribute A_{n+1} is expected to take on value 1. The algorithms proposed in the literature for mining the association rules are based on the concepts pf a support and confidence, and are well known (cf. Agrawal and Srikant [1] and Mannila *et al.*[32]). Moreover, these algorithms may be easily adopted for non-binary valued data and more sophisticated rules can be sought.

In particular, *fuzzy association rules* may be considered:

$$A_1 \text{ IS } R_1 \wedge A_2 \text{ IS } R_2 \wedge \ldots \wedge A_n \text{ IS } R_n \longrightarrow A_{n+1} \text{ IS } S \tag{9}$$

where R_i is a linguistic term defined in the domain of the attribute A_i, i.e. a qualifier fuzzy predicate in terms of linguistic summaries (cf. Section 2) and S is another linguistic term corresponding to the summarizer. The confidence of the rule may be interpreted in terms of linguistic quantifiers employed in the definition of a linguistic summary. Thus, a fuzzy association rule may be treated as a special case of a linguistic summary of type defined by (4). The structure of the fuzzy predicates R_i and S is to some extent fixed but due to that efficient algorithms for rule generation may be employed. These algorithms are easily adopted to fuzzy association rules. Usually, the first step is a preprocessing of original, crisp data. Values of all attributes considered are replaced with linguistic terms best matching them. Additionally, a degree of this matching may be optionally recorded and later taken into account. Then, each combination of attribute and linguistic term may be considered as a Boolean attribute and original algorithms, such as Apriori [1], may be applied. They, basically, boil down to an efficient counting of support for all conjunctions of Boolean attributes, i.e., so-called itemsets (in fact, the essence of these algorithms is to count support for as small a subset of itemsets as possible). In case of fuzzy association rules attributes may be treated strictly as Boolean attributes - they may appear or not in particular tuples - or interpreted in terms of fuzzy logic as in linguistic summaries. In the latter case they appear in a tuple to a degree and the support counting should take that into account. In our context we employ basically the approach by Lee and Lee-Kwang [30] and Au and Chan [2]. Hu *et al.* [13] who simplify the fuzzy association rules sought by assuming a single specific attribute (class) in the consequent. Kacprzyk, Yager and Zadrożny [17, 28, 25, 24, 23] advocated the use of fuzzy association rules for mining linguistic summaries in the framework of flexible querying interface. Chen *et al.* [4] investigated the issue of generalized fuzzy rules where a fuzzy taxonomy of linguistic terms is taken into account. Kacprzyk and Zadrożny [29] proposed to use more flexible aggregation operators instead of conjunction, but still in context of fuzzy association rules.More information on fuzzy association rules, from various perspectives, may be found later in this volume.

3.3 Other approaches to linguistic data summaries

Among some other approaches to the derivation of fuzzy linguistic summaries, we can mention the following ones. George and Srikanth [9], [10] use a genetic algorithm to mine linguistic summaries in which the summarizer is a conjunction of atomic fuzzy predicates. Then, they search for two linguistic summaries: the most specific generalization and the most general specification, assuming a dictionary of linguistic quantifiers and linguistic values over domains of all attributes. Kacprzyk and Strykowski [14, 15] have also implemented the mining of linguistic summaries using genetic algorithms. In their approach, the fitting function is a combination of a wide array of indices a degree of imprecision (fuzziness), a degree of covering, a degree of appropriateness, a length of a summary, etc. (cf. also Kacprzyk and Yager [16]).

Rasmussen and Yager [34, 35] propose an extension, SummarySQL, to SQL to cover linguistic summaries. Actually, they do not address the mining linguistic summaries but merely their verification. The SummarySQL may also be used to verify a kind of fuzzy gradual rules (cf. Dubois and Prade [5]) and fuzzy functional dependencies.

Raschia and Mouaddib [33] consider the problem of mining hierarchies of summaries, and their understanding of summaries is slightly different than here as it is a conjunction of atomic fuzzy predicates (each referring to just one attribute). However, these predicates are not defined by just one linguistic value but possibly by fuzzy sets of linguistic values (i.e., fuzzy sets of higher levels are considered). The mining of summaries (a whole hierarchy of summaries) is based on a concept formation (conceptual clustering) process.

4 Examples of Linguistic Summaries and Possible Extensions

Finally, to show the essence of our approach, and provide a convining example that Zadeh's computing with words and perception paradigm does work, and his conceot of a protoform is constructive and valuable, we we will briefly present an implementation of a system for deriving linguistic database summaries for a computer retailer. Basically, we will deal with its sales database, and will only show some examples of linguistic summaries for some interesting (for the user!) choices of relations between attributes.

The basic structure of the database in question is shown in Table 2.

Table 2. The basic structure of the database

Attribute name	Attribute type	Description
Date	Date	Date of sale
Time	Time	Time of sale transaction
Name	Test	Name of the product
Amount (number)	Numeric	Number of products sold in the transaction
Price	Numeric	Unit price
Commission	Numeric	Commission (in %) on sale
Value	Numeric	Value = amount (number) × price, of the product
Discount	Numeric	Discount (in %) for transaction
Group	Test	Product group to which the product belongs
Transaction value	Numeric	Value of the whole transaction
Total sale to customer	Numeric	Total value of sales to the customer in fiscal year
Purchasing frequency	Numeric	Number of purchases by customer in fiscal year
Town	Test	Town where the customer lives or is based

Linguistic summaries are generated using a genetic algorithm [14, 15]. We will now give a couple of examples of resulting summaries. First, suppose that we are interested in a relation between the commission and the type of goods sold. The best linguistic summaries obtained are as shown in Table 3.

As we can see, the results can be very helpful, for instance while negotiating commissions for various products sold.

Next, suppose that we are interested in relations between the groups of products and times of sale. The best results obtained are as in Table 4.

Notice that in this case the summaries are much less obvious than in the former case expressing relations between the group of product and commission. But, again, they provide very useful information.

Finally, let us show in Table 5 some of the obtained linguistic summaries expressing relations between the attributes: size of customer, regularity of customer (purchasing frequency), date of sale, time of sale, commission, group of product and day of sale.

Notice that the linguistic summaries obtained do provide much of relevant and useful information, and can help the decision maker make decisions. It should be stressed that in the construction of the data mining paradigm

Table 3. Linguistic summaries expressing relations between the group of products and commission

Summary
About 1/3 of sales of network elements is with a high commission
About 1/2 of sales of computers is with a medium commission
Much sales of accessories is with a high commission
Much sales of components is with a low commission
About 1/2 of sales of software is with a low commission
About 1/3 of sales of computers is with a low commission
A few sales of components is without commission
A few sales of computers is with a high commission
Very few sales of printers is with a high commission

Table 4. Linguistic summaries expressing relations between the groups of products and times of sale

Summary
About 1/3 of sales of computers is by the end of year
About 1/2 of sales in autumn is of accessories
About 1/3 of sales of network elements is in the beginning of year
Very few sales of network elements is by the end of year
Very few sales of software is in the beginning of year
About 1/2 of sales in the beginning of year is of accessories
About 1/3 of sales in the summer is of accessories
About 1/3 of sales of peripherals is in the spring period
About 1/3 of sales of software is by the end of year
About 1/3 of sales of network elements is in the spring period
About 1/3 of sales in the summer period is of components
Very few sales of network elements is in the autumn period
A few sales of software is in the summer period

presented we do not want to replace the decision maker but just to provide him or her with a help (support). This is clearly an example of the promising philosophy of decision support, i.e. to maintain users autonomy and just to provide a support for decision making, and by no means to replace the user.

The system for deriving linguistic summaries developed and implemented for a computer retailer has been found useful by the user who has indicated its human friendliness, and ease of calibration and adaptation to new tasks (summaries involving new attributes of interest) and users (of a variable preparation, knowledge, flexibility, etc.). However, after some time of intensive use, the user has expressed his intention to go beyond data from the own database of a company, and use some external data We have extended the class of linguistic summaries handled by the system to include those that take into

Table 5. Linguistic summaries expressing relations between the attributes: size of customer, regularity of customer (purchasing frequency), date of sale, time of sale, commission, group of product and day of sale

Summary
Much sales on Saturday is about noon with a low commission
Much sales on Saturday is about noon for bigger customers
Much sales on Saturday is about noon
Much sales on Saturday is about noon for regular customers
A few sales for regular customers is with a low commission
A few sales for small customers is with a low commission
A few sales for one-time customers is with a low commission
Much sales for small customers is for nonregular customers

account data easily (freely) available from Internet sources, more specifically data on weather conditions as, first, they have an impact on the operation, and are easily and inexpensively available from the Internet.

For instance, if we are interested in relations between group of products, time of sale, temperature, precipitacion, and type of customers, the best linguistic summaries (of both our "internal" data from the sales database, and external meteorological data from an Internet service) are as shown in Table 6.

Table 6. Linguistic summaries expressing relations between the attributes: group of products, time of sale, temperature, precipitacion, and type of customers

Summary
Very few sales of software in hot days to individual customers
About 1/2 of sales of accessories in rainy days on weekends by the end of the year
About 1/3 of sales of computers in rainy days to individual customers

Notice that the use of external data gives a new quality to possible linguistic summaries. It can be viewed as providing a greater adaptivity to varying conditions because the use of free or inexpensive data sources from the Internet makes it possible to easily and quickly adapt the form and contents of summaries to varying needs and interests. And this all is practically at no additional price and effort.

5 Concluding Remarks

We show how Zadeh's idea of computing with words and perceptions, based on his concept of a precisiated natural language (PNL), can lead to a new direction in the use of natural language in data mining, linguistic data(base) summaries. We emphasize the relevance of Zadeh's another idea, that of a protoform, and show that various types of linguistic data summaries may be viewed as items in a hierarchy of protoforms of summaries. We briefly present an implementation for a sales database of a computer retailer as a convincing example that these tools and techniques are implemenatnle and functional. These summaries involve both data from an internat database of the company and data downloaded from external databases via the Internet.

References

1. R. Agrawal and R. Srikant. Fast algorithms for mining association rules. Proceedings of the 20th International Conference on Very Large Databases, Santiago de Chile, 1994.
2. W.-H. Au and K.C.C. Chan. FARM: A data mining system for discovering fuzzy association rules. Proceedings of the 8th IEEE International Conference on Fuzzy Systems, pp. 1217 - 1222, Seoul, Korea, 1999.
3. P. Bosc, D. Dubois, O. Pivert, H. Prade and M. de Calmes. Fuzzy summarization of data using fuzzy cardinalities. Proceedings of IPMU 2002, pp. 1553 - 1559, Annecy, France, 2002.
4. G. Chen, Q. Wei and E. Kerre. Fuzzy data mining: discovery of fuzzy generalized association rules. In G. Bordogna and G. Pasi (Eds.): Recent Issues on Fuzzy Databases, pp. 45 - 66. Springer-Verlag, Heidelberg and New York, 2000.
5. D. Dubois, H. Fargier and H. Prade. Beyond min aggregation in multicriteria decision: (ordered) weighted min, discri-min,leximin. In R.R. Yager and J. Kacprzyk (Eds.): The Ordered Weighted Averaging Operators. Theory and Applications, pp. 181 - 192, Kluwer, Boston, 1997.
6. D. Dubois and H. Prade. Gradual rules in approximate reasoning. Information Sciences, 61, 103 - 122, 1992.
7. D. Dubois and H. Prade. Fuzzy sets in approximate reasoning, Part 1: Inference with possibility distributions. Fuzzy Sets and Systems, 40, 143 - 202, 1991.
8. P. Bosc, D. Dubois and H. Prade. Fuzzy functional dependencies – an overview and a critical discussion. Proceedings of 3rd IEEE International Conference on Fuzzy Systems, pp. 325 - 330, Orlando, USA, 1994.
9. R. George and Srikanth R. Data summarization using genetic algorithms and fuzzy logic. In F. Herrera and J.L. Verdegay (Eds.): Genetic Algorithms and Soft Computing, pp. 599 - 611, Springer-Verlag, Heidelberg, 1996.
10. R. George and R. Srikanth. A soft computing approach to intensional answering in databases. Information Sciences, 92, 313 - 328, 1996.
11. I. Glöckner. Fuzzy quantifiers, multiple variable binding, and branching quantification. In T.Bilgi c *et al.* IFSA 2003. LNAI 2715, pp. 135 - 142, Springer-Verlag, Berlin and Heidelberg, 2003.
12. P. Hájek, M. Holeňa. Formal logics of discovery and hypothesis formation by machine. Theoretical Computer Science, 292, 345 357, 2003.
13. Y.-Ch. Hu, R.-Sh. Chen and G.-H. Tzeng. Mining fuzzy association rules for classification problems. Computers and Industrial Engineering, 43, 735 750, 2002.

14. J. Kacprzyk and P. Strykowski. Linguistic data summaries for intelligent decision support. In R. Felix (Ed.): Fuzzy Decision Analysis and Recognition Technology for Management, Planning and Optimization - Proceedings of EFDAN'99, pp. 3 - 12, Dortmund, Germany, 1999.
15. J. Kacprzyk and P. Strykowski. Linguitic summaries of sales data at a computer retailer: a case study. Proceedings of IFSA'99, pp. 29 - 33, Taipei, Taiwan R.O.C, vol. 1, 1999.
16. J. Kacprzyk and R.R. Yager. Linguistic summaries of data using fuzzy logic. International Journal of General Systems, 30, 33 - 154, 2001.
17. J. Kacprzyk, R.R. Yager and S. Zadrożny. A fuzzy logic based approach to linguistic summaries of databases. International Journal of Applied Mathematics and Computer Science, 10, 813 - 834, 2000.
18. J. Kacprzyk, R.R. Yager and S. Zadrożny. Fuzzy linguistic summaries of databases for an efficient business data analysis and decision support. In W. Abramowicz and J. Zurada (Eds.): Knowledge Discovery for Business Information Systems, pp. 129-152, Kluwer, Boston, 2001.
19. J. Kacprzyk and S. Zadrożny. FQUERY for Access: fuzzy querying for a Windows-based DBMS. In P. Bosc and J. Kacprzyk (Eds.): Fuzziness in Database Management Systems, pp. 415-433, Springer-Verlag, Heidelberg, 1995.
20. J. Kacprzyk and S. Zadrożny. Protoforms of linguistic data summaries: towards more general natural-language-based data mining tools. In A. Abraham, J. Ruiz-del-Solar, M. Koeppen (Eds.): Soft Computing Systems, pp. 417 - 425, IOS Press, Amsterdam, 2002.
21. J. Kacprzyk and S. Zadrożny. Data Mining via Linguistic Summaries of Data: An Interactive Approach. In T. Yamakawa and G. Matsumoto (Eds.): Methodologies for the Conception, Design and Application of Soft Computing. Proc. of IIZUKA98, pp. 668 - 671, Iizuka, Japan, 1998.
22. J. Kacprzyk and S. Zadrożny. The paradigm of computing with words in intelligent database querying. In L.A. Zadeh and J. Kacprzyk (Eds.): Computing with Words in Information/Intelligent Systems. Part 2. Foundations, pp. 382 - 398, Springer–Verlag, Heidelberg and New York, 1999.
23. J. Kacprzyk and S. Zadrożny. Computing with words: towards a new generation of linguistic querying and summarization of databases. In P. Sinčak and J. Vaščak (Eds.): Quo Vadis Computational Intelligence?, pp. 144 - 175, Springer-Verlag, Heidelberg and New York, 2000.
24. J. Kacprzyk and S. Zadrożny. On a fuzzy querying and data mining interface, Kybernetika, 36, 657 - 670, 2000.
25. J. Kacprzyk J. and S. Zadrożny. On combining intelligent querying and data mining using fuzzy logic concepts. In G. Bordogna and G. Pasi (Eds.): Recent Research Issues on the Management of Fuzziness in Databases, pp. 67 - 81, Springer–Verlag, Heidelberg and New York, 2000.
26. J. Kacprzyk and S. Zadrożny. Data mining via linguistic summaries of databases: an interactive approach. In L. Ding (Ed.): A New Paradigm of Knowledge Engineering by Soft Computing, pp. 325-345, World Scientific, Singapore, 2001.
27. J. Kacprzyk and S. Zadrożny. Computing with words in intelligent database querying: standalone and Internet-based applications. Information Sciences, 134, 71 - 109, 2001.
28. J. Kacprzyk and S. Zadrożny. On linguistic approaches in flexible querying and mining of association rules. In H.L. Larsen, J. Kacprzyk, S. Zadrożny, T. An-

dreasen and H. Christiansen (Eds.): Flexible Query Answering Systems. Recent Advances, pp. 475 - 484, Springer-Verlag, Heidelberg and New York, 2001.
29. J. Kacprzyk and S. Zadrożny. Linguistic summarization of data sets using association rules. Proceedings of The IEEE International Conference on Fuzzy Systems, pp. 702 - 707, St. Louis, USA, 2003.
30. J.-H. Lee and H. Lee-Kwang. An extension of association rules using fuzzy sets. Proceedings of the Seventh IFSA World Congress, pp. 399 - 402, Prague, Czech Republic, 1997.
31. Y. Liu and E.E. Kerre. An overview of fuzzy quantifiers. (I). Interpretations. Fuzzy Sets and Systems, 95, 1 - 21, 1998.
32. H. Mannila, H. Toivonen and A.I. Verkamo. Efficient algorithms for discovering association rules. In U.M. Fayyad and R. Uthurusamy (Eds.): Proceedings of the AAAI Workshop on Knowledge Discovery in Databases, pp. 181 - 192, Seattle, USA, 1994.
33. G. Raschia and N. Mouaddib. SAINTETIQ: a fuzzy set-based approach to database summarization. Fuzzy Sets and Systems, 129, 137 - 162, 2002.
34. D. Rasmussen and R.R. Yager. Fuzzy query language for hypothesis evaluation. In Andreasen T., H. Christiansen and H. L. Larsen (Eds.): Flexible Query Answering Systems, pp. 23 - 43, Kluwer, Boston, 1997.
35. D. Rasmussen and R.R. Yager. Finding fuzzy and gradual functional dep

Enhancing the Power of Search Engines and Navigations Based on Conceptual Model: Web Intelligence

Masoud Nikravesh[(1)], Tomohiro Takagi[(2)], Masanori Tajima[(2)], Akiyoshi Shinmura[(2)], Ryosuke Ohgaya[(2)], Koji Taniguchi[(2)], Kazuyosi Kawahara[(2)], Kouta Fukano[(2)], and Akiko Aizawa[(3)]

[(1)]BISC Program, EECS Department-CS Division
University of California, Berkeley, CA 94720
Nikravesh@cs.berkeley.edu
[(2)] Dept. of Computer Science, Meiji University.
[(3)]National Institute of Informatics

Abstract: Retrieving relevant information is a crucial component of cased-based reasoning systems for Internet applications such as search engines. The task is to use user-defined queries to retrieve useful information according to certain measures. Even though techniques exist for locating exact matches, finding relevant partial matches might be a problem. It may not be also easy to specify query requests precisely and completely - resulting in a situation known as a fuzzy-querying. It is usually not a problem for small domains, but for large repositories such as World Wide Web, a request specification becomes a bottleneck. Thus, a flexible retrieval algorithm is required, allowing for imprecise or fuzzy query specification or search. In this chapter, first we will present the role of the fuzzy logic in the Internet. Then we will present an intelligent model that can mine the Internet to conceptually match and rank homepages based on predefined linguistic formulations and rules defined by experts or based on a set of known homepages. The Fuzzy Conceptual Matching (FCM) model will be used for intelligent information and knowledge retrieval through conceptual matching of both text and images (here defined as "Concept"). The FCM can also be used for constructing fuzzy ontology or terms related to the context of the query and search to resolve the ambiguity. This model can be used to calculate conceptually the degree of match to the object or query. We will also present the integration of our technology into commercial search engines such as Google ™ and Yahoo! as a framework that can be used to integrate our model into any other commercial search engines, or development of the next generation of search engines.

1 Introduction

World Wide Web search engines have become the most heavily-used online services, with millions of searches performed each day. Their popularity is due, in part, to their ease of use. The central tasks for the most of the search engines can be summarize as 1) query or user information request- do what I mean and not what I say!, 2) model for the Internet, Web representation-web page collection, documents, text, images, music, etc, and 3) ranking or matching function-degree of relevance, recall, precision, similarity, etc.

Design of any new intelligent search engine should be at least based on two main motivations (Zadeh 2001a and 2002):

i· The web environment is, for the most part, unstructured and imprecise. To deal with information in the web environment what is needed is a logic that supports modes of reasoning which are approximate rather than exact. While searches may retrieve thousands of hits, finding decision-relevant and query-relevant information in an imprecise environment is a challenging problem, which has to be addressed.

ii· Another, and less obvious, is deduction in an unstructured and imprecise environment given the huge stream of complex information.

During the recent years, applications of fuzzy logic and the Internet from Web data mining to intelligent search engine and agents for Internet applications have greatly increased (Nikravesh and Azvine, 2001). Martin (2001) concluded that semantic web includes many aspects, which require fuzzy knowledge representation and reasoning. This includes the fuzzification and matching of concepts. In addition, it is concluded that fuzzy logic can be used in making useful, human-understandable, deduction from semi-structured information available in the web. It is also presented issues related to knowledge representation focusing on the process of fuzzy matching within graph structure. This includes knowledge representation based on conceptual graphs and Fril++. Baldwin and Morton (1985) studied the use of fuzzy logic in conceptual graph framework. Ho (1994) also used fuzzy conceptual graph to be implemented in the machine-learning framework. Baldwin (2001) presented the basic concept of fuzzy Bayesian Nets for user modeling, message filtering and data mining. For message filtering the protoype model representation has been used. Given a context, prototypes represent different types of people and can be modeled using fuzzy rules, fuzzy decision tree, fuzzy Bayesian Net or a fuzzy conceptual graph. In their study, fuzzy set has been used for better generalization. It has been also concluded that the new approach has many applications. For example, it can be used for personalization of web pages, intelligent filtering of the Emails, providing TV programs, books or movie and video of interest. Cao (2001) presented the fuzzy conceptual graphs for the semantic web. It is concluded that the use of conceptual graph and fuzzy logic is complementary for the semantic web. While conceptual graph provide a structure for natural language sentence, fuzzy logic provide a methodology for computing with words. It has been concluded that fuzzy conceptual graphs is suitable language for knowledge representation to be used by Semantic web. Takagi and Tajima (2001) presented the conceptual matching of text notes to be used by search engines. An new search engine proposed which conceptually matches keywords and the web pages. Conceptual fuzzy set has been used for context-dependent keyword expansion. A new structure for search engine has been proposed which can resolve the context-dependent word ambiguity using fuzzy conceptual matching technique. Berenji (2001) used Fuzzy Reinforcement Learning (FRL) for text data mining and Internet search engine. Choi (2001) presented a new technique, which integrates document index with perception index. The techniques can be used for refinement of fuzzy queries on the Internet. It has been concluded that the use of perception index in commercial search engine provides a framework to handle fuzzy terms (perception-based), which is further step toward a human-friendly, natural language-based interface for the Internet. Sanchez (2001) presented the concept of Internet-based fuzzy Telerobotic for the WWW. The system receives the information from human and has the capability for fuzzy reasoning. It has be proposed to use fuzzy applets such as fuzzy logic propositions in the form of fuzzy rules that can be used for smart data base search. Bautista and Kraft (2001) presented an approach to use fuzzy logic for user profiling in Web retrieval applications. The technique can be used to expand the queries and knowledge extraction related to a group of users with common interest. Fuzzy representation of terms based on linguistic qualifiers has been used for their study. In addition, fuzzy clustering of the user profiles can be used to construct fuzzy rules and inferences in order to modify queries. The result can be used for knowledge extraction from user profiles for marketing purposes. Yager (2001) introduced fuzzy aggregation methods for intelligent search. It is concluded that the new technique can increase the expressiveness in the queries. Widyantoro and Yen (2001) proposed the use of fuzzy ontology in search engines. Fuzzy ontology of term relations can be built automatically from a collection of documents. The proposed fuzzy ontology can be used for query refinement and to suggest narrower and broader terms suggestions during user search activity. Presser (2001) introduced fuzzy logic for rule-based personalization and can be implemented for personalization of newsletters. It is concluded that the use of fuzzy logic provide better flexibility and better interpretation which helps in keeping the knowledge bases easy to maintain. Zhang et al. (2001a) presented granular fuzzy technique for web search engine to increase Internet search speed and the Internet quality of service. The techniques can be used for personalized fuzzy web search engine, the personalized granular web search agent. While current fuzzy search engines uses keywords, the proposed technique provide a framework to not only use traditional fuzzy-key-word but also fuzzy-user-preference-based search algorithm. It is concluded that the proposed model reduces web search redundancy, increase web search relevancy, and decrease user's web search time. Zhang et al. (2001b) proposed fuzzy neural web agents based on granular neural network, which discovers fuzzy rules for stock prediction. Fuzzy logic can be used for web mining. Pal et al. (2002) presented issues related to web mining using soft computing framework. The main tasks of web mining based on fuzzy logic include information retrieval and generalization. Krisnapuram et al. (1999) used fuzzy c medoids and triimed medoids for clustering of web documents. Joshi and Krisnapuram (1998) used fuzzy clustering for web log data mining. Sharestani (2001) presented the use of fuzzy logic for network intruder detection. It is concluded that fuzzy logic can be used for approximate reasoning and handling detection of intruders through approximate matching; fuzzy rule and summarizing the audit log data. Serrano (2001) presented a web-based intelligent assistance. The model is an agent-based system which uses a knowledge-based model of the e-business that provide advise to user through intelligent reasoning and dialogue evolution. The main advantage of this system is based on the human-computer understanding and expression capabilities, which generate the right information in the right time.

In our perspective, one can use clarification dialog, user profile, context, and ontology, into an integrated frame work to design a more intelligent search engine. The model will be used for intelligent information and knowledge retrieval through conceptual matching of text. The selected query doesn't need to match the decision criteria exactly, which gives the system a more human-like behavior. The model can also be used for constructing ontology or terms related to the context of search or query to resolve the ambiguity. The new model can execute conceptual matching dealing with context-dependent word ambiguity and produce results in a format that permits the user to interact dynamically to customize and personalized its search strategy.

It is also possible to automate ontology generation and document indexing using the terms similarity based on Conceptual-Latent Semantic Indexing Technique (CLSI). Often time it is hard to find the "right" term and even in some cases the term does not exist.

The ontology is automatically constructed from text document collection and can be used for query refinement. It is also possible to generate conceptual documents similarity map that can be used for intelligent search engine based on CLSI, personalization and user profiling. The user profile is automatically constructed from text document collection and can be used for query refinement and provide suggestions and for ranking the information based on pre-existence user profile.

Given the ambiguity and imprecision of the "concept" in the Internet, which may be described by both textual and image information, the use of Fuzzy Conceptual Matching (FCM) is a necessity for search engines. In the FCM approach, the "concept" is defined by a series of keywords with different weights depending on the importance of each keyword. Ambiguity in concepts can be defined by a set of imprecise concepts. Each imprecise concept in fact can be defined by a set of fuzzy concepts. The fuzzy concepts can then be related to a set of imprecise words given the context. Imprecise words can then be translated into precise words given the ontology and ambiguity resolution through clarification dialog. By constructing the ontology and fine-tuning the strength of links (weights), we could construct a fuzzy set to integrate piecewise the imprecise concepts and precise words to define the ambiguous concept.

Currently on the Internet there exists a host of illegal web sites which specialize in the distribution of commercial software and music. This chapter proposes a method to distinguish illegal web sites from legal ones not only by using tf-idf values but also to recognize the purpose/meaning of the web sites. It is achieved by describing what are considered to be illegal sites and by judging whether the objective web sites match the description of illegality. Conceptual fuzzy sets (CFSs) are used to describe the concept of illegal web sites. First, we introduced the usefulness of CFSs in overcoming those problems, and propose the realization of CFSs using RBF-like networks. In a CFS, the meaning of a concept is represented by the distribution of the activation values of the other nodes. Because the distribution changes depend on which labels are activated as a result of the conditions, the activations show a context-dependent meaning. Next, we proposed the architecture of the filtering system. Finally, we compared the proposed method with the tf-idf method with the support vector machine. The e-measures as a total evaluation indicate that the proposed system showed better results as compared to the tf-idf method with the support vector machine.

In addition, we propose a menu navigation system which conceptually matches input keywords and paths. For conceptual matching, we use conceptual fuzzy sets (CFSs) based on radial basis function (RBF) networks. In a CFS, the meaning of a concept is represented by the distribution of the activation values of the other concepts. To expand input keywords, the propagation of activation values is carried out recursively. The proposed system recommends users paths to appropriate categories. We use 3D user interface to navigate users.

2 Fuzzy Conceptual Model and Search Engine

The Conceptual Fuzzy Set (CFS) model will be used for intelligent information and knowledge retrieval through conceptual matching of both text and images (here defined as "Concept"). The CFS can also be used for constructing fuzzy ontology or terms related to the context of search or query to resolve the ambiguity. It is intended to combine the expert knowledge with soft computing tool. Expert knowledge needs to be partially converted into artificial intelligence that can better handle the huge information stream. In addition, sophisticated management workflow needs to be designed to make optimal use of this information. In this Chapter, we present the foundation of CFS-Based Intelligent Model and its applications to both information filtering and design of navigation.

2.1 Search Engine based on Conceptual Matching of Text Notes

Information retrieval in the Internet is generally done by using keyword matching, which requires that for words to match, they must be the same or synonyms. But essentially, not only the information that matches the keywords exactly, but also information related in meaning to the input keywords should be retrieved. The following reasons are why fuzzy sets are essential for information retrieval.

First, a fuzzy set is defined by enumerating its elements and the degree of membership of each element. It is useful for retrieving information which includes not only the keyword, but also elements of the fuzzy set labeled by the input keyword. For example, a search engine may use baseball, diving, skiing, etc., as kinds of sports, when a user inputs "sports" as the keyword.

Second, the same word can have various meanings. Several words are used concurrently in usual sentences, but each word has multiple possible meanings (region), so we suppose an appropriate context which suits all regions of meaning of all words (**Figure 1**). At the same time, the context determines the meaning of each word.

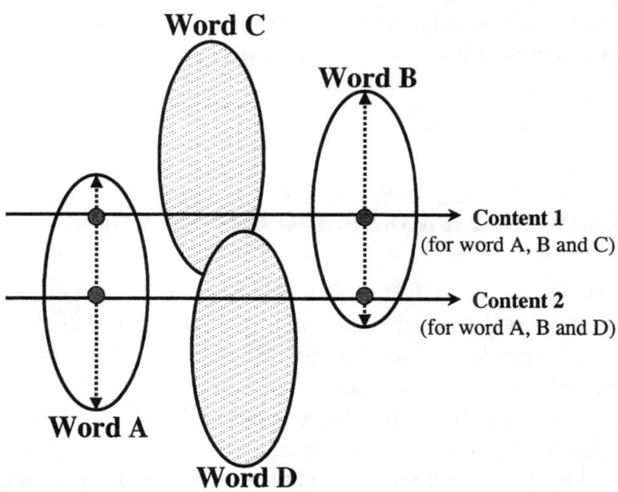

Figure 1. Meanings of words determined by a context.

For example, "sports" may mean "diving" or "sailing" when it is used with "marine," and may mean "baseball" or "basketball" when used with "TV programs." That is, each possibility distribution of meaning is considered as a fuzzy set itself. For information retrieval, keyword expansion that considers context is necessary, because simple expansion of a possible region causes a flood of words. For example, even if the user intends "marine sports," the set of expanded keywords includes unnecessary sports such as "baseball." However, an ordinary fuzzy set does not provide us the method to deal with context-dependent word ambiguity. To overcome this problem, we previously proposed using conceptual fuzzy sets (CFSs) (Takagi et al. 1995, 1996, 1999a and 1999b), which conform to Wittgenstein's concept, to represent the meanings of concepts.

In this section, we propose a search engine which conceptually matches input keywords and Web pages. The conceptual matching is attained by context dependent keyword expansion using conceptual fuzzy sets. We describe the necessity of conceptual fuzzy sets for information retrieval in Section 2.1.1, and propose the use of conceptual fuzzy sets using Hopfield Networks in section 2.1.2. Section 2.1.3 proposes the search engine which can execute conceptual matching and deal with context-dependent word ambiguity. In Section 2.1.4, we show two simulations of retrieving actual Web pages comparing the proposed method with the ordinary TF-IDF method. In section 2.1.5, we provide the summary.

2.1.1 Fuzzy Sets and Context Dependent Word Ambiguity

In this section we will present the context dependent word ambiguity and how to resolve the issue.

2.1.1.1 Conceptual Fuzzy Sets (Takagi et al. 1995, 1996, 1999a and 1999b)

Let's think about the meaning of "heavy." A person weighting 100kg would usually be considered heavy. But there is no clear boundary between "heavy" and "not heavy." Fuzzy sets are generally used to indicate these regions. That is, we have a problem of specificity.

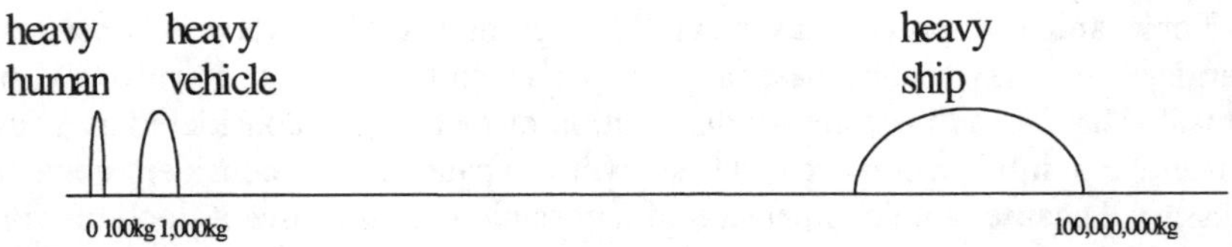

Figure 2. The meaning of "heavy."

Let's think about it some more. For a vehicle, "heavy" might be several thousand kgs. For a ship, it might be more than ten thousand tons. Therefore, the item "heavy" being judged affects the vagueness of the meaning of "heavy" much more than the non-specificity of the amount when the item is already determined as shown in **Figure 2.** Moreover, "work" can be heavy, "traffic" can be heavy, and "smoking" can be heavy. So the meaning of "heavy" changes with the context, which results in the vagueness of the meaning.

That is, the main cause of vagueness is ambiguity in the language, not specificity. Ordinary fuzzy set theory has not dealt with the context dependent meaning representation concerning language ambiguity. However, as we mentioned in the Introduction, a fuzzy set is defined by enumerating its elements and the degree of membership of each element, we can use it to express word ambiguity by enumerating all possible meanings of a word, then estimating the degrees of compatibilities between the word and the meanings. Fuzzy set theory should therefore deal with language ambiguity as the main cause of vagueness.

To overcome this problem, we previously proposed using conceptual fuzzy sets. Although several works have been published, we will explain CFSs for understanding the following section. According to Wittgenstein (1953), the meaning of a concept can be represented by the totality of its uses. In this spirit, conceptual fuzzy sets, in which a concept is represented by the distribution of the activation concepts, are proposed.

The label of a fuzzy set represents the name of a concept, and a fuzzy set represents the meaning of the concept. Therefore, the shape of a fuzzy set is determined by the meaning of the label, which depends on the situation (Figure 3).

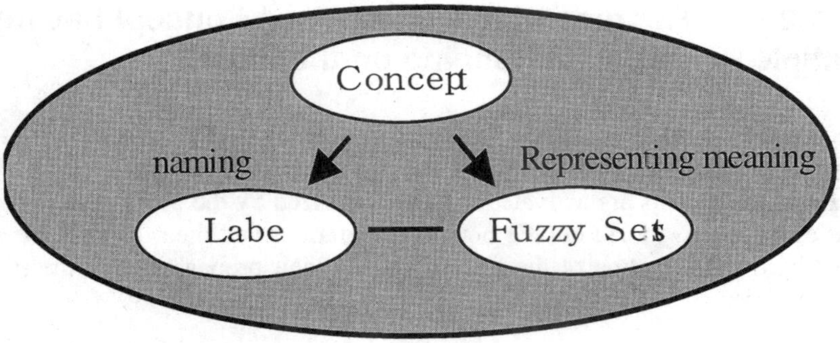

Figure 3. A fuzzy set as a meaning representation.

According to the theory of "meaning representation from use" proposed by Wittgenstein) the various meanings of a label (word) can be represented by other labels (words), and we can assign grades of activation showing the degree of compatibility between labels.

A CFS achieves this by using distributions of activations. A CFS is realized as an associative memory in which a node represents a concept and a link represents the strength of the relation between two (connected) concepts. The activation values agreeing with the grades of membership are determined through this associative memory. In a CFS, the meaning of a concept is represented by the distribution of the activation values of the other nodes. The distribution evolves from the activation of the node representing the concept of interest. The image of a CFS is shown in **Figure 4**.

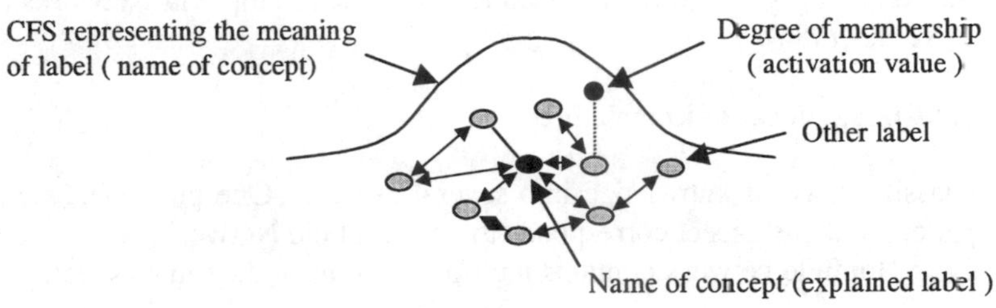

Figure 4. A conceptual fuzzy set represented by associative memories.

2.1.1.2 CFS Representing a Composed Concept having Multiple Meanings Depending on the Situation

Because the distribution changes depending on which labels are activated as a result of the conditions the activations show a context-dependent meaning. When more than two labels are activated, a CFS is realized by the overlapping propagations of their activations. In CFS notation, operations and their controls are all realized by the distribution of the activations and their propagations in the associative memories.

We can say that the distribution determined by the activation of a label agrees with the region of thought corresponding to the word expressing its meaning. The distribution (meaning of a label), that is a figure of a fuzzy set, changes depending on considered aspects that reflect the context.

2.1.2 Creating of Conceptual Fuzzy Sets

Previously we used bidirectional associative memories (BAMs) (Kasko 1987 and 1992) to generate CFSs, because of the clarity of the constraints used for their utilization. In this paper, we use Hopfield Networks, whose output can be also used with a continuous value, to overcome the limitation of BAMs that are a layered neural network. We do so because in a correlative neural network, relationships between concepts may not be a layered structure.

The following shows how to construct CFSs using Hopfield Networks (Hopfield 1982 and 1984).

Memorizing pieces of knowledge:

1. Classify piece of knowledge into several aspects. One piece becomes one pattern and one aspect corresponds to one Hopfield Network.
2. Each Hopfield network contains multiple patterns in the same aspect.

Generating CFSs:

1. Recollect patterns which include a node of interest in each Hopfield Network.
2. Sum all recollected patterns and normalize the activation values.

Figure 5 shows the image of memorized patterns and a generated CFS.

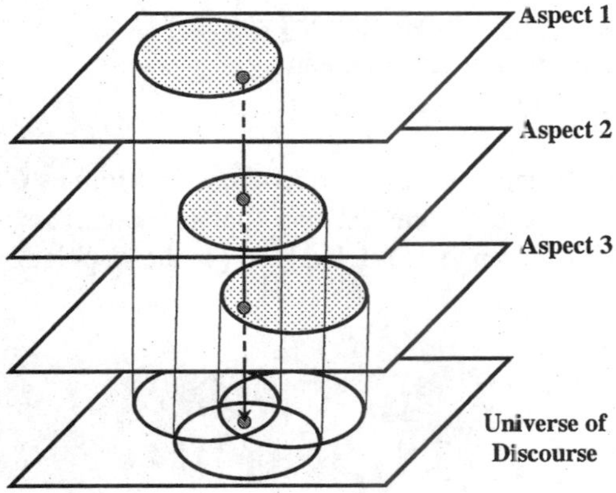

Figure 5. Image of memorized patterns and a generated CFS.

Logic-based paradigms for knowledge representation use symbolic processing both for concept representation and inference. Their underlying assumption is that a concept can be defined precisely. This causes an exponential increase in the number of cases, because we have to memorize all cases according to every possible context. In contrast, knowledge representation using CFSs memorizes knowledge about generalizations instead of exact cases (**Figure 5**). The following is an example to compare the proposed knowledge representation with ordinary logic-based paradigms. It shows that context-dependent meanings are generated by activating several keywords.

Example:

Let's think about the meaning of "heavy" again. The subject may be a human, a cat, or an elephant. Moreover, the age of the subject may be a baby or an adult, which also influences the meaning of "heavy." Therefore, since the number of cases increases exponentially as:

(cat, human, elephant......) *
(baby, adult, ...) * ....,

it is impossible to know how heavy the subjects are in all cases. On the other hand, using CFSs, which create meaning by overlapping activations, number of cases to be memorized becomes:

(cat, human, elephant......) +
(baby, adult.......) +...........,

and increases linearly.

Let's generate CFSs in these contexts. Assume the universe of discourse is "weight," from 0-1000 kg. Aspects and memorized patterns are as follows.

(Aspect)	(Memorized pattern)
kind	cat, human, elephant
age	baby, adult

<Step1> Memorize patterns such as those in **Figures 6** and **7** for each aspect. For example, the pattern "cat" shows that it memorizes its usual heavy weight within the activation range of [-1,1]. [-1,1] is the bi-polar expression of [0,1].

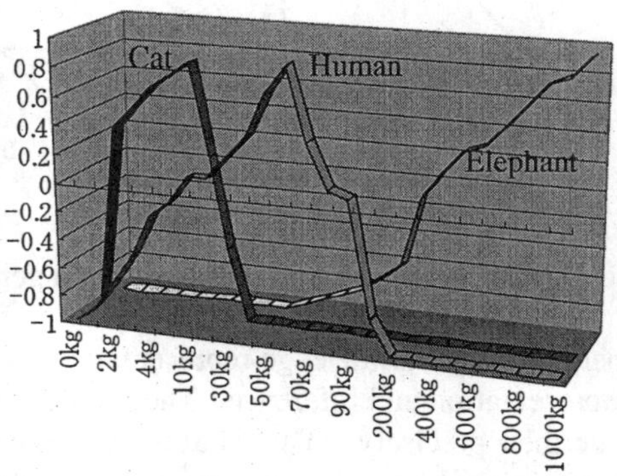

Figure 6. Memorized patterns of "heavy cat," "heavy human," and "heavy elephant"

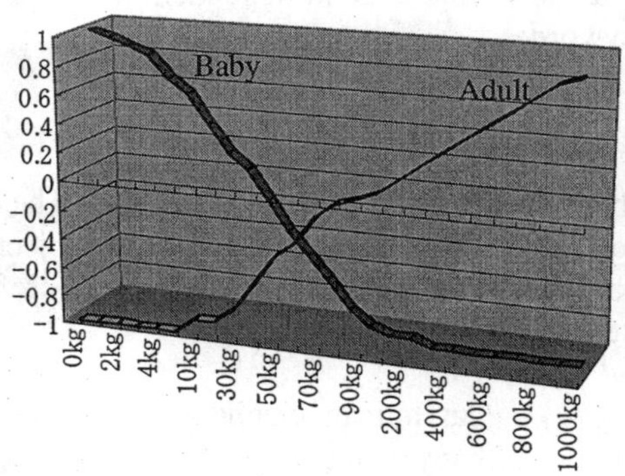

Figure 7. Memorized patterns of "heavy baby," and "heavy adult"

<Step2>When keywords are input, the input values of the neurons corresponding to these keywords are set as 1 and the input values of the other neurons are set as -1, and each Hopfield Network recollects the patterns.

<Step3> Finally, the activations of nodes in all aspects are summed up, and they are normalized in the range of [-1,1]. The normalized outputs become the final outputs result.

Figure 8 shows the ability of CFSs to generate context-dependent meanings of "heavy human" in the case of "adult" and "baby." We can recognize that both fuzzy sets have different shapes even when considering the same word "human." **Figure 9** compares the difference between the case of "human" and "elephant." Here, [a + b] means that the activation is started from the concepts in nodes "a "(ex: adult) and "b" (ex: elephant).

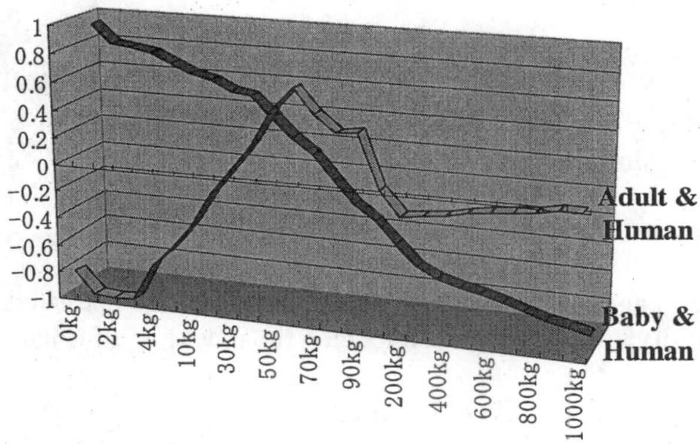

Figure 8. Example of outputs "heavy human."

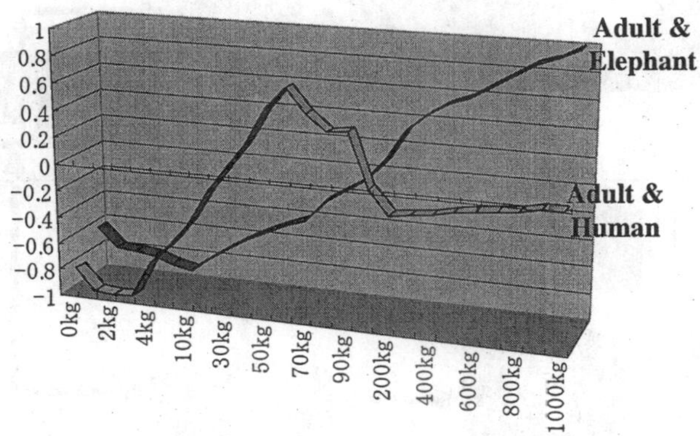

Figure 9. Example of outputs "heavy adult."

2.1.3 Conceptual Matching in a Search Engine Using CFS

In this section, we will focus on the use of CFS in search engines.

2.1.3.1 Scheme of Search Engine (Kobayashi and Takeda 2000, Quarino and Vetere 1999)

Usually, search engines work as follows: (We may want to add information about the search engines, my slides)

Index collecting of Web pages:

An indexer extracts words from Web pages, analyzes them, and stores the result as indexing information.

Retrieving information:

The Web pages, which include input keywords, are extracted. The pages are assigned priority and are sorted referencing the indexing information above.

As we mentioned earlier, information retrieval is generally done by using keyword matching, which requires words to match and is different from conceptual matching.

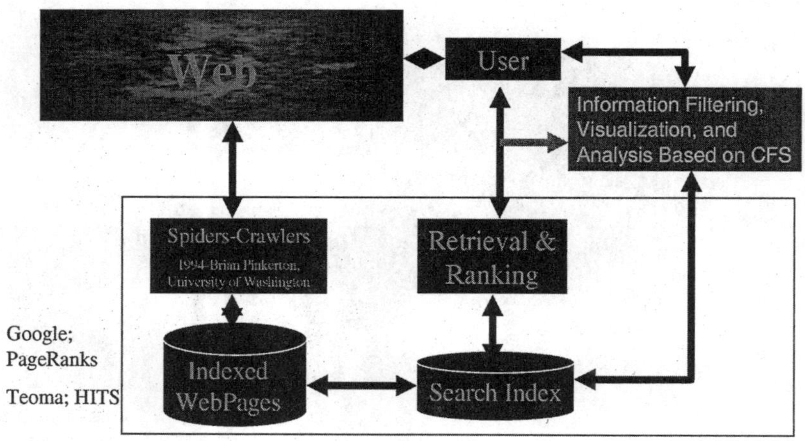

Figure 10. Search Engines Architecture

2.1.3.2 Conceptual Matching

We propose a search engine system which conceptually matches input keywords and Web pages according to the following idea.

1. Expand input keywords to the elements of CFSs.
2. Evaluate the matching degree between the set of expanded keywords and the set of words included in each Web page.
3. Sort the Web pages and display them according to the matching degrees.

The following shows the process in the proposed search engine.

Index collecting of Web pages:

1. Extract nouns and adjectives, and count the frequency of each word.
2. Calculate an evaluation of each word using the TF-IDF method for each Web page.
3. Store the evaluation into a lexicon.

Retrieve information:

1. A user inputs keywords into a browser, which transfers the keywords to a CFS unit.
2. Propagation of activation occurs from input keywords in the CFS unit. The meanings of the keywords are represented in other expanded words using conceptual fuzzy sets, and the activation value of each word is stored into the lexicon.
3. Matching is executed in the following process for each Web page. Obtain the final evaluation of each word by multiplying the evaluation by the TF-IDF method and the activation value. Sum up the final evaluations of all words and attach the result to each Web page as a matching degree.
4. The matched Web pages are sorted according to the matching degrees, and their addresses are returned to the browser with their matching degrees.

2.1.4 Simulations and Evaluations

Let's think about the case where we are searching for Web pages of places to visit using certain keywords, we indexed 200 actual Web pages, and compared the search result of the following two matching methods.

1. TF-IDF method
2. our proposed method (**Figure 11** using CFS)

Evaluation 1:

If the CFS unit has knowledge in fuzzy sets about places, and if a user inputs "famous resort" as a keyword, relating name of places are added as expanded keywords with their activation degrees agreeing with membership degrees.

Famous resort = 0.95/gold coast + 0.95/the Cote d'Azur + 0.91/Fiji + ..

Table 2.a shows the result when "famous resort" and "the Mediterranean Sea" are input as keywords. It consists of names of places and activation values, and shows the extended keywords generated by the activation of the above two keywords.

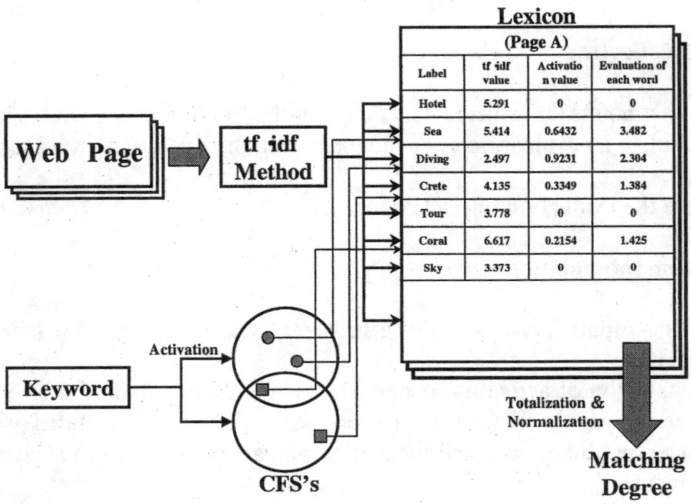

Figure 11. Scheme of the proposed search engine.

Table 2.a Extended keywords.

Ranking	Word	Activation Value
1	The Côte d'Azur	1.0000
2	The Mediterranean Sea	0.9773
3	Famous Resort	0.9187
4	Crete	0.8482
5	Capri	0.6445
6	Anguilla	0.6445
7	Santorini	0.6445
8	Taormina	0.6445
9	Sicily	0.4802
10	Gold Coast	0.0748

Next, abstracts of the retrieved Web pages are listed. Note that, no Web pages were matched by the simple TF-IDF method starting with the keyword input of "famous resort and the Mediterranean Sea (**Figure 12.a**).

Taormina: Ranking 1, Matching degree 1.0

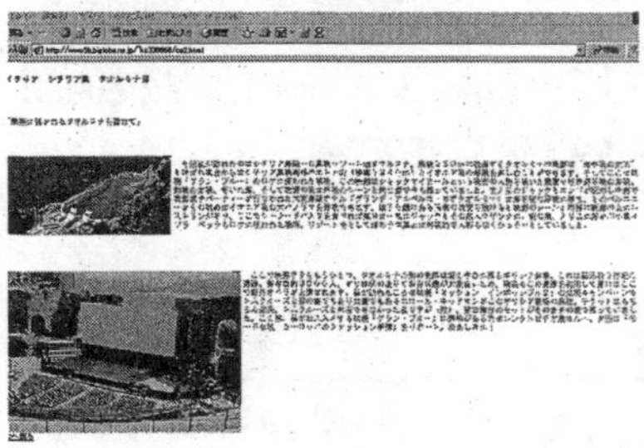

The greatest high-class resort on the island of Sicily. It is located 250 meters above sea level and is known as the "Mediterranean Queen." It has superb views of Mount Etna and the Ionian Sea.

Crete island: Ranking 2, Matching degree 0.83

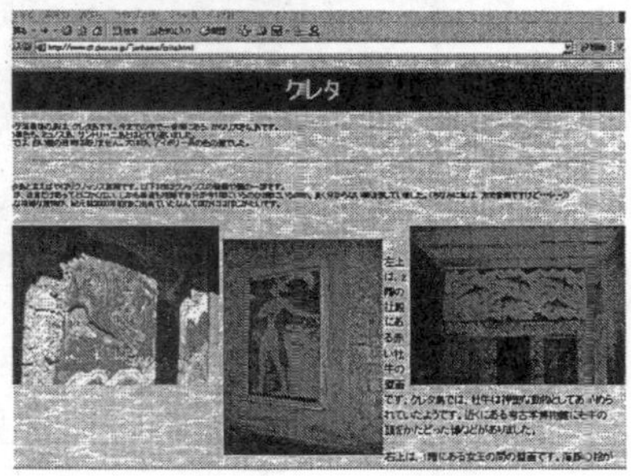

It is a big island and located in the south. The scenery is different from Mykonos and Santorini islands.

Cote d'Azur: Ranking 3, Matching degree 0.53

Deep-blue coast.

The above results show that our proposed method effectively retrieves information relating to input keywords even when there are no matches with the input keyword itself.

Figure 12.a.

Evaluation 2:

If the CFS unit memorizes knowledge about "vacation" and "sports" such as,

vacation = 1.0/vacance + 0.6/sea + 0.6/sandy beach + 0.6/the South Pacific + ..

sports = 1.0/spots + 0.6/diving + 0.6/trekking + 0.6/golf + ..

then a ranked list of Web pages appears. **Table 2** shows the extended key-words generated by the activation of "vacation" and "sports."

Table 2.b Extended keywords.

Ranking	Word	Activation Value
1	Diving	0.6079
1	Surfing	0.6079
3	Sports	0.5000
3	Vacation	0.5000
5	Golf	0.3039
5	Rock Climbing	0.3039
5	Baseball	0.3039
5	Sea	0.3039
5	Paradise	0.3039
5	Sandy Bearch	0.3039

Next, abstracts of the retrieved Web pages are listed. In contrast, no Web pages were matched by the TF-IDF method using "vacation and sports" (**Figure 12.b**).

Tahiti island: Ranking 1, Matching degree 1.00

Fun for jet skiing and surfing. Diving is also enjoyable.

Boracay island: Ranking 2, Matching degree 0.91

Diving boats and cruising boats come and go. Vacationers have to climb on the boat from the water because there are no piers. It exemplifies resort life.

Genuine diving! Even snorkeling and a glass-bottomed boat allow glimpses of its mystery.

Rangiroa island: Ranking 3, Matching degree 0.84

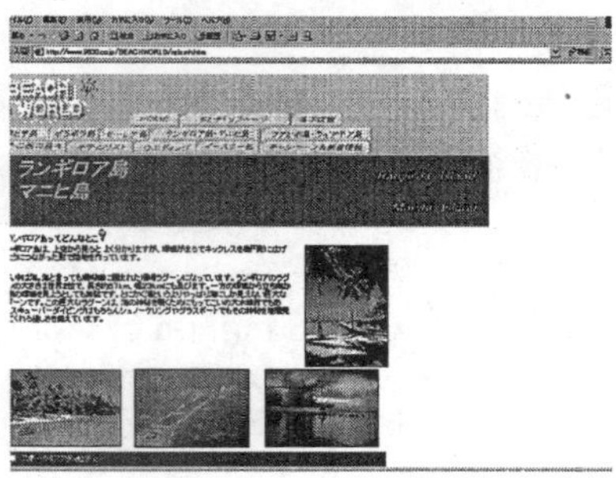

Figure 12.b.

From the results, we demonstrate the effectiveness of our proposed method. Unlike the first case, the Web pages were not retrieved by place names, but by the activities corresponding to the context of "vacation sports." Jet skiing and surfing were suggested by the CFS as a relevant sports, but baseball was not.

We show that pertinent Web pages can be retrieved independently of the key ward, because even though the region "sport" can encompass a huge number of different activities.

2.1.5. General Observations and Summaries

First, we showed the necessity and also the problems of applying fuzzy sets to information retrieval. Next, we introduced using conceptual fuzzy sets in overcoming those problems, and proposed the realization of conceptual fuzzy sets using Hopfield Networks. Based on above, we proposed the architecture of the search engine which can execute conceptual matching dealing with context-dependent word ambiguity. Finally, we evaluated our proposed method through two simulations of retrieving actual web pages, and compare the proposed method with the ordinary TF-IDF method. We showed that our method could correlate seemingly unrelated input keywords and produce matching Web pages, whereas the simple TF-IDF method could not.

3. Exposure of Illegal Web Sites Using : Conceptual Fuzzy Sets-Based Information Filtering System

Currently, about 1,600 million or more web pages exist on the Internet. People can obtain necessary information from this huge network quickly and easily. On the other hand, various problems have arisen. For example, there are the adult sites, the criminal sites which illegally distribute software (Warez) and music (MP3) and the criminal promotion sites which promote illegal behavior such as making a bomb etc. Therefore, one of the technologies needed currently is the filtering of web sites.

Some software are put to practical use as an internet filter. However, typical software simply match the web sites with illegal URL lists. This approach does not take their contents into consideration at all. These methods lack updating capabilities due to the drastic increase of illegal sites. Additionally, some software eliminates web sites that contain illegal words. They eliminate web sites by calculating confidence of the illegal words in documents by the TF-IDF method. The content-based approach using the TF-IDF method may eliminate any sites that contain harmful words. For example, news sites, which contain harmful words, may be eliminated. In this paper, we propose a filtering system that performs semantic analysis of a web document using conceptual fuzzy sets (CFSs). Our approach concerns not only the appearance of words in a document but also the meanings of words to recognize the harmful nature of a document.

We describe the construction of CFSs using Radial Basis Function (RBF)-like networks in section 3.1. Section 3.2 proposes the filtering system that deals with the semantics of words. Section 3.3 compares our proposed method with the TF-IDF method, and shows the result of filtering simulations of actual web pages. We will conclude the paper in section 3.4.

3.1 Construction of CFSs based on RBF networks

In this section, we will present the construction of CFSs based on RBF networks.

3.1.1 Construction of the network

In the CFSs, words may have a synonymous, antonymous, hypernymous and hyponymous relationship to other words. These relationships are too complicated to be represented in a hierarchical structure. Therefore, we use RBF-like networks to generate CFSs. The image of CFSs is shown in **Figure 13.**

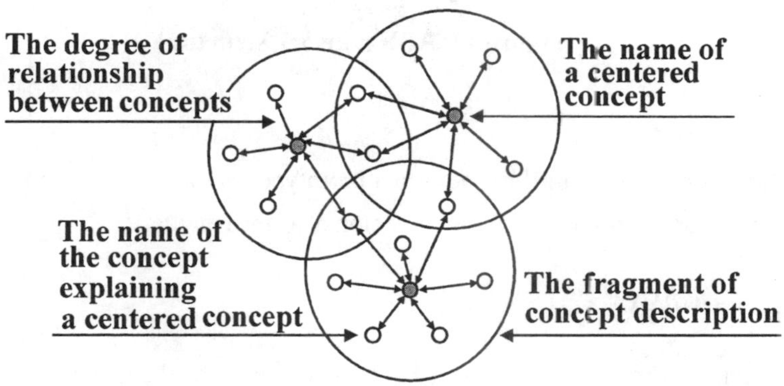

Figure 13. CFSs based on RBF networks

White surrounding concepts explain the centered gray concept. The strength of the links between concepts reflect their degrees of relationship. The centered concept and its connected concepts constitute a fragment of concept description. A CFS is generated by overlapping the fragments of the activated concept desctiption. A CFS expresses the meaning of a concept by the activation values of other concepts in these fragments.

3.1.2 Generation of CFSs

To generate CFSs, concepts are activated using the RBF networks as follows. In general, RBF networks have a structure shown in **Figure 14.**

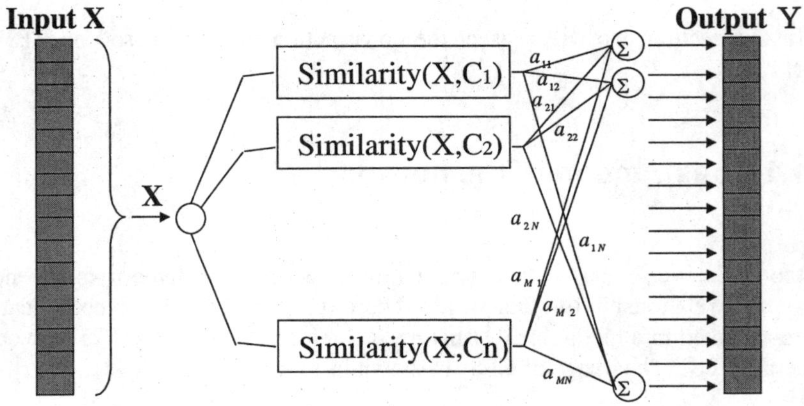

Figure 14. RBF network structure

1) The degree of relationship between a prototype vector c_i (i-th fragment of the concept description) and an input vector x is measured as,

$$\phi(dist(x, c_i))$$

and dist means the distance. Function φ translates the distance to the activation value of the prototype vector. Usually the distance is calculated by Euclidean distance.

$$dist(x, c_i) = \| x - c_i \|.$$

2) The activation values of prototype vectors are weighted with degrees of rela-

tionship a_{ij}, and propagate to the relating nodes. So the activation value propagated to j-th node from i-th prototype vector c_i becomes,

$$a_{ij} \times \phi(dist(x, c_i))$$

3) Each node in output layer sums up values translated from all prototype vectors as,

$$\sum_{i=1}^{M} a_{ij} \times \phi(dist(x, c_i))$$

3.2 System description

We developed a system filtering web pages using conceptual fuzzy sets based on RBF networks.

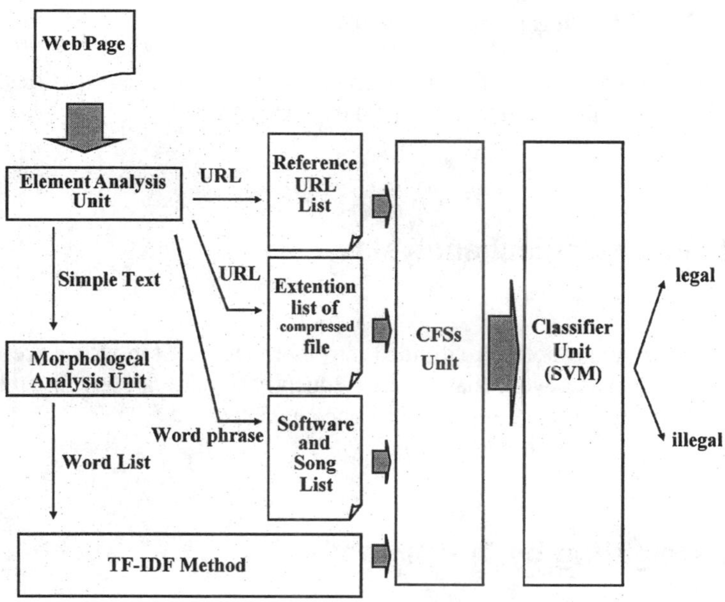

Figure 15. Filtering system using CFSs

3.2.1. Element analysis

In the element analysis unit following two rates and a matching degree are calculated and transferred to the CFSs unit.

- **Rate of links:**

$$\text{Rate of links to compressed files} = \frac{\text{Number of links to compressed files}}{\text{Total number of links}}$$

$$\text{Rate of links to major underground sites} = \frac{\text{Number of links to major illegal sites}}{\text{Total number of links}}$$

We deal with the links to compressed files and the links to major underground sites. It can be conjectured that web pages containing these links have high illegality.

- **Matching degree with name lists:**

The system matches words in the web site with the list of music titles and software names to evaluate tendency toward illegality.

3.2.2 Morphological analysis

The extracted text notes are divided into morphemes. Stop words are excluded. A stop word means a word that occurs frequently despite not having an important meaning and a word that consist of just one character.

3.2.3 Weighting by TF-IDF method

Weights of the words are obtained using general TF-IDF method.

3.2.4 CFSs Unit

A word vector, which consists of TF-IDF values, the link rates and the matching degree, is inputed into the CFSs unit. Propagation of activations occurs from input word vector throughout fragments of the concept descriptions and then abstract concepts "warez", "MP3" and "Emulator" are recognized. CFSs units consists of fragments of the concept descriptions, such as "warez", "MP3", etc as shown in **Figure 16.**

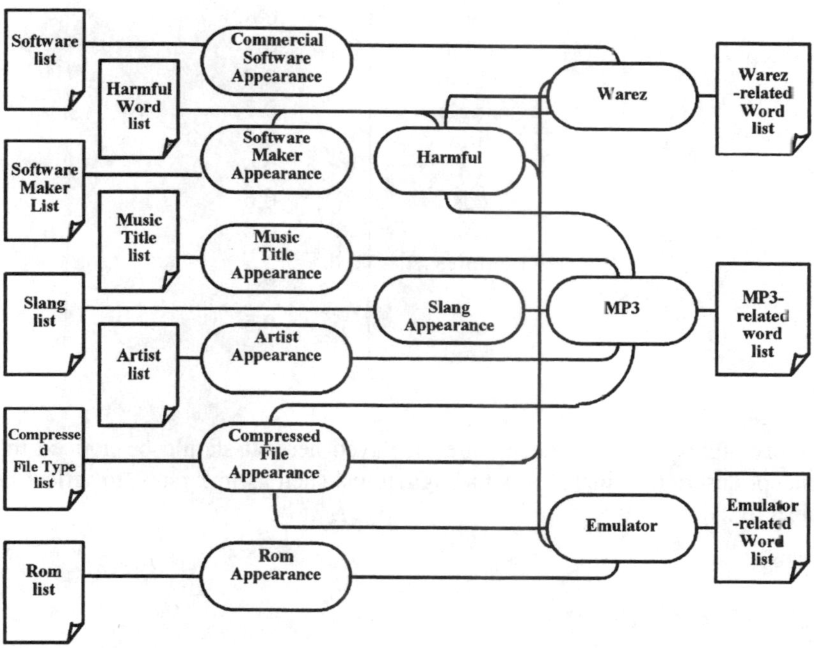

Figure 16. Concept description in the CFS unit

For example, the concept description of "MP3" is defined as shown in **Table 3.**

Table 3. Concept description of "MP3"

Concepts	Relationship
mp3	1
mp3s	0.9
mp3z	0.9
<song title>	0.8
<artist name>	0.8
napster	0.8
winmx	0.8
artist	0.8
song	0.8
music	0.7
album	0.7
single	0.7
title	0.6
cd	0.6
cds	0.6
cdz	0.6
<compressed file>	0.5
zip	0.5
band	0.5

Only highly ranked concepts are displayed here. It should be noticed that the concept description includes some signatures, such as link rates, to reflect human subjectivity.

3.2.5 Classifier Unit

We used the Support Vector Machine (SVM) as a classifier. The classifier unit inputs the activation values of words in the CFSs unit and distinguishes illegal sites from legal ones.

3.3 Evaluations Procedure

In this study, we randomly selected 300 actual web sites as samples for evaluation, and we compared the proposed method with the support vector machine. The samples contained 85 illegal sites. We assumed seven types of the illegal sites shown in **Table 4.** This classification is not based on the law strictly but on common sense. We evaluated the system by filtering Warez, Emulator and MP3 sites in this study.

Table 4. Seven types of illegal sites

Group	A classification criterion
Warez	Illegal distribution and sale of commercial softwares
Emulater	Illegal distribution of softwares, such as consumer games and video games
MP3	Disutribution of music data which infringe on copyright
Adult	Dirty depictions and expressions
Hack & Crack	Distribution of illegal hacking and cracking softwares Instruction of technical know-how
Drug & Gun	Sale of drugs and guns Introduction of acquisition route
Killing	Expressions about murder, violete depiction, etc.

3.3.1 Results

The results of the experiments are shown in **Table 5** and **Table 6.**

Table 5. The classification results

	proposed system		TF-IDF method	
	success	failure	success	failure
illegal document	81	4	74	11
legal document	214	1	215	0
all document	295	5	289	11

Table 6. Comparison by E measure

		300 documents	
		proposed system	TF-IDF method
illegal document	precision	0.9878	1.0000
	recall	0.9529	0.8706
	E measure	0.0299	0.0692
ligal document	precision	0.9817	0.9556
	recall	0.9953	1.0000
	E measure	0.0115	0.0227

Table 6 compares the experimental results from the viewpoint of precision, recall and E measure. The proposed method exceeded simple TF-IDF method in finding illegal sites, although it is inferior in the case of legal sites.

Here is an example (in **Figure 17**) which is distinguished correctly by the proposed system but not by simple TF-IDF method.

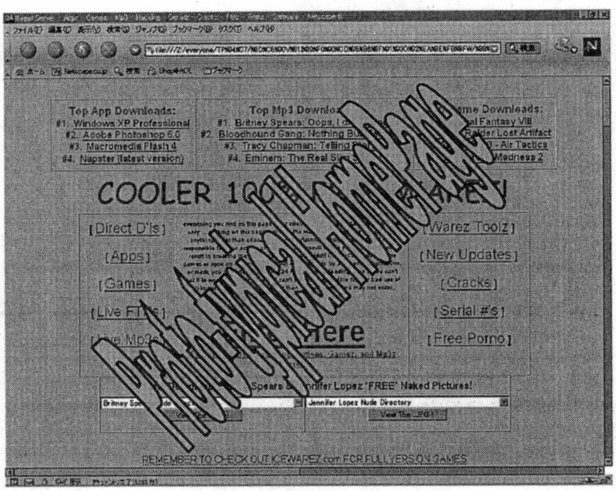

Figure 17. Page evaluated by the proposed system.

Highly ranked 20 words of TF-IDF value are shown in **Table 7**.

Table 7. Word vector of Web site

Ranking	Word	TF-IDF value
1	1spc	0.10751
2	crusader	0.10146
3	rate	0.09140
4	jpg	0.05966
5	flt	0.05759
6	com1	0.05553
7	apps	0.04492
8	port	0.03896
9	tag	0.03660
10	jennifer	0.02922
11	britney	0.02885
12	pub	0.02820
13	nov	0.02722
14	razor1911	0.02687
15	3spcs	0.02687
16	echelon	0.02687
17	wolfenstein	0.02598
18	systran	0.02564
19	simak	0.02304
20	crackin	0.02304

The reason the TF-IDF method failed could be due to the fact that this Web page is very large. It also contains a legal part in the latter half, and thus the TF-IDF values indicating illegality were decreased. Another reason is interpretation of link rates. **Table 8** indicates highly ranked 10 words used as the input of CFSs Unit.

Table 8. Input values to CFSs

Ranking	Word	Input-value (TF-IDF value)
1	<software list>	0.13333
2	apps	0.04492
3	iso	0.02192
4	warez	0.01934
5	psx	0.01568
6	mp3	0.01424
7	game	0.01280
8	ftp	0.01276
9	album	0.01074
10	get	0.01052

Table 9. Output values of CFSs

Ranking	Word	Activation value
1	<software list>	0.14441
2	apps	0.05646
3	warez	0.03673
4	iso	0.02884
5	game	0.02341
6	gamez	0.02130
7	mp3	0.02020
8	psx	0.01845
9	appz	0.01798
10	serialz	0.01787

Because the famous software names appeared frequently, activation values of the concepts besides "warez", such as gamez, appz and serialz, arose. (**Table 9**). It is considered that CFSs can recognize that the concept "warez" fusing word frequency and link information reflecting human subjectivity.

3.4 General Observations and Summaries

In this section, we applied the CFSs using RBF networks. Moreover, we proposed a system which is capable of filtering harmful web sites. We showed that the semantic interpretation of the concept by CFSs exceeded the TF-IDF method which is based on the superficial statistical information.

However, the proposed system has been evaluated using the limited number of target web documents. In our future work, we need to strengthen the conceptual descriptions and generalizations of CFSs that can be used in the entire Internet.

3.6 Fuzzy-TF.IDF

The use of Fuzzy-tf-idf is an alternative to the use of the conventional tf-idf. In this case, the original tf-idf weighting values will be replaced by a fuzzy set rather than original crisp value. To reconstruct such value both ontology and similarity measure can be used. To develop ontology and similarity one can used the conventional Latent Semantic Indexing (LSI) or Fuzzy-LSI (Nikravesh and Azvine 2002). The fuzzy-LSI (**Figure 18**), fuzzy-TF-IDF, and CFS can be used through an integrated system to develop fuzzy conceptual model for intelligent search engine. One can use clarification dialog, user profile, context, and ontology, into a integrated frame work to address some of the issues related to search engines were described earlier. In our perspective, we define this framework as ***Fuzzy Conceptual Matching based on Human Mental Model*** (**Figure19**).

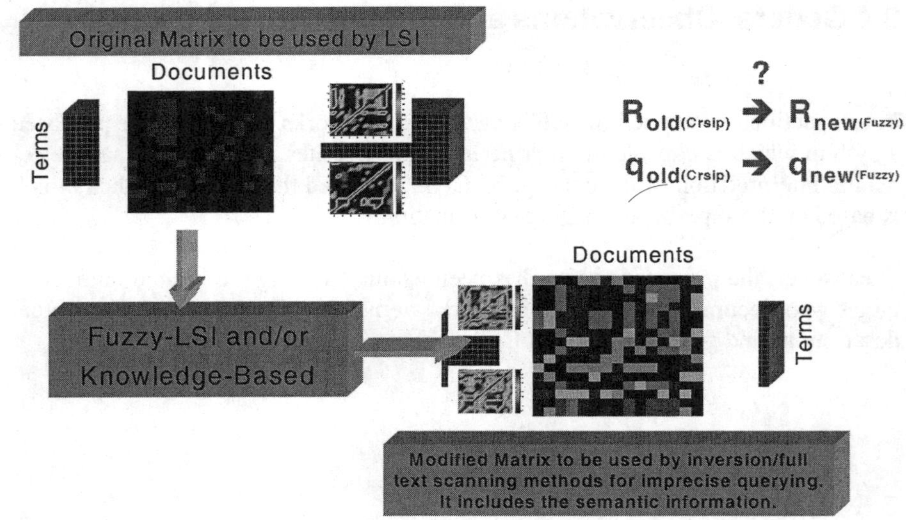

Figure 18. Fuzzy-Latent Semantic Indexing-Based Conceptual Technique

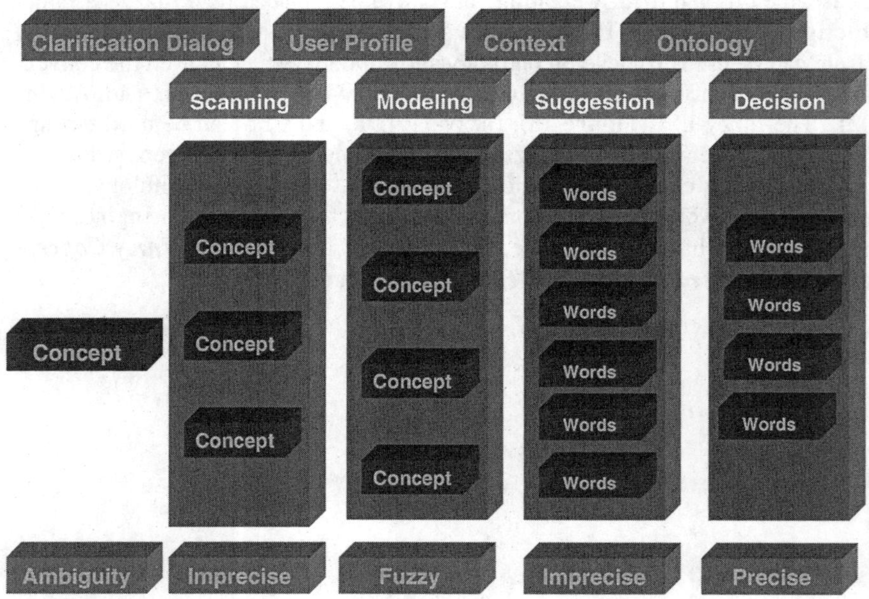

Figure 19. Fuzzy Conceptual Matching and Human Mental Model

4. Conceptual Fuzzy Sets-Based Navigation System for Yahoo!

Humans have a remarkable capability (perception) to perform a wide variety of physical and mental tasks without any measurements or computations. Familiar examples of such tasks are: playing golf, assessing wine, recognizing distorted speech, and summarizing a story. The question is whether a special type information retrieval processing strategy can be designed that build in perception (Zadeh 2001b and 1999).

World Wide Web search engines have become the most heavily-used online services, with millions of searches performed each day. Their popularity is due, in part, to their ease of use. The central tasks for the most of the search engines can be summarize as 1) query or user information request- do what I mean and not what I say!, 2) model for the Internet, Web representation-web page collection, documents, text, images, music, etc, and 3) ranking or matching function-degree of relevance, recall, precision, similarity, etc. Already explosive amount of users on the Internet is estimated over hundreds of millions. For example over 30 million new users visiting Google™ every month. While the number of pages available on the Internet almost double every year, the main issue will be the size of the internet when we include multimedia information as part of the Web and also when the databases connected to the pages to be considered as part of an integrated Internet and Intranet structure. Databases are now considered as backbone of most of the E-commerce and B2B and business and sharing information through Net between different databases (Internet-Based Distributed Database) both by user or clients are one of the main interest and trend in the future. In addition, the estimated user of wireless devices is estimated 1 billion within 2003 and 95 % of all wireless devices will be Internet enabled within 2005.

Courtois and Berry (Martin P. Courtois and Michael W. Berry, ONLINE, May 1999-Copyright © Online Inc.) published a very interesting paper "Results Ranking in Web Search Engines". In their work for each search (Altavista, Excite, HotBot, Infoseek, and Lycos), the following topics were selected: credit card fraud, quantity theory of money, liberation tigers, evolutionary psychology, French and Indian war, classical Greek philosophy, Beowulf criticism, abstract expressionism, tilt up concrete, latent semantic indexing, fm synthesis, pyloric stenosis, and the first 20 and 100 items were downloaded using the search engine. Three criteria 1) All Terms, 2) Proximity, and 3) Location were used as a major for testing the relevancy ranking (For all five search engines; Mean Hit: %15, Proximity: %21.4, and Location: %50.2). The effectiveness of the classification is defined based on the precision and recall. Effectiveness is a measure of the system ability to satisfy the user in terms of the relevance of documents retrieved. In probability theory, precision is defined as conditional probability, as the probability that if a random document is classified under selected terms or category, this decision is correct. Precision is defined as portion of the retrieved documents that are relevant with respect to all retrieved documents; number of the relevant documents retrieved divided by all documents retrieved. Recall is defined as the conditional probability and as the probability if a random document should be classified under selected terms or category, this decision is taken. Recall is defined as portion of the relevant retrieved documents that are relevant with respect to all relevant documents exists; number of the relevant documents retrieved divided by all relevant documents. The performance of each request is usually given by precision-recall curve. The overall performance of a system is based on a series of query request. Therefore, the performance of a system is represented by a precision-recall curve, which is an average of the entire precision-recall curve for that set of query request. To improve the performance of a system one can use different mathematical model for aggregation operator for ($A \cap B$) such as fuzzy logic. This will sift the curve to a higher or lower value. However, this may be a matter of scale change and may not change the actual performance of the system. We call this improvement, virtual improvement. However, one can shit the curve to the next level, by using a more intelligent model that for example have deductive capability or may resolve the ambiguity.

Many search engines support Boolean operators, field searching, and other advanced techniques such as fuzzy logic in variety of definition and in a very primitive ways. While searches may retrieve thousands of hits, finding relevant partial matches and query relevant information with deductive capabilities might be a problem. What is also important to mention for search engines is query-relevant information rather than generic information. Therefore, the query needs to be refined to capture the user's perception. However, to design such a system is not trivial, however, Q/A systems information can be used as a first step to build a knowledge based to capture some of the common user's perceptions. Given the concept of the perception, new machineries and tools need to be developed. Therefore, we envision that non-classical techniques such as fuzzy logic based-clustering methodology based on perception, fuzzy similarity, fuzzy aggregation, and Fuzzy-LSI for automatic information retrieval and search with partial matches are required.

4.1 Navigation System for Yahoo!

Many search engines such as *Yahoo!* classify a large number of web sites into their own large hierarchical categories (directories). Although category menus are provided for users, the users don't commonly know the hierarchical structure nor do they understand which item (categories) on the menus to select to find documents they want.

In this section, we propose a navigation system which conceptually matches input keywords with all paths from a root category to leaf categories. Input keywords don't always match words on category menus directly. The proposed system conceptually matches keywords with paths by taking the meaning of a path into consideration and by expanding keywords. For conceptual matching, we use CFSs based on RBF networks.

We describe the CFSs based on RBF networks in section 4.2 and our navigation system in section 4.3. In section 4.4, we present our experiments and results.

4.2 Conceptual fuzzy sets

For conceptual matching, we use conceptual fuzzy sets (CFSs) based on radial basis function (RBF) networks. In a CFS, the meaning of a concept is represented by the distribution of the activation values of the other concepts. To expand input keywords, the propagation of activation values is carried out recursively.

4.2.1 RBF-based CFSs

In the CFSs, relationships between words may be synonymous, antonymous, hypernymous and hyponymous . These relationships are too complicated to be represented in a hierarchical structure. Therefore, we use RBF-like networks to generate CFSs. **Figure 4** shows the image of CFSs based on RBF networks.
Each node corresponds to one concept and input/output is represented as the activation values of nodes.

RBF networks originally learn the prototype vectors and the weights between nodes from data. In this section, however, we generate CFSs using a concept base which we made manually.

4.2.2 Concept base

A concept base is a knowledge base which stores words represented in other words with their degrees of relationship. Although word co-occurrence measures are widely used to calculate the degrees of relationship between words, we use an original method based on some rules found in Japanese language dictionaries.

The rules are:

1. A word that has highly relative meaning to a headword usually appears first.

2. If a single word represents the meaning of a headword, it is a synonym or it has strong relationship to the headword.

We add some words that are relative to headwords because the networks are too sparse if we construct them only with a Japanese language dictionary. Then the degrees of relationships between words are calculated with the rules above. **Table 10** shows some examples of headwords and their relative words with the degrees of relationships.

Table 10. Example of degrees of relationship

headword	magazine		book	
words explaining the headword	magazine	1.0	book	1.0
	book	0.8	publication	0.84
	bookstore	0.72	magazine	0.72
	publication	0.6	journal	0.7
	newspaper	0.5	bookstore	0.64
	information	0.4	dictionary	0.6

We generate the CFSs considering each vector of a headword as a prototype vector of RBF networks, and each degree of relationship between words as a weight between output and corresponding unit.

4.3 Navigation system

In this section, we describe our navigation system which conceptually matches input keywords and all paths. The system consists of a CFS unit, a path base, a matching unit and a user interface. **Figure 20** shows the architecture of the system.

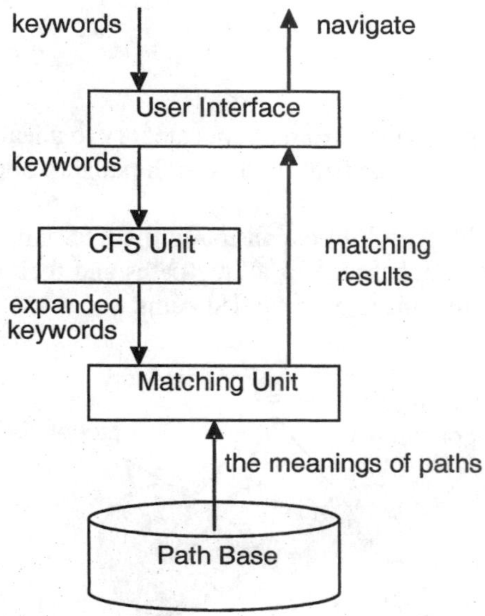

Figure 20. Navigation System Architecture

4.3.1 CFS unit

When keywords are inputed into the CFS unit, propagation of activation values occurs from the keywords. It results in the distribution of the activation values of the other words and represents the concept of the keywords. The propagation is carried out recursively several times to associate with relative concepts. This recursive propagation enables the association of concepts which are connected indirectly with input keywords (**Figure 21**).

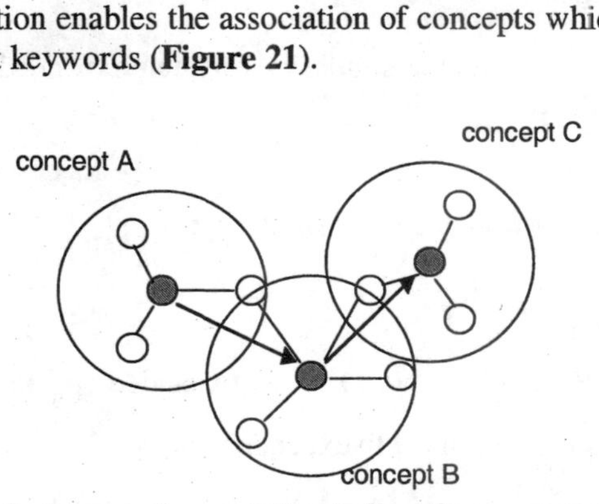

Figure 3. Image of association indirectly connected concepts

4.3.2. Path base

A path is a sequence of category labels from a root category to a leaf category. We take the meaning of a path into consideration to search paths to appropriate categories.

The meaning of a path is the result of expansion in the CFS unit from the category labels in the path. The path base stores all the paths and their corresponding meanings. **Figure 22** shows the image of expanded paths.

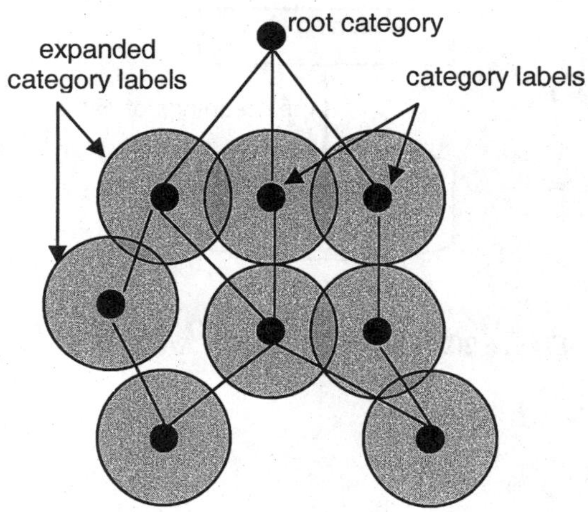

Fiaure 4. Imaae of exoanded

4.3.3 Matching unit

The matching unit calculates the similarity between expanded input keywords and each expanded path. We

use the cosine measure to calculate the similarity:

$$Sim(C_k, C_{p_i}) = \sum_{j=0}^{N} w_{kj} \times w_{p_i j}$$

where $C_k = (w_{k1}, w_{k2}, \ldots, w_{kN})$ is expanded keywords, and $C_{p_i} = (w_{p_i 1}, w_{p_i 2}, \ldots, w_{p_i N})$ is the i th expanded path.

4.3.4 User interface

The user interface displays recommended paths according to matching degrees to navigate the user.

4.4 Experiments

In this section, we evaluate the effectiveness of our navigation system using test data shown in **Table 11. Table 12** shows some examples of paths. They referred *Yahoo! JAPAN.*

First, we evaluated how many times the propagation of activation values should be carried out in the CFS unit. Second, we actually search the paths and evaluate the results.

Table 11. Test data

the number of paths	213
the number of words	803
the number of headwords	214

Table 3. Example of paths

Business and Economy > Cooperatives > Architecture > Software
Computer and Internet > Software > Internet
Entertainment > Music > Composition
Government > Law > Tax
Science > Physics > Libraries

4.4.1 Determine the repeat number of propagation

Figure 23 shows the changes of activation values of some words with "personal computer" and "book" as input to the CFS unit.

The activation value of the word "magazine" gets higher as the propagation is carried out and is at the peak in the third to fifth iteration. The word "magazine", which highly relates to "book", is associated by iterating propagation of activation values in CFS unit. The activation value of the word "information" is also at the peak in the third to fifth iteration. Although the word "information" is not connected directly to "personal computer" nor "book", the iteration of the propagation of activation values enables the association of the word.

4.4.2 Search the paths

We assume that a user is interested in books about personal computers, and then he/she inputs "personal computer" and "book" as keywords. We fixed the repeat number of propagation in the CFS unit on three times and searched the paths with these keywords. The result is shown in **Table 13.**

The top ranked path leads to the category which is assigned to web sites of online computer bookstores. The system could search the path that seems to be the best for the input keywords.

Table 13. Matching results

Business and Economy > Cooperatives > Books > Booksellers > Computers	Similarity = 0.951106
Business and Economy > Cooperatives > Books > Booksellers > Magazines	Similarity = 0.945896
Business and Economy > Cooperatives > Books > Booksellers > Movies	Similarity = 0.918033
Business and Economy > Cooperatives > Books > Booksellers > Architecture	Similarity = 0.9093
Business and Economy > Cooperatives > Books > Booksellers > Literature	Similarity = 0.904156

Note that the first item in the best path is "Business and Economy", which may be unexpected for him/her to have to click on to reach the computer bookstores. Our system could recommend such a path that lets the user find categories he/she wants.

However, all the top five paths in the search result lead to categories about books. The reason of this may be that the concept base includes too many words about books.

4.4.3 3D user interface

We have developed a 3D user interface to navigate users effectively. The interface displays hierarchical structure in the same manner as Cone Tree (Robertson et al 90 and 91). **Figure 24** is a screenshot of the user interface. Users can easily understand their position in large categorical hierarchy and the system can prevent them from getting lost. And for users who want to get more detail, functions such as rotation and zoom are also provided.

Paths with high similarities to input keywords are highlighted and the system can help users to reach appropriate categories.

4.5 General Observations and Summaries

We proposed a navigation system which conceptually matches input keywords and paths using CFSs based on RBF networks. Taking the meaning of a path into consideration and propagating activations of concepts recursively in CFS unit to associate relative words with input keywords enabled the system to search the path leading to an appropriate category.

However, the following are some problems which require further study:

- The scale of the system is small.
- The associations in CFS unit are affected by un-uniformity of the concept base.
- The number of propagation of activation values in CFS unit is empirical.

In this study, we used the cosine measure to calculate the similarity. In the future work, we intent to use other similarity measures (as shown in **Table 14** and **Table 15**) especially perception-based and fuzzy-based similarity measures (Nikravesh 2002, Nikravesh et al. 2002).

Table 14. Five commonly used measure of similarity and association in IR

Simple matching Coefficiet : $|X \cap Y|$

Dice's Coefficiet : $2\frac{|X \cap Y|}{|X|+|Y|}$

Jaccard's Coefficiet : $\frac{|X \cap Y|}{|X \cup Y|}$

Cosine Coeffciet : $\frac{|X \cap Y|}{|X|^{1/2} \times |Y|^{1/2}}$

Overlap Coefficiet : $\frac{|X \cap Y|}{\min(|X|,|Y|)}$

Disimilarity Coefficeint : $\frac{|X \Delta Y|}{|X|+|Y|} = 1 - Dice's$ Coefficient

$|X \Delta Y| = |X \cup Y| - |X \cap Y|$

Table 15. Term-Document Matrix Representation (R) and Similarity Measure (For definition of the terms used in this table refer to Ref. (Zobel and Moffat).

$$Similarity\text{ - }measure: S_{q,d} = f(w_{q,t}, w_{d,t}, W_q, W_d, \tau_{q,d}, C, w_t, r_{d,t})$$

$$Term\text{ - }weights: w_t = f(N, f_t, f^m, f_{d,t}, F_t, n_t, s_t)$$

$$Document - term\text{ - }weights: w_{d,t} = f(r_{d,t}, w_t)$$

$$\text{Re}lative\text{ - }term\text{ - }frequencies: r_{d,t} = f(t, \tau_d, f_{d,t}, f_d^m, W_d)$$

$$Document\text{ - }lengths\ and\ query\text{ - }lengths: W_d = f(t, \tau_d, w_{d,t}, f_d, s, W_d')$$

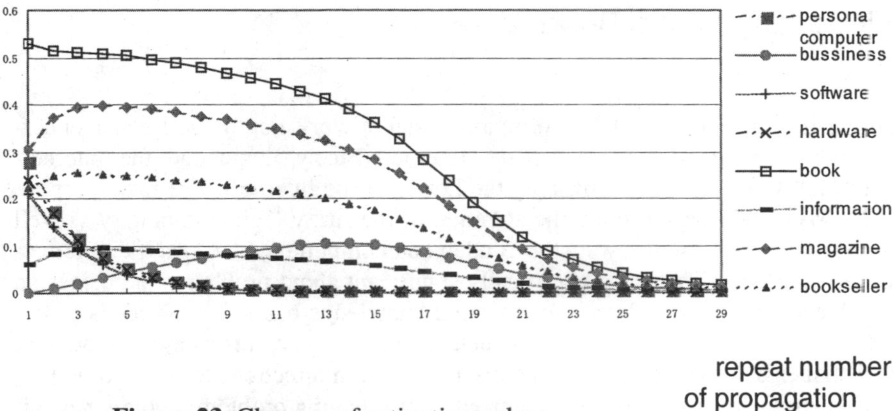

Figure 23. Changes of activation values

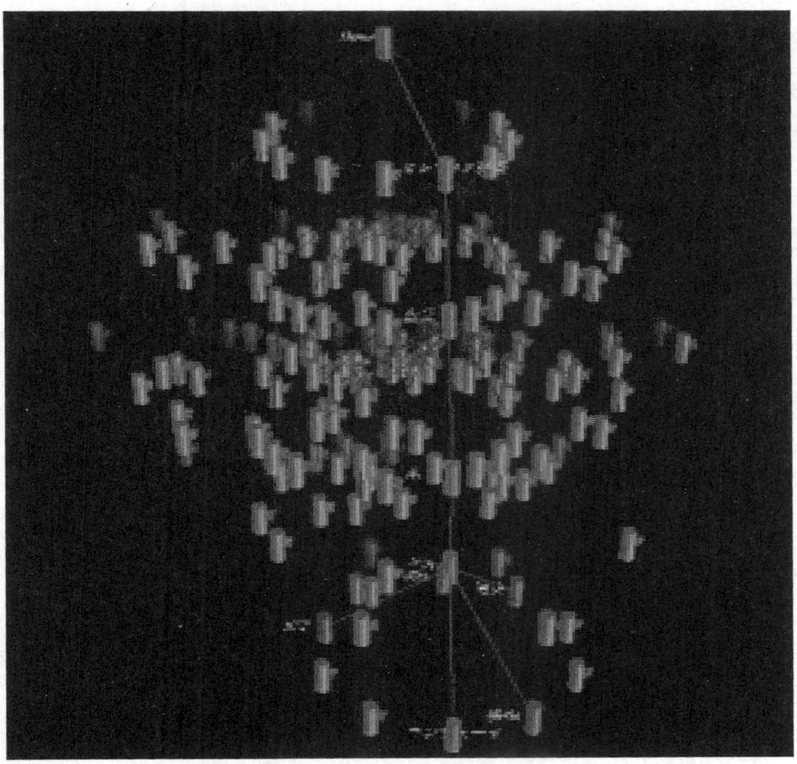

Figure 24. 3D user interface

5 Challenges and Road Ahead

During the August 2001, BISC program hosted a workshop toward better understanding of the issues related to the Internet (Fuzzy Logic and the Internet-FLINT2001, Toward the Enhancing the Power of the Internet). The main purpose of the Workshop was to draw the attention of the fuzzy logic community as well as the Internet community to the fundamental importance of specific Internet-related problems. This issue is critically significant about problems that center on search and deduction in large, unstructured knowledge bases. The Workshop provided a unique opportunity for the academic and corporate communities to address new challenges, share solutions, and discuss research directions for the future. Following are the areas that were recognized as challenging problems and the new direction toward the next generation of the search engines and Internet. We summarize the challenges and the road ahead into four categories as follows:

- ***Search Engine and Queries:***
    - Deductive Capabilities
    - Customization and Specialization
    - Metadata and Profiling
    - Semantic Web
    - Imprecise-Querying
    - Automatic Parallelism via Database Technology
    - Approximate Reasoning
    - Ontology
    - *Ambiguity Resolution through Clarification Dialog; Definition/Meaning & Specificity*User Friendly
    - Multimedia
    - Databases
    - Interaction
- ***Internet and the Academia:***
    - Ambiguity and Conceptual and Ontology
    - Aggregation and Imprecision Query
    - Meaning and structure Understanding
    - Dynamic Knowledge
    - Perception, Emotion, and Intelligent Behavior
    - Content-Based
    - Escape from Vector SpaceDeductive Capabilities
    - Imprecise-Querying
    - *Ambiguity Resolution through Clarification Dialog*

    - *Precisiated Natural Languages (PNL)*
- ***Internet and the Industry:***
    - XML=>Semantic Web
    - Workflow
    - Mobile E-Commerce
    - CRM
    - Resource Allocation
    - Intent
    - Ambiguity Resolution
    - Interaction
    - Reliability
    - Monitoring
    - Personalization and Navigation
    - Decision Support
    - Document Soul
    - Approximate Reasoning
    - Imprecise QueryContextual Categorization
- ***Fuzzy Logic and Internet; Fundamental Research:***
    - Computing with Words (CW)
    - Computational Theory of Perception (CTP)
    - Precisiated Natural Languages (PNL)

The potential areas and applications of Fuzzy Logic for the Internet include:

- ***Potential Areas:***
    - Search Engines
    - Retrieving Information
    - Database Querying
    - Ontology
    - Content Management
    - Recognition Technology
    - Data Mining
    - Summarization
    - Information Aggregation and Fusion
    - E-Commerce
    - Intelligent Agents
    - Customization and Personalization
- ***Potential Applications:***

- Search Engines and Web Crawlers
- Agent Technology (i.e., Web-Based Collaborative and Distributed Agents)
- Adaptive and Evolutionary techniques for dynamic environment (i.e. Evolutionary search engine and text retrieval, Dynamic learning and adaptation of the Web Databases, etc)
- Fuzzy Queries in Multimedia Database Systems
- Query Based on User Profile
- Information Retrievals
- Summary of Documents
- Information Fusion Such as Medical Records, Research Papers, News, etc
- Files and Folder Organizer
- Data Management for Mobile Applications and eBusiness Mobile Solutions over the Web
- Matching People, Interests, Products, etc
- Association Rule Mining for Terms-Documents and Text Mining
- E-mail Notification
- Web-Based Calendar Manager
- Web-Based Telephony
- Web-Based Call Centre
- Workgroup Messages
- E-Mail and Web-Mail
- Web-Based Personal Info
- Internet related issues such as Information overload and load balancing, Wireless Internet-coding and D-coding (Encryption), Security such as Web security and Wireless/Embedded Web Security, Web-based Fraud detection and prediction, Recognition, issues related to E-commerce and E-bussiness, etc.

6 Conclusions

Intelligent search engines with growing complexity and technological challenges are currently being developed. This requires new technology in terms of understanding, development, engineering design and visualization. While the technological expertise of each component becomes increasingly complex, there is a need for better integration of each component into a global model adequately capturing the imprecision and deduction capabilities. In addition, intelligent models can mine the Internet to conceptually match and rank homepages based on predefined linguistic formulations and rules defined by experts or based on a set of known homepages. The FCM model can be used as a framework for intelligent information and knowledge retrieval through conceptual matching of both text and images (here defined as "Concept"). The FCM can also be used for constructing fuzzy ontology or terms related to th context of the query and search to resolve the ambiguity. This model can be used to calculate conceptually the degree of match to the object or query.

In this work, we proposed a search engine which conceptually matches input keywords and web pages. The conceptual matching is realized by context-dependent keyword expansion using conceptual fuzzy sets. First, we show the necessity and also the problems of applying fuzzy sets to information retrieval. Next, we introduce the usefulness of conceptual fuzzy sets in overcoming those problems, and propose the realization of conceptual fuzzy sets using Hopfield Networks. We also propose the architecture of the search engine which can execute conceptual matching dealing with context-dependent word ambiguity. Finally, we evaluate our proposed method through two simulations of retrieving actual web pages,and compare the proposed method with the ordinary TF-IDF method. We show that our method can correlate seemingly unrelated input keywords and produce matching Web pages, whereas the TF-IDF method cannot.

Currently on the Internet there exists a host of illegal web sites which specialize in the distribution of commercial software and music. This section proposes a method to distinguish illegal web sites from legal ones not only by using tf-idf values but also to recognize the purpose/meaning of the web sites. It is achieved by describing what are considered to be illegal sites and by judging whether the objective web sites match the description of illegality. Conceptual fuzzy sets (CFSs) are used to describe the concept of illegal web sites. First, we introduced the usefulness of CFSs in overcoming those problems, and propose the realization of CFSs using RBF-like networks. In a CFS, the meaning of a concept is represented by the distribution of the activation values of the other nodes. Because the distribution changes depend on which labels are activated as a result of the conditions, the activations show a context-dependent meaning. Next, we proposed the architecture of the filtering system. Additionally, we compared the proposed method with the tf-idf method with the support vector machine. The e-measures as a total evaluation indicate that the proposed system showed better results as compared to the tf-idf method with the support vector machine.

Finaly, we proposed a menu navigation system which conceptually matches input keywords and paths. For conceptual matching, we used conceptual fuzzy sets (CFSs) based on radial basis function (RBF) networks. In a CFS, the meaning of a concept is represented by the distribution of the activation values of the other concepts. To expand input keywords, the propagation of activation values is carried out recursively. The proposed system recommends users paths to appropriate categories. We used 3D user interface to navigate users.

7. Future Works

7.1 TIKManD (Tool for Intelligent Knowledge Management and Discovery)

In the future work, we intent to develop and deploy an intelligent computer system is called *"**TIKManD (Tool for Intelligent Knowledge Management and Discovery)**"*.

The system can mine Internet homepages, Emails, Chat Lines, and/or authorized wire tapping information (which may include Multi-Lingual information) to recognize, conceptually match, and rank potential terrorist and criminal activities (both common and unusual) by the type and seriousness of the activities. This will be done automatically or semi-automatically based on predefined linguistic formulations and rules defined by experts or based on a set of known terrorist activities given the information provided through law enforcement databases (text and voices) and huge number of "tips" received immediately after the attack. Conceptual Fuzzy Set (CFS) model will be used for intelligent information and knowledge retrieval through conceptual matching of text, images and voice (here defined as "Concept"). The CFS can be also used for constructing fuzzy ontology or terms relating the context of the investigation (Terrorism or other criminal activities) to resolve the ambiguity. This model can be used to calculate conceptually the degree of match to the object or query. In addition, the ranking can be used for intelligently allocating resources given the degree of match between objectives and resources available (Nikravesh et al. 2002, Nikravesh 2002).

The use of the Conceptual Fuzzy Set (CFS) is a necessity, given the ambiguity and imprecision of the "concept" in law enforcement databases and information related to terrorism, which may be described by Multi-Lingual textual, images and voice information. In the CFS approach, the "concept" is defined by a series of keywords with different weights depending on the importance of each keyword. Ambiguity in concepts can be defined by a set of imprecise concepts. Each imprecise concept in fact can be defined by a set of fuzzy concepts. The fuzzy concepts can then be related to a set of imprecise words given the context. Imprecise words can then be translated into precise words given the ontology and ambiguity resolution through clarification dialog. By constructing the ontology and fine-tuning the strength of links (weights), we could construct a fuzzy set to integrate piece-wise the imprecise concepts and precise words to define the ambiguous concept.

7.2 Web Intelligence: Google™ and Yahoo! Concept-Based Search Engine

There are two type of search engine that we are interested and are dominating the Internet. First, the most popular search engines that are mainly for unstructured data such as Google ™ and Teoma which are based on the concept of Authorities and Hubs. Second, search engines that are task spcifics such as 1) Yahoo!: manually-pre-classified, 2) NorthernLight: Classification, 3) Vivisimo: Clustering, 4) Self-organizing Map: Clustering + Visualization and 5) AskJeeves: Natural Languages-Based Search; Human Expert.

Google uses the PageRank and Teoma uses HITS (Ding et al. 2001) for the Ranking. **Figure 25** shows the Authorities and Hubs concept and the possibility of comparing two homepages.

Figures 26 shows the possible model for similarity analysis is called "fuzzy Conceptual Similarity". **Figure 27** shows the matrix representation of Fuzzy Conceptual Similarity model. **Figure 28** shows the evolution of the Term-Document matrix. **Figure 29** shows the structure of the Concept-based Google ™ search engine for Multi-Media Retrieval . Finally, **Figure 30** shows the structure of the Concept-Based Intelligent Decision Analysis. To develop such models, state-of-the-art computational intelligence techniques are needed. These include and are not limited to:

- Latent-Semantic Indexing and SVD for preprocessing,
- Radial-Basis Function Network to develop concepts,
- Support Vector Machine (SVM) for supervised classification,
- fuzzy/neuro-fuzzy clustering for unsupervised classification based on both conventional learning techniques and Genetic and Reinforcement learning,
- non-linear aggregation operators for data/text fusion,
- automatic recognition using fuzzy measures and a fuzzy integral approach

- self organization map and graph theory for building community and clusters,

- both genetic algorithm and reinforcement learning to learn the preferences,

- fuzzy-integration-based aggregation technique and hybrid fuzzy logic-genetic algorithm for decision analysis, resource allocation, multi-criteria decision-making and multi-attribute optimization.

- text analysis: next generation of the Text, Image Retrieval and concept recognition based on soft computing technique and in particular Conceptual Search Model (CSM). This includes

    - Understanding textual content by retrieval of relevant texts or paragraphs using CSM followed by clustering analysis.
    - Hierarchical model for CSM
    - Integration of Text and Images based on CSM
    - CSM Scalability, and
    - The use of CSM for development of

        - Ontology
        - Query Refinement and Ambiguity Resolution
        - Clarification Dialog
        - Personalization-User Profiling

Acknowledgement

Funding for this research was provided by the British Telecommunication (BT) and the BISC Program of UC Berkeley.

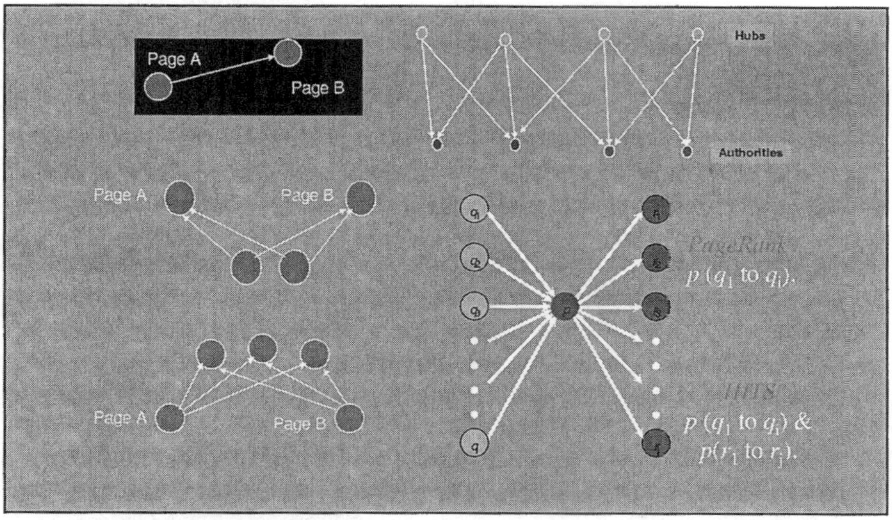

Figure 25. Similarity of web pages.

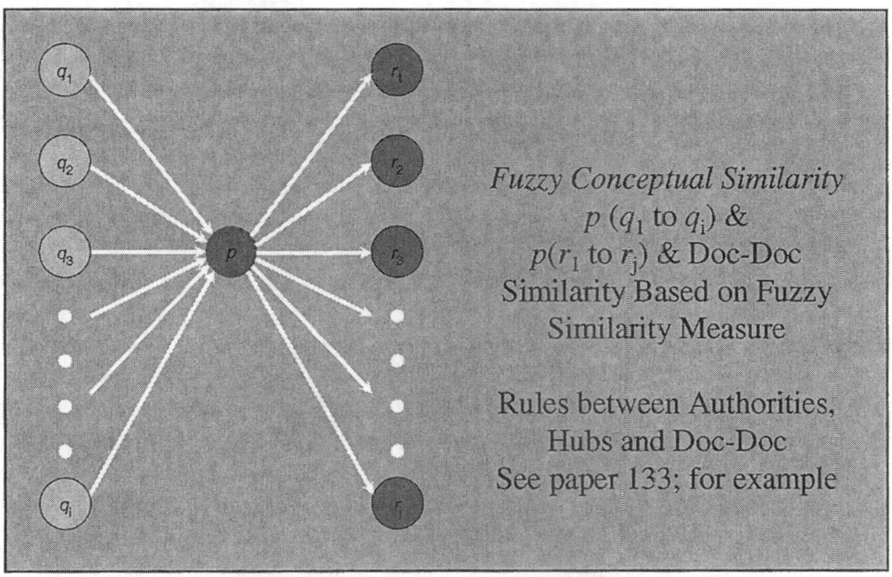

Figure 26. Fuzzy Conceptual Similarity

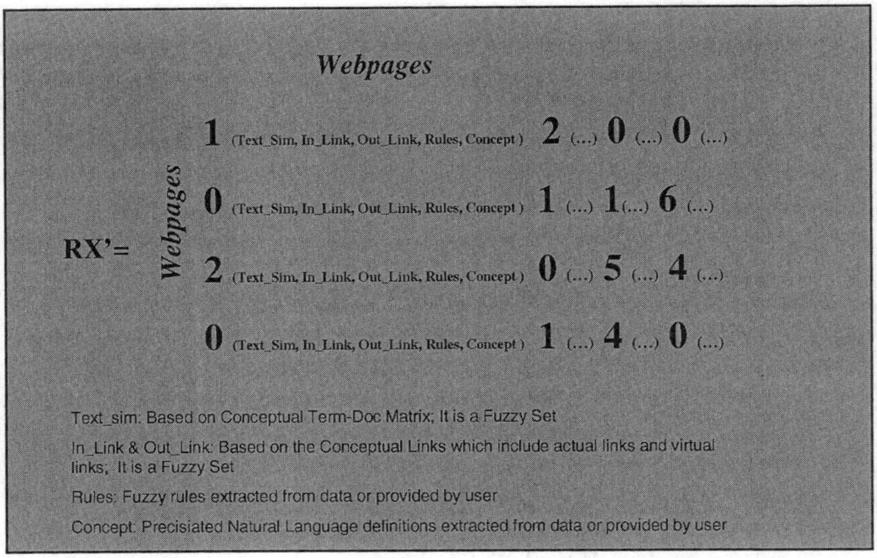

Figure 27. Matrix representation of Fuzzy Conceptual Similarity model

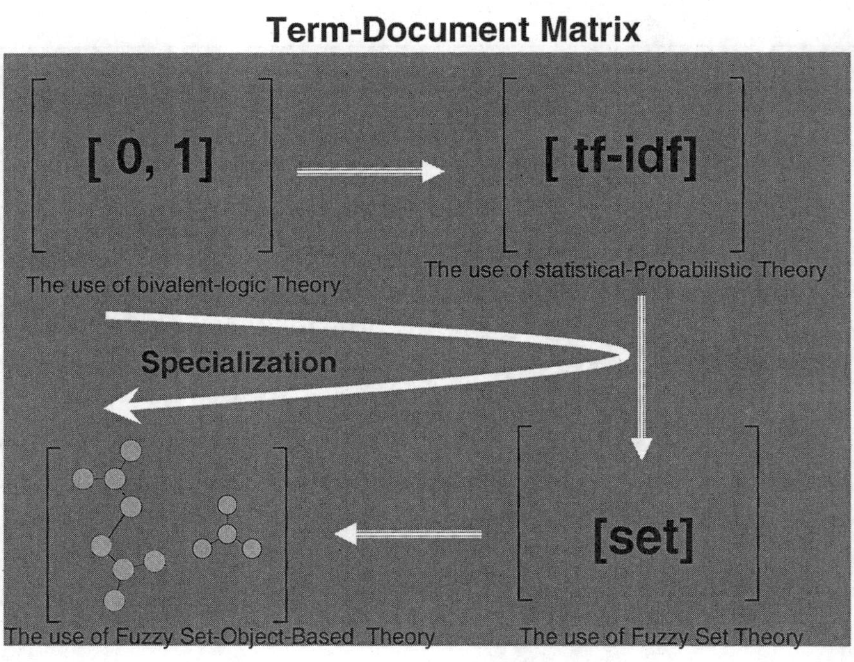

Figure 28. Evolution of Term-Document Matrix representation

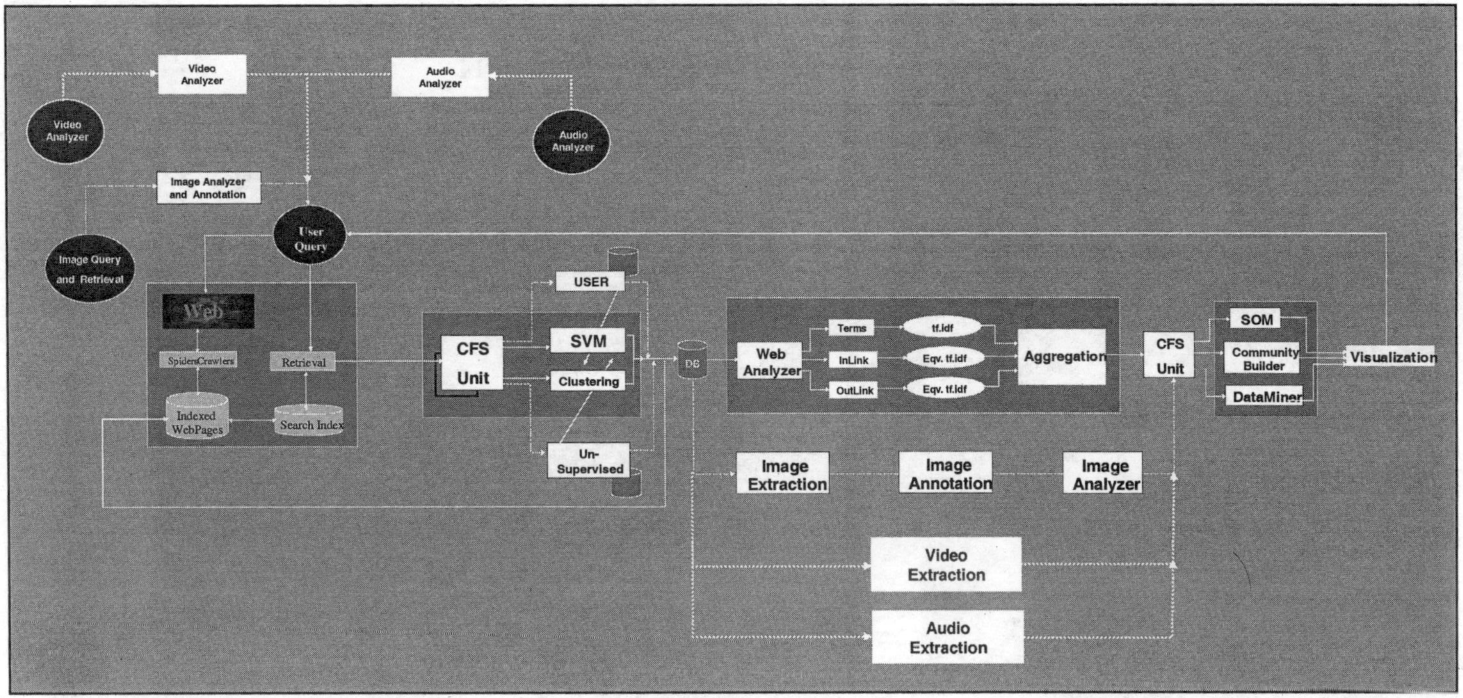

Figure 29. Concept-Based Google™ Search Engine for Multi-Media Retrieval

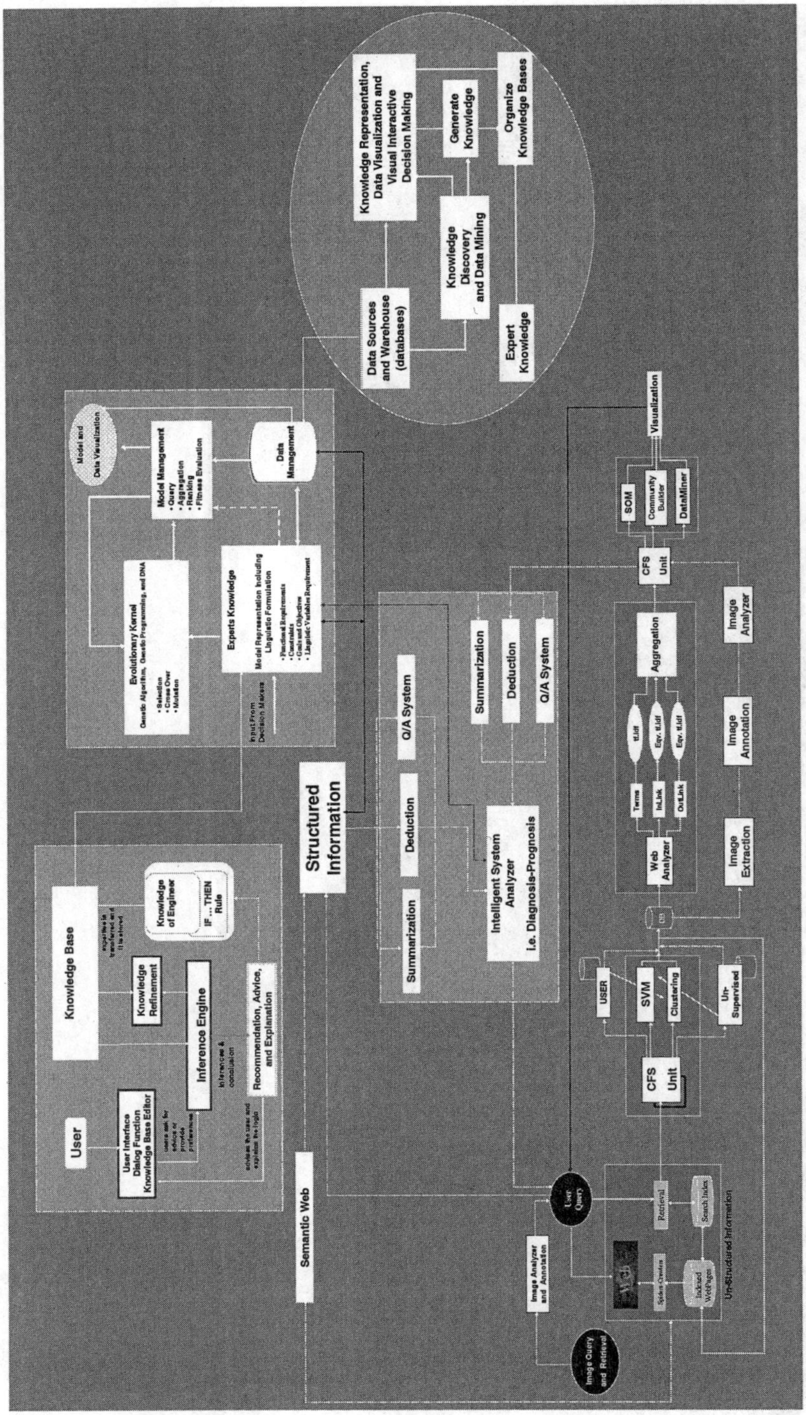

Figure 30. concept-Based Intelligent Decision Analysis

References

J. Baldwin, Future directions for fuzzy theory with applications to intelligent agents, in M. Nikravesh and B. Azvine, FLINT 2001, New Directions in Enhancing the Power of the Internet, UC Berkeley Electronics Research Laboratory, Memorandum No. UCB/ERL M01/28, August 200.

J. F. Baldwin and S. K. Morton, conceptual Graphs and Fuzzy Qualifiers in Natural Languages Interfaces, 1985, University of Bristol.

M. J. M. Batista et al., User Profiles and Fuzzy Logic in Web Retrieval, in M. Nikravesh and B. Azvine, FLINT 2001, New Directions in Enhancing the Power of the Internet, UC Berkeley Electronics Research Laboratory, Memorandum No. UCB/ERL M01/28, August 2001.

H. Beremji, Fuzzy Reinforcement Learning and the Internet with Applications in Power Management or wireless Networks, in M. Nikravesh and B. Azvine, FLINT 2001, New Directions in Enhancing the Power of the Internet, UC Berkeley Electronics Research Laboratory, Memorandum No. UCB/ERL M01/28, August 2001.

T.H. Cao, Fuzzy Conceptual Graphs for the Semantic Web, in M. Nikravesh and B. Azvine, FLINT 2001, New Directions in Enhancing the Power of the Internet, UC Berkeley Electronics Research Laboratory, Memorandum No. UCB/ERL M01/28, August 2001.

D. Y. Choi, Integration of Document Index with Perception Index and Its Application to Fuzzy Query on the Internet, in M. Nikravesh and B. Azvine, FLINT 2001, New Directions in Enhancing the Power of the Internet, UC Berkeley Electronics Research Laboratory, Memorandum No. UCB/ERL M01/28, August 2001.

Chris Ding, Xiaofeng He, Parry Husbands, Hongyuan Zha and Horst D. Simon, PageRank, HITS and a Unified Framework for Link Analysis. LBNL Tech Report 50007. Nov 2001. Proc. of 25th ACM SIGIR Conf. pp.353 354, 2002 (poster), Tampere, Finland

N. Guarino, C. Masalo, G. Vetere, "OntoSeek : content-based access to the Web", IEEE Intelligent Systems, Vol.14, pp.70-80 (1999)

K.H.L. Ho, Learning Fuzzy Concepts by Example with Fuzzy Conceptual Graphs. In 1st Australian Conceptual Structures Workshop, 1994. Armidale, Australia.

J. J. Hopfield, "Neural networks and physical systems with emergent collective computational abilities", Proceedings of the National Academy of Sciences U.S.A., Vol.79, pp.2554-2558 (1982)

J. J. Hopfield, "Neurons with graded response have collective computational properties like those of two-state neurons, Proceedings of the National Academy of Sciences U.S.A., Vol.81, pp.3088-3092 (1984)

A. Joshi and R. Krishnapuram, Robust Fuzzy Clustering Methods to Support Web Mining, in Proc Workshop in Data Mining and Knowledge Discovery, SIGMOD, pp. 15-1 to 15-8, 1998.

M. Kobayashi, K. Takeda, "Information retrieval on the web", ACM Computing Survey, Vol.32, pp.144-173 (2000)

B. Kosko, "Adaptive Bi-directional Associative Memories," Applied Optics, Vol. 26, No. 23, 4947-4960 (1987).

B. Kosko, "Neural Network and Fuzzy Systems," Prentice Hall (1992).

R. Krishnapuram et al., A Fuzzy Relative of the K-medoids Algorithm with application to document and Snippet Clustering , in Proceedings of IEEE Intel. Conf. Fuzzy Systems-FUZZIEEE 99, Korea, 1999.

T. P. Martin, Searching and smushing on the Semantic Web – Challenges for Soft Computing, in M. Nikravesh and B. Azvine, FLINT 2001, New Directions in Enhancing the Power of the Internet, UC Berkeley Electronics Research Laboratory, Memorandum No. UCB/ERL M01/28, August 2001.

M. Nikravesh and B. Azvine, FLINT 2001, New Directions in Enhancing the Power of the Internet, UC Berkeley Electronics Research Laboratory, Memorandum No. UCB/ERL M01/28, August 2001.

M. Nikravesh, Fuzzy Logic and Internet: Perception Based Information Processing and Retrieval, Berkeley Initiative in Soft Computing, Report No. 2001-2-SI-BT, September 2001a.

M. Nikravesh, BISC and The New Millennium, Perception-based Information Processing, Berkeley Initiative in Soft Computing, Report No. 2001-1-SI, September 2001b.

M. Nikravesh, V. Loia,, and B. Azvine, Fuzzy logic and the Internet (FLINT), Internet, World Wide Web, and Search Engines, to be appeared in International Journal of Soft Computing-Special Issue in fuzzy logic and the Internet , 2002

M. Nikravesh, Fuzzy Conceptual-Based Search Engine using Conceptual Semantic Indexing, NAFIPS-FLINT 2002, June 27-29, New Orleans, LA, USA

M. Nikravesh and B. Azvin, Fuzzy Queries, Search, and Decision Support System, to be appeared in International Journal of Soft Computing-Special Issue in fuzzy logic and the Internet , 2002

M. Nikravesh, V. Loia,, and B. Azvine, Fuzzy logic and the Internet (FLINT), Internet, World Wide Web, and Search Engines, to be appeared in International Journal of Soft Computing-Special Issue in fuzzy logic and the Internet , 2002

M. Nikravesh, Fuzzy Conceptual-Based Search Engine using Conceptual Semantic Indexing, NAFIPS-FLINT 2002, June 27-29, New Orleans, LA, USA

S. K. Pal, V. Talwar, and P. Mitra, Web Mining in Soft Computing Framework: Relevance, State of the Art and Future Directions, to be published in IEEE Transcations on Neural Networks, 2002.

G. Presser, Fuzzy Personalization, in M. Nikravesh and B. Azvine, FLINT 2001, New Directions in Enhancing the Power of the Internet, UC Berkeley Electronics Research Laboratory, Memorandum No. UCB/ERL M01/28, August 2001.

George G. Robertson, Stuart K. Card, and Jock D. Mackinlay, "Information Visualization Using 3D Interactive Animation", *Communications of the ACM*, Vol.36 No.4, pp.57-71, 1990.

George G. Robertson, Jock D. Machinlay, and Stuart K. Card, "Cone Trees: Animated 3D Visualizations of Hierarchical Information", *Proceedings of CHI '91*, pp.189-194.

E. Sanchez, Fuzzy logic e-motion, in M. Nikravesh and B. Azvine, FLINT 2001, New Directions in Enhancing the Power of the Internet, UC Berkeley Electronics Research Laboratory, Memorandum No. UCB/ERL M01/28, August 2001.

A. M. G. Serrano, Dialogue-based Approach to Intelligent Assistance on the Web, in M. Nikravesh and B. Azvine, FLINT 2001, New Directions in Enhancing the Power of the Internet, UC Berkeley Electronics Research Laboratory, Memorandum No. UCB/ERL M01/28, August 2001.

S. Shahrestani, Fuzzy Logic and Network Intrusion Detection, in M. Nikravesh and B. Azvine, FLINT 2001, New Directions in Enhancing the Power of the Internet, UC Berkeley Electronics Research Laboratory, Memorandum No. UCB/ERL M01/28, August 2001.

T. Takagi and M. Tajima, Proposal of a Search Engine based on Conceptual Matching of Text Notes, in M. Nikravesh and B. Azvine, FLINT 2001, New Directions in Enhancing the Power of the Internet, UC Berkeley Electronics Research Laboratory, Memorandum No. UCB/ERL M01/28, August 2001.

T. Takagi, A. Imura, H. Ushida, and T. Yamaguchi, "Conceptual Fuzzy Sets as a Meaning Representation and their Inductive Construction," International Journal of Intelligent Systems, Vol. 10, 929-945 (1995).

T. Takagi, A. Imura, H. Ushida, and T. Yamaguchi, "Multilayered Reasoning by Means of Conceptual Fuzzy Sets," International Journal of Intelligent Systems, Vol. 11, 97-111 (1996).

T. Takagi, S. Kasuya, M. Mukaidono, T. Yamaguchi, and T. Kokubo, "Realization of Sound-scape Agent by the Fusion of Conceptual Fuzzy Sets and Ontology," 8th International Conference on Fuzzy Systems FUZZ-IEEE'99, II, 801-806 (1999).

T. Takagi, S. Kasuya, M. Mukaidono, and T. Yamaguchi, "Conceptual Matching and its Applications to Selection of TV Programs and BGMs," IEEE International Conference on Systems, Man, and Cybernetics SMC'99, III, 269-273 (1999).

Wittgenstein, "Philosophical Investigations," Basil Blackwell, Oxford (1953).

R. Yager, Aggregation Methods for Intelligent Search and Information Fusion, in M. Nikravesh and B. Azvine, FLINT 2001, New Directions in Enhancing the Power of the Internet, UC Berkeley Electronics Research Laboratory, Memorandum No. UCB/ERL M01/28, August 2001.

John Yen, Incorporating Fuzzy Ontology of Terms Relations in a Search Engine, in M. Nikravesh and B. Azvine, FLINT 2001, New Directions in Enhancing the Power of the Internet, UC Berkeley Electronics Research Laboratory, Memorandum No. UCB/ERL M01/28, August 2001.

L. A.Zadeh, The problem of deduction in an environment of imprecision, uncertainty, and partial truth, in M. Nikravesh and B. Azvine, FLINT 2001, New Directions in Enhancing the Power of the Internet, UC Berkeley Electronics Research Laboratory, Memorandum No. UCB/ERL M01/28, August 2001 [2001a].

L.A. Zadeh, A Prototype-Centered Approach to Adding Deduction Capability to Search Engines -- The Concept of Protoform, BISC Seminar, Feb 7, 2002, UC Berkeley, 2002.

L. A. Zadeh, " A new direction in AI – Toward a computational theory of perceptions, AI Magazine 22(1): Spring 2001b, 73-84

L.A. Zadeh, From Computing with Numbers to Computing with Words-From Manipulation of Measurements to Manipulation of Perceptions, IEEE Trans. On Circuit and Systems-I Fundamental Theory and Applications, 45(1), Jan 1999, 105-119.

Y. Zhang et al., Granular Fuzzy Web Search Agents, in M. Nikravesh and B. Azvine, FLINT 2001, New Directions in Enhancing the Power of the Internet, UC Berkeley Electronics Research Laboratory, Memorandum No. UCB/ERL M01/28, August 2001.

Y. Zhang et al., Fuzzy Neural Web Agents for Stock Prediction, in M. Nikravesh and B. Azvine, FLINT 2001, New Directions in Enhancing the Power of the Internet, UC Berkeley Electronics Research Laboratory, Memorandum No. UCB/ERL M01/28, August 2001.

J.Zobel and A. Moffat, Exploring the Similarity Space, http://www.cs.mu.oz.au/~alistair/exploring/.

Soft Computing for Perception-Based Decision Processing and Analysis: Web-Based BISC-DSS

Masoud Nikravesh and Souad Bensafi
BISC Program, Computer Sciences Division, EECS Department
University of California, Berkeley, CA 94720, USA
Email: nikravesh@cs.berkeley.edu
Tel: (510) 643-4522
Fax: (510) 642-5775
URL: http://www-bisc.cs.berkeley.edu

Abstract: Searching a database records and ranking the results based on multi-criteria queries is central for many database applications used within organizations in finance, business, industrial and other fields. For Example, the process of ranking (scoring) has been used to make billions of financing decisions each year serving an industry worth hundreds of billion of dollars. To a lesser extent, ranking has also been used to process hundreds of millions of applications by U.S. Universities resulting in over 15 million college admissions in the year 2000 for a total revenue of over $250 billion. College admissions are expected to reach over 17 million by the year 2010 for total revenue of over $280 billion. In this paper, we will introduce fuzzy query and fuzzy aggregation as an alternative for ranking and predicting the risk for credit scoring and university admissions, which currently utilize an imprecise and subjective process. In addition we will introduce the BISC Decision Support System. The main key features of the BISC Decision Support System for the internet applications are 1) to use intelligently the vast amounts of important data in organizations in an optimum way as a decision support system and 2) To share intelligently and securely company's data internally and with business partners and customers that can be process quickly by end users. The model consists of five major parts: the Fuzzy Search Engine (FSE), the Application Templates, the User Interface, the database and the Evolutionary Computing (EC).

1 Introduction

Most of the available systems 'software' are modeled using crisp logic and queries, which results in rigid systems with imprecise and subjective process and results. In this chapter, we introduce fuzzy querying and ranking as a flexible tool allowing approximation where the selected objects do not need to match exactly the decision criteria resembling natural human behavior.

The model consists of five major parts: the Fuzzy Search Engine (FSE), the Application Templates, the User Interface, the database and the Evolutionary Computing (EC). We developed the software with many essential key features. The system is designed as generic system that can run different application domains. To this end, the Application Template module provides all needed information for a certain application as object attributes and properties, and serve as a guideline structure for building a new application.

The Fuzzy Search Engine (FSE) is the core module of the system. It has been developed to be generic so that it would fit any application with minimal changes. The main FSE components are the membership functions, similarity functions and aggregators. Administrator can also change the membership function to be used to do searches.

Through the user interface a user can enter and save his/her profile, input criteria for a new query, run different queries and display results. The user can manipulate manually the result by eliminating what he/she disproof and the ranking according to his/her preferences.

This process is monitored and learned by the Evolutionary Computing (EC) module recording and saving user preferences to be used as basic queries for that particular user. We present our approach with three important applications: ranking (scoring) which has been used to make financing decisions concerning credit cards, cars and mortgage loans; the process of college admissions where hundreds of thousands of applications are processed yearly by U.S. Universities; and date matching as one of the most popular internet programs. However, the software is generic software for much more diverse applications and to be delivered as stand alone software to both academia and businesses.

Consider walking into a car dealer and leaving with an old used car paying a high interest rate of around 15% to 23% and your colleague leaves the dealer with a luxury car paying only a 1.9% interest rate. Consider walking into a real estate agency and finding yourself ineligible for a loan to buy your dream house. Also consider getting denied admission to your college of choice but your classmate gets accepted to the top school in his dream major. Welcome to the world of ranking, which is used both for deciding college admissions and determining credit risk. In the credit rating world, FICO (Fair Isaac Company) either makes you or breaks you, or can at least prevent you from getting the best rate possible (Fair Isaac). Admissions ranking can either grant you a better educational opportunity or stop you from fulfilling your dream.

When you apply for credit, whether it's a new credit card, a car loan, a student loan, or a mortgage, about 40 pieces of information from your credit card report are fed into a model. That model provides a numerical score designed to predict your risk as a borrower. When you apply for university or college admission, more than 20 pieces of information from your application are fed into the model. That model provides a numerical score designed to predict your success rate and risk as a student. In this paper, we will introduce fuzzy query and fuzzy aggregation as an alternative for ranking and predicting risk in areas which currently utilize an imprecise and subjective process.

The areas we will consider include: credit scoring (***Table 1***), credit card ranking (***Table 2***), and university admissions (***Table 3***). Fuzzy query and ranking is robust, provides better insight and a bigger picture, contains more intelligence about an underlying pattern in data and is capable of flexible querying and intelligent searching (Nikravesh, 2001a). This greater insight makes it easy for users to evaluate the results related to the stated criterion and makes a decision faster with improved confidence. It is also very useful for multiple criteria or when users want to vary each criterion independently with different degrees of confidence or weighting factor (Nikravesh, 2001b).

2 Fuzzy Query and Ranking

In the case of crisp queries, we can make multi-criterion decision and ranking where we use the functions AND and OR to aggregate the predicates. In the extended Boolean model or fuzzy logic, one can interpret the AND as a fuzzy-MIN function and the OR as a fuzzy-MAX function. Fuzzy querying and ranking is a very flexible tool in which linguistic concepts can be used in the queries and ranking in a very natural form. In addition, the selected objects do not need to match the decision criteria exactly, which gives the system a more human-like behavior.

2.1 Measure of Association and Fuzzy Similarity

As in crisp query and ranking, an important concept in fuzzy query and ranking applications is the measure of association or similarity between two objects in consideration. For example, in a fuzzy query application, a measure of similarity between two a query and a document, or between two documents, provides a basis for determining the optimal response from the system. In fuzzy ranking applications, a measure of similarity between a new object and a known preferred (or non-preferred) object can be used to define the relative goodness of the new object. Most of the measures of fuzzy association and similarity are simply extensions from their crisp counterparts. However, because of the use of perception

Table 1. Variables, Granulation and Information used to create the Credit Rating System Model.

Variable	Granulation
AOA: Amount owed on accounts is too high. 01	AOA={'Too Low'; 'Low'; 'Average'; 'High'; 'Too High';'Extremely High'; 'Not Care'};
LDA: Level of Delinquency on accounts. 02	LDA = {'Too Low'; 'Low'; 'Average'; 'High'; 'Too High';'Extremely High'; 'Not Care'};
BRA: Too few bank revolving accounts.03	BRA ={'Too Few'; 'Few'; 'Some'; 'Many'; 'Too Many'; 'Not Care'};
BorNRA: Too many bank or national revolving accounts. 04	BorNRA = {'Too Few'; 'Few'; 'Some'; 'Many'; 'Too Many'; 'Not Care'};
RILI: lack of recent installment loan information: 04	RILI = {'Lacking'; 'Not Enough'; 'Enough';'Not Care'};
ACB: Too many accounts with balances. 05	ACB = {'Too Few'; 'Few'; 'Some'; 'Many'; 'Too Many'; 'Not Care'};
CFA: Too many Consumer finance accounts. 06	CFA = {'Too Few'; 'Few'; 'Some'; 'Many'; 'Too Many'; 'Not Care'};
APH: Account payment history too new to rate.07	APH = {'Too New'; 'New'; 'Kind of New'; 'Established'; 'Well Established'; 'Not Care'};
RI: Too many recent inquiries in the last 12 months.08	RI = {'Too Few'; 'Few'; 'Some'; 'Many'; 'Too Many'; 'Not Care'};
AOinL12M: Too many accounts opened in the last 12 months. 09	AOinL12M = {'Too Few'; 'Few'; 'Some'; 'Many'; 'Too Many'; 'Not Care'};
PBtoCLRI: Proportion of balances to credit limits is too high on revolving accounts. 10	PBtoCLRI= {'Too Low'; 'Low'; 'Average'; 'High'; 'Too High';'Extremely High'; 'Not Care'};
AORI: Amount owed on revolving accounts is too high.11	AORI = {'Too Low'; 'Low'; 'Average'; 'High'; 'Too High';'Extremely High'; 'Not Care'};
LRCH: Length of revolving credit history is too short.12	LRCH ={'Too Short'; 'Short'; 'Average'; 'Long'; 'Too Long'; 'Not Care'};
TD: Time since delinquency is too recent or unknown.13	TD = {'Too Recent'; 'Recent'; 'No Recent';'Unkown'; 'Not Care'};
LCH: Length of credit history is too short.14	LCH= {'Too Short'; 'Short'; 'Average'; 'Long'; 'Too Long'; 'Not Care'};
LRBRI: Lack of recent bank revolving information.15	LRBRI = {'Lacking'; 'Not Enough'; 'Enough';'Not Care'};
LRRAI: Lack of recent revolving account information. 16	LRRAI= {'Lacking'; 'Not Enough'; 'Enough';'Not Care'};
RNMBI: No recent non-mortgage balance information.17	RNMBI= {'Too Recent'; 'Recent'; 'No Recent';'Unkown'; 'Not Care'};
NAwD: Number of accounts with delinquency.18	NAwD = {'Too Few'; 'Few'; 'Some'; 'Many'; 'Too Many'; 'Not Care'};
ACPasA: Too few accounts currently paid as agreed.19	ACPasA = {'Too Few'; 'Few'; 'Some'; 'Many'; 'Too Many';'Not Care'};
TDPRorC: Time since derogatory public record or collection.20	TDPRorC ={'Too Short'; 'Short'; 'Average'; 'Long'; 'Too Long';'Not Care'};
APDonA: Amount past due on accounts.21	APDonA = {'Too Low'; 'Low'; 'Average'; 'High'; 'Too High';'Extremely High'; 'Not Care'};
SDDPRorC: Serious delinquency, derogatory public record, or collection.22	SDDPRorC= {'Not Serous'; 'Serious'; 'Very Serious'; 'Extremely Serious'; 'Not Care'};
BorNRAwB: Too many bank or national revolving accounts with balances.23	BorNRAwB = {'Too Few'; 'Few'; 'Some'; 'Many'; 'Too Many'; 'Not Care'};
RB: No recent revolving balances.24	RB = {'Too Recent'; 'Recent'; 'No Recent'; 'Not Care'};
LILH: Length of installment loan history 25	LILH= {'Too Short'; 'Short'; 'Average'; 'Long'; 'Too Long';'Not Care'};
NRA: Number of revolving accounts.26	NRA = {'Too Low'; 'Low'; 'Average'; 'High'; 'Too High';'Extremely High'; 'Not Care'};
BNRorORA: Number of bank revolving or other revolving accounts.26	BNRorORA ={'Too Low'; 'Low'; 'Average'; 'High'; 'Too High';'Extremely High'; 'Not Care'};
ACPasA: Too few accounts currently paid as agreed. 27	ACPasA = {'Too Few'; 'Few'; 'Some'; 'Many'; 'Too Many'; 'Not Care'};
NofEA: Number of established accounts.28	NofEA = {'Too Low'; 'Low'; 'Average'; 'High'; 'Too High';'Extremely High'; 'Not Care'};
DofLI: Date of last inquiry too recent.29	DofLI = {'Too Recent'; 'Recent'; 'No Recent'; 'Not Care'};
BB: No recent bankcard balances.29	BB = {'Too Recent'; 'Recent'; 'No Recent'; 'Not Care'};
TRAO: Time since most recent account opening too short.30	TRAO = {'Too Short'; 'Short'; 'Average'; 'Long'; 'Too Long';'Not Care'};
AwRPI: Too few accounts with recent payment information.31	AwRPI = {'Too Few'; 'Few'; 'Some'; 'Many'; 'Too Many'; 'Not Care'};
AOonDA: Amount owed on delinquent accounts. 31	AOonDA = {'Too Low'; 'Low'; 'Average'; 'High'; 'Too High';'Extremely High'; 'Not Care'};
LofrILI: Lack of recent installment loan information.32	LofrILI = {'Lacking'; 'Not Enough'; 'Enough';'Not Care'};
PofLBtoLA: Proportion of loan balances to loan amounts is too high. 33	PofLBtoLA = {'Too Low'; 'Low'; 'Average'; 'High'; 'Too High';'Extremely High'; 'Not Care'};
LTOILE: Length of time open installment loans have been established * 36	LTOILE = {'Too Short'; 'Short'; 'Average'; 'Long'; 'Too Long';'Not Care'};
NFCAERLFH: Number of finance company accounts established relative to length of finance history 37	NFCAERLFH = {'Too Low'; 'Low'; 'Average'; 'High'; 'Too High';'Extremely High'; 'Not Care'};
SDPRCF: Serious delinquency and public record or collection filed X 38	SDPRCF = {'Not Serous'; 'Serious'; 'Very Serious'; 'Extremely Serious'; 'Not Care'};
SD: Serious delinquency X 39	SD = {'Not Serous'; 'Serious'; 'Very Serious'; 'Extremely Serious'; 'Not Care'};
DPRCF: Derogatory public record or collection filed X 40	DPRCF = {'Not Serous'; 'Serious'; 'Very Serious'; 'Extremely Serious'; 'Not Care'};
LRHFALFA: Lack of recent history on finance accounts, or lack of finance accounts * 99	LRHFALFA = {'Lacking'; 'Not Enough'; 'Enough';'Not Care'};
LRIALAL: Lack of recent information on auto loan, or lack of auto loans * 98	LRIALAL = {'Lacking'; 'Not Enough'; 'Enough';'Not Care'};

Table 2. Variables, Granulation and Information used to create the Credit Card Ranking System Model.

```
% Vis: Vissaa
% VisG: Vissaa Gold
% VisP: Vissaa Platinum
% MSCS: Masters Cards
% MSCSG: Masters Cards Gold
% MSCSP: Masters Cards Platinum
% Amaexs: Americana Experesses
% APR: Annual Percentage Rate
% APRC: Cash Advance APR
% AF: Annual Fee
% GP: Grace Periods
% CAF: Cash Advance Fee
% IIR: Introductory Interest Rate
% RBP: Rebate Programs
% FVR: Fix vs. Variable Rate
% GF: General Fee
% CF: Consumer Feedback
% RI: Reputation of Issuer
% FF: Frequet Flyer
% CA: Card Acceptability
% RCF: Return Check Fee
% LPF: Late Payment Fee
% SI: Security Interest
% DO: Dispute Option
% CS: Customer Service
% SPP: Special Payment Plan
% PP: Partner Programs
% IYR: Itemize Annual Report
```

```
CARDName= {'Vissaa'; 'Vissaa Gold'; 'Vissaa Platinum'; 'Masters Cards'; 'Masters Cards Gold'; ...
    'Masters Cards Platinum'; 'Americana Experesses'; 'Not Care'};
APR={'Very Low'; 'Low'; 'Average'; 'High'; 'Very High';'Extremely High'; 'Not Care'};
APRC={'Very Low'; 'Low'; 'Average'; 'High'; 'Very High';'Extremely High'; 'Not Care'};
AF= {'Very Low'; 'Low'; 'Average'; 'High'; 'Very High';'Extremely High'; 'Not Care'};
GP={'Extremely Short'; 'Very Short'; 'Short'; 'Medium'; 'Long'; 'Very Long'; 'Not Care'};
CAF= {'Very Low'; 'Low'; 'Average'; 'High'; 'Very High';'Extremely High'; 'Not Care'};
IIR= {'Very Low'; 'Low'; 'Average'; 'High'; 'Very High';'Extremely High'; 'Not Care'};
RBP= {'No Rebate'; 'Some Rebate'; 'Good Rebate'; 'Great Rebate'; 'Not Care'};
FVR= {'Fix Rate'; 'Not Quite Fix'; 'Not Quite Vaiable'; 'Variable'; 'Not Care'};
GF= {'Very Low'; 'Low'; 'Average'; 'High'; 'Very High';'Extremely High'; 'Not Care'};
CF= {'Vey Bad'; 'Bad'; 'Not Bad'; 'Average'; 'Good'; 'Great'; 'Not Care'};
RI= {'Vey Bad'; 'Bad'; 'Not Bad'; 'Average'; 'Good'; 'Great'; 'Not Care'};
FF={'No Frequent Flyer'; 'Some Frequent Flyer'; 'Good Frequent Flyer'; 'Great Frequent Flyer'; 'Not Care'};
CA= {'Very Low'; 'Low'; 'Average'; 'High'; 'Very High';'Extremely High'; 'Not Care'};
RCF = {'Very Low'; 'Low'; 'Average'; 'High'; 'Very High';'Extremely High'; 'Not Care'};
LPF = {'Very Low'; 'Low'; 'Average'; 'High'; 'Very High';'Extremely High'; 'Not Care'};
SI = {'Very Low'; 'Low'; 'Average'; 'High'; 'Very High';'Extremely High'; 'Not Care'};
DO={'No Disbute'; 'Some Disbute'; 'Good Disbute'; 'Great Disbute'; 'Not Care'};
CS = {'Vey Bad'; 'Bad'; 'Not Bad'; 'Average'; 'Good'; 'Great'; 'Not Care'};
SPP= {'No Option'; 'Some Option'; 'Good Option'; 'Great Option'; 'Not Care'};
PP={'No Partner'; 'Some Partner'; 'Good Partner'; 'Great Partner'; 'Not Care'};
IYR ={'Yes', 'No'};
```

Table 3. Variables, Granulation and Information used to create the University Admission System Model.

```
% AP : Advanced Placement
% IBHL : International Bacculaureat Higher Level (IBHL)
% HW: Honors and Awards
% GPA: 12th Grade Courses GPA
% CP: Course pattern
% GPAP: Pattern of Grades through time
% SAT II
% SAT I
% CAoSI: Creative Achievement or Sustained Intellectual
% AAaO: Academic Achievement and Outreach
% ClaCV: Contribution to the intellectual and cultural vitality
% DPBaE: Diversity in the Personal Background and Experience
% Leadership
% Motivation
% Concern: Concern for Community and others
% AAA: Achievements; Art or Athletics
% Employment
% IMajor: Interest in the Major
```

```
EthnicName = {'American'; 'Chinese'; 'French'; 'Greek'; 'Indian'; 'Irish'; 'Italian'; 'Japanese'; 'Mediterranean ';'Persian';
'Spanish'; 'Taiwanese', 'Not Care'};
Residency={'California Resident'; 'US Resident'; 'International', 'NotCare' };
Sex={'Male'; 'Female, 'Not Care' };
Minority={'No'; 'Yes'; 'Not Care'};
HW= {'Few'; 'Some'; 'Lot'; 'Not Care'};
AAA= {'Kind of Active'; 'Active'; 'Exceptional'; 'Not Care'};
CP={'Less Than Required'; 'Required'; 'Recommended'; 'Above Recommendation' };
Concern={'Kind of Concern'; 'Concern'; 'Very Concern'; 'Enthusiast'};
Motivation={'Kind of Motivated'; 'Motivated'; 'Highly Motivated'; 'Enthusiast'};
IMajor={'Kind of Interested'; 'Interested'; ' Very Interested'; 'Enthusiast'};
AP= {'Very Low'; 'Low'; 'Medium'; 'High'; 'Very High' };
IBHL= {'Very Low'; 'Low'; 'Medium'; 'High'; 'Very High' };
SATI={'Very Low'; 'Low'; 'Medium'; 'High'; 'Very High' };
SATII={'Very Low'; 'Low'; 'Medium'; 'High'; 'Very High' };
GPA= {'Very Low'; 'Low'; 'Medium'; 'High'; 'Very High' };
Employment={'Few'; 'Average'; 'Kind High'; 'High'; 'Lot'};
CAoSAI= {' Low'; 'Kind Low'; 'Average'; 'Kind of High'; 'High'; 'Exceptional' };
AAaO={' Low'; 'Kind Low'; 'Average'; 'Kind of High'; 'High'; 'Exceptional'};
ClaCV={' Low'; 'Kind Low'; 'Average'; 'Kind of High'; 'High'; 'Exceptional'};
DPBaE={' Low Diversity'; 'Kind Low Diversity'; 'Diverse'; 'Kind of High Diversity'; 'High Diversity'; 'Exceptional'};
Leadership={' Low'; 'Kind Low'; 'Average'; 'Kind of High'; 'High'; 'Exceptional' };
```

based and fuzzy information, the computation in the fuzzy domain can be more powerful and more complex. This section gives a brief overview of various measures of fuzzy association and similarity and various types of aggregation operators involved, along with the description of a simple procedure of utilizing these tools in real applications.

Various definitions of similarity exist in the classical, crisp domain, and many of them can be easily extended to the fuzzy domain. However, unlike in the crisp case, in the fuzzy case the similarity is defined on two fuzzy sets. Suppose we have two fuzzy sets *A* and *B* with membership functions $\mu_A(x)$ and $\mu_B(x)$, respectively. Table 4 lists a number of commonly used fuzzy similarity measures between *A* and *B*. The arithmetic operators involved in the fuzzy similarity measures can be treated using their usual definitions while the union and the intersection operators need to be treated specially. It is important for these operator pairs to have the following properties: (1) conservation, (2) monotonicity, (3) commutativity, and (4) associativity (cf. Table 5 for the definitions of these properties). It can be verified that the triangular norm (T-norm) and triangular co-norm (T-conorm) (Nikravesh, 2001b; Bonissone and Decker, 1986; Mizumoto, 1989; Fagin, 1998 and 1999) conform to these properties and can be applied here. A detailed survey of some commonly used T-norm and T-conorm pairs will be provided shortly along with other aggregation operators.

Table 4. Measures of Association

Simple Matching Coefficient :	$\lvert A \cap B \rvert$
Dice's Coefficient :	$2\dfrac{\lvert A \cap B \rvert}{\lvert A \rvert + \lvert B \rvert}$
Jaccard's Coefficient :	$\dfrac{\lvert A \cap B \rvert}{\lvert A \cup B \rvert}$
Cosine Coefficient :	$\dfrac{\lvert A \cap B \rvert}{\lvert A \rvert^{1/2} \cdot \lvert B \rvert^{1/2}}$
Overlap Coefficient :	$\dfrac{\lvert A \cap B \rvert}{min(\lvert A \rvert, \lvert B \rvert)}$
Disimilarity Coefficient :	$\dfrac{\lvert A \Delta B \rvert}{\lvert A \rvert + \lvert B \rvert} =$
1 − Dice's Coefficient :	$\lvert A \Delta B \rvert = \lvert A \cup B \rvert - \lvert A \cap B \rvert$

While any of the five fuzzy similarity measures can be used in an application, they have different properties. The Simple Matching Coefficient essentially generalizes the inner product and is thus sensitive to the vector length. The Cosine Coefficient is a simple extension to the Simple Matching Coefficient but normalized with respect to the vector lengths. The Overlap Coefficient computes the degree of overlap (the size of intersection) normalized to the size of the smaller of the two fuzzy sets. The Jaccard's Coefficient is an extension to the Overlap Coefficient by using a different normalization. The Dice's Coefficient is yet another extension to the Overlap Coefficient, and both the Jaccard's and Dice's Coefficients are frequently used in traditional information retrieval applications.

In the definition of all five similarity metrics, appropriate aggregation operator pairs are substituted in place of the fuzzy intersection ($\cap$) and fuzzy union operators ($\cup$). As discussed previously, a number of different T-norm and T-conorm pairs are good candidates for this application. There exist many different types of T-norm and T-conorm pairs (Mizumoto, 1989), and they are all functions from $[0,1]x[0,1] \rightarrow [0,1]$ and conform to the list of properties in Table 5. Table 6 shows a number of commonly used T-norm and T-conorm pairs that we consider here. Note that each pair of T-norm and T-conorm satisfies the DeMorgan's law: $\sim T(x,y) = S(\sim x, \sim y)$ where "~" is the negation operator defined by $\sim x = 1\text{-}x$.

The minimum and the maximum are the simplest T-norm and T-conorm pair. It can be verified that the minimum is the largest T-norm in the sense that $T(x,y) \leq min(x,y)$ for any T-norm operator T. Similarly, the maximum is the smallest T-conorm. Both the minimum and the maximum are idempotent since $min(x,x)=x$ and $max(x,x)=x$ for any x.

Contrary to the minimum the drastic product produces as small a T-norm value as possible without a violation of the properties in Table 5. Similarly, the drastic sum produces as large a T-conorm value as possible. Thus, the value produced by any other T-norm (T-conorm) operator must lie between the minimum (maximum) and the drastic product (drastic sum).

The bounded difference and its dual, the bounded sum, are sometimes referred to as the Lukasiewicz T-norm and T-conorm. It is important to note that they conform to the law of excluded middle of classific bivalent logic, i.e. $T(x,\sim x)=0$ and $S(x,\sim x)=1$.

The algebraic product and algebraic sum have intuitive interpretations in the probabilistic domain as being the probability of the intersection and the union of two independent events, respectively. In addition, they are smooth functions that are continuously differentiable.

Besides the fixed T-norm and T-conorm pairs described above, there are also a number of parametric T-norm and T-conorm pairs that contain a free parameter for adjusting the behavior (such as softness) of the operators. A commonly used pair due to Hamacher is defined as: $T(x,y) = xy / [p+(1-p)(x+y-xy)]$ and $S(x,y) = [x+y-xy-(1-p)xy]/[1-(1-p)xy]$ where $p \geq 0$ is free parameter. In particular, we obtain the Hamacher product/sum and the Einstein product/sum (cf. Table 6) by setting p to 0 and 2, respectively.

So far we have introduced several different types of fuzzy association/similarity metrics involving a variety T-norm and T-conorm pairs. An appropriate similarity metric can be selected to compute the distance between two objects according to the requirements of a particular application. In most practical applications we may have to consider more than one attribute when comparing two objects. For example, computing the similarity between two students' academic achievements may require separate comparisons for different subjects, e.g. sciences, mathematics, humanities, etc. Thus, it is useful to have a principled manner for aggregating partial similarity scores between two objects computed on individual attributes. We call such a function an aggregation operator (or simply an aggregator) and define it as a function $f: [0,1]x...x[0,1] \rightarrow [0,1]$.

As for the similarity metric, there are a variety of aggregation operators to choose from, depending on the nature of a particular application (Detyniecki M, 2000). Given our discussion of the T-norm and T-conorm operators, it should not be surprising that many T-norms and T-conorms can be used as aggregation operators. In particular, the associative property (cf. Table 5) of T-norms and T-conorms make them applicable in aggregating more than two values. Intuitively, T-norm aggregators have a minimum-like (or conjunctive) behavior while T-conorms have a maximum-like (or disjunctive) behavior, and these behaviors should be taken into account in selecting an appropriate aggregator to use.

Table 5. Properties of aggregation operators for triangular norms and triangular co-norms.

• *Conservation* $t(0,0) = 0; t(x,1) = t(1,x) = x$	• *Conservation* $s(1,) = 1; s(x,0) = s(0,x) = x$
• *Monotonicity* $t(x_1,x_2) \leq t(x'_1,x'_2)$ *if* $x_1 \leq x'_1$ *and* $x_2 \leq x'_2$	• *Monotonicity* $s(x_1,x_2) \leq s(x'_1,x'_2)$ *if* $x_1 \leq x'_1$ *and* $x_2 \leq x'_2$
• *Commutativity* $t(x_1,x_2) = t(x_2,x_1)$	• *Commutativity* $s(x_1,x_2) = s(x_2,x_1)$
• *Associativity* $t(t(x_1,x_2),x_3) = t(x_1,t(x_2,x_3))$	• *Associativity* $s(s(x_1,x_2),x_3) = s(x_1,s(x_2,x_3))$

Table 6. Triangular norm/triangular co-norm pairs.

Minimum : $t(x_1, x_2) = \min\{x_1, x_2\}$
Maximum : $s(x_1, x_2) = \max\{x_1, x_2\}$

Drastic Product : $t(x_1, x_2) = \begin{cases} \min\{x_1, x_2\} & \text{if } \max\{x_1, x_2\} = 1 \\ 0 & \text{othewise} \end{cases}$

Drastic sum : $s(x_1, x_2) = \begin{cases} \max\{x_1, x_2\} & \text{if } \min\{x_1, x_2\} = 0 \\ 1 & \text{othewise} \end{cases}$

Bounded difference : $t(x_1, x_2) = \max\{0, x_1 + x_2 - 1\}$
Boubded sum : $s(x_1, x_2) = \min\{1, x_1 + x_2\}$

Einstein product : $t(x_1, x_2) = (x_1 \cdot x_2)/(2 - (x_1 + x_2 - x_1 \cdot x_2))$
Einstein sum : $s(x_1, x_2) = (x_1 + x_2)/(1 + x_1 \cdot x_2)$

Algebraic product : $t(x_1, x_2) = x_1 \cdot x_2$
algebraic sum : $s(x_1, x_2) = x_1 + x_2 - x_1 \cdot x_2$

Hamacher product : $t(x_1, x_2) = (x_1 \cdot x_2)/(x_1 + x_2 - x_1 \cdot x_2)$
Hamacher sum : $s(x_1, x_2) = (x_1 + x_2 - 2x_1 \cdot x_2)/(1 - x_1 \cdot x_2)$

Table 7. Fuzzy-Min and Fuzzy-Max Operators.

Conjunction rule : $\mu_{A \wedge B}(x) = \min\{\mu_A(x), \mu_B(x)\}$

Disjunction rule : $\mu_{A \vee B}(x) = \max\{\mu_A(x), \mu_B(x)\}$

Negation rule : $\mu_{\sim A}(x) = 1 - \mu_A(x)$

$\mu_{A \wedge A}(x) = \mu_A(x)$

$\mu_{A \wedge (B \vee C)}(x) = \mu_{(A \wedge B)}(x) \vee \mu_{(A \wedge C)}(x)$

If : $\mu_A(x) \leq \mu_A(x')$ **AND** $\mu_B(x) \leq \mu_B(x')$
Then: $\mu_{A \wedge B}(x) \leq \mu_{A \wedge B}(x')$

If Query (A) and Query (B) are equivalent:
$\mu_A(x) = \mu_B(x)$

One of the simplest aggregation operators is the arithmetic mean: $f(x_1,...,x_N) = (x_1+...+x_N)/N$. This simple averaging operator is often considered as the most unbiased aggregator when no further information is available about an application. It is also most applicable when different attributes all have relatively even importance or relevance to the overall aggregated result.

A simple extension of the arithmetic mean, the linearly weighted mean, attaches different weights to the attributes, and is defined by: $f(x_1,...,x_N) = (w_1x_1+...+w_Nx_N)/N$ where $w_1,...,w_N \geq 0$ are linear weights assigned to different attributes and the weights add up to one. The weights can be interpreted as the relative importance or relevance of the attributes and can be specified using domain knowledge or from simple linear regression.

Extension to the arithmetic mean also includes the geometric mean: $f(x_1,...,x_N) = (x_1...x_N)^{1/n}$ which is equivalent to taking the arithmetic mean in the logarithmic domain (with an appropriate exponential scaling), and the harmonic mean: $f(x_1,...,x_N) = n/(1/x_1+...+1/x_N)$ which is particularly appropriate when the x_i's are rates (e.g. units/time). Both geometric mean and harmonic mean also have their weighted versions.

Another family of non-linear aggregation operator involves ordering of the aggregated values. This family includes the median, the k-order statistic, and more generally the ordered weighted average. For N values in ascending order the median is taken to be the $(N+1)/2$'th value if N is odd or the average of the $N/2$ and $N/2+1$'th value if N is even. The k-order statistics generalizes the median operator to take the k'th value, thus including median, minimum, and maximum as special cases. The ordered weighted average (OWA), first introduced by Yager (1988), generalizes both the k-order statistic and the arithmetic mean and is defined as: $f(x_1,...,x_N) = w_1x_{\sigma(1)}+...+w_Nx_{\sigma(N)}$ where w's are non-negative and add up to one, and $x_{\sigma(i)}$ denotes the i'th value of x's in ascending order. By using appropriate weights OWA provide a compromise between the conjunctive behavior of the arithmetic mean and the disjunctive behavior of the k-order statistic.

Finally, it is of interest to include in our discussion a family of aggregators based on fuzzy measures and fuzzy integrals since they subsume most of the aggregators described above. The concept of fuzzy measure was originally introduced by Sugeno (Sugeno, 1974) in the early 1970's in order to extend the classical (probability) measure through relaxation of the additivity property. A formal definition of the fuzzy measure is as follows:

Definition 1. Fuzzy measure: Let X be a non-empty finite set and Ω a Boolean algebra (i.e. a family of subsets of X closed under union and complementation, including the empty set) defined on X. A fuzzy measure, g, is a set function $g:\Omega \rightarrow [0,1]$ defined on Ω, which satisfies the following properties: (1) Boundary conditions: $g(\phi)=0$, $g(X)=1$. (2) Monotonicity: If $A \subseteq B$, then $g(A) \leq g(B)$. (3) Continuity: If $F_n \in \Omega$ for $1 \leq n < \infty$ and the sequence $\{F_n\}$ is monotonic (in the sense of inclusion), then $lim_{n \to \infty} g(F_n) = g(lim_{n \to \infty} F_n)$. And (X, Ω, g) is said to be a fuzzy measure space.

To aggregate values with respect to specific fuzzy measures a technique based on the concept of the fuzzy integral can be applied. There are actually several forms of fuzzy integral; for brevity let us focus on only the discrete Choquet integral proposed by Murofushi and Sugeno (1989).

Definition 4 (Choquet) Fuzzy integral: Let (X, Ω, g) be a fuzzy measure space, with $X = \{x_1, \cdots, x_N\}$. Let $h: X \to [0,1]$ be a measurable function. Assume without loss of generality that $0 \leq h(x_1) \leq \cdots \leq h(x_N) \leq 1$, and $A_i = \{x_i, x_{i+1}, \cdots, x_N\}$. The Choquet integral of h with respect to the fuzzy measure g is defined by

$$\int_C h \circ g = \sum_{i=1}^{N} [h(x_i) - h(x_{i-1})] g(A_i) \tag{1}$$

where $h(x_0) = 0$.

An interesting property of the (Choquet) fuzzy integral is that if g is a probability measure, the fuzzy integral is equivalent to the classical Lebesgue integral and simply computes the expectation of h with respect to g in the usual probability framework. The fuzzy integral is a form of averaging operator in the sense that the value of a fuzzy integral is between the minimum and maximum values of the h function to be integrated. It can be verified that most of the aggregation operators we have described so far, including the minimum, maximum, median, arithmetic mean, weighted average, k-order statistic, ordered-weighted average, are all special cases of the Choquet fuzzy integral. A distinct advantage of the fuzzy integral as a weighted operator is that, using an appropriate fuzzy measure, the weights represent not only the importance or relevance of individual information sources but also the interactions (redundancy and synergy) among any subset of the sources. However, the representational power of fuzzy integrals and fuzzy measures comes at the expense of having a greater number of free parameters to specify. For N attributes a full specification of fuzzy measures requires 2^N-2 numbers. Alternatives such as using a decomposable k-additive fuzzy measure have been proposed to trade off the number of parameters and the representational power (Grabisch, 1996). Further description of these alternatives, as well as techniques for specifying and learning fuzzy measures, are beyond the scope of this paper and interested readers can refer to (Grabisch et al., 2000).

Having introduced a variety of tools that are required to evaluate fuzzy association/similarity between two objects, a simple algorithm in pseudo code is provided below to illustrate how these machineries can be used in a practical implementation.

Input: two objects A and B
  A: N discrete attributes
    For the ith attribute, A^i is an array of length M^i, where M^i is the number of possible linguistic values of the ith attribute.
    i.e. each A_j^i, i in 1,…,N and j in 1,…,M^i, gives the degree of A's i^{th} attribute having j^{th} linguistic value.
  B: similar to A with the same dimensions.

Other parameters:
  AggregatorType
  SimilarityType
  TNormType
  OptionalWeights

Output: An aggregated similarity score between A and B

Algorithm:
  For each i=1 to N
    SAB^i = ComputeSimilarity(A^i, B^i ,SimilarityType, TNormType)
  End
  Return Aggregate(SAB, AggregatorType, OptionalWeights)

  Sub ComputeSimilarity(X, Y, SimilarityType, TNormType)
    Switch SimilarityType:
    Case SimpleMatchingCoefficient:
      Return $|X \cap Y|$

    Case CosineCoefficient:
      Return $|X \cap Y| / (|X|^{½} |Y|^{½})$

    Case OverlapCoefficient:
      Return $|X \cap Y| / \min(|X|, |Y|)$

    Case Jaccard's Coefficient:
      Return $|X \cap Y| / (|X \cup Y|)$

    Case Dice's Coefficient:
      Return $2|X \cap Y| / (|X| + |Y|)$

    …
  End

  Sub Aggregate(S, AggregatorType, OptionalWeights)

```
        Switch AggregatorType:
        Case Min:
                Return min(S)
        Case Max:
                Return max(S)
        Case Mean:
                Return mean(S)
        Case Median:
                Return median(S)
        Case WeightedAverage:
                Return WeightedAverage(S, OptionalWeights)
        Case OrderedWeightedAverage:
                Return OrderedWeightedAverage(S, OptionalWeights)
        Case ChoquetIntegral:
                Return ChoquetIntegral(S, OptionalWeights)
        Case SugenoIntegral:
                Return SugenoIntegral(S, OptionalWeights)
        ...
End
```

This algorithm takes as input two objects, each with *N* discrete attributes. Similarity scores between the two objects are first computed with respect to each attribute separately, using a specified similarity metric and T-norm/conorm pair. As described previously, the computation of a similarity score with respect to an attribute involves a pair wise application of the T-norm or T-conorm operators on the possible values of the attribute, followed by other usual arithmetic operation specified in the similarity metric. Finally, an aggregation operator with appropriate weights is used to combine the similarity measures obtained with respect to different attributes.

In many situations, the controlling parameters, including the similarity metric, the type of T-norm/conorm, the type of aggregation operator and associated weights, can all be specified based on the domain knowledge of a particular application. However, in some other cases, it may be difficult to specify a priori an optimal set of parameters. In those cases, various machine learning methods can be employed to automatically "discover" a suitable set of parameters using a supervised or unsupervised approach. For example, the Genetic Algorithm (GA) and DNA-based computing, as described in later sections, can be quite effective.

2.2 Precisions and Recall Measure

Table 8 and ***Figure 1*** show the definition of precision, recall and their relationship. Given a user's criteria, the data provided for modeling, and the strategy defined in ***Figure 2***, the recall/precision relationship has been optimized. Therefore, a user will get better precision and recall in fuzzy or imprecise situations.

Table 8. Measures of Precision, Recall and several other relevant attributes.

$$Precision: P = \frac{|A \cap B|}{|B|}$$

$$Recall: R = \frac{|A \cap B|}{|A|}$$

$$Fallout: F = \frac{|\overline{A} \cap B|}{|\overline{A}|}$$

$$Generality: G = \frac{|A|}{N}$$

$Retrieved/Relevent: A \cap B$

$Retrieved/Non-Relevent: \overline{A} \cap B$

$Not-Retrieved/Relevent: A \cap \overline{B}$

$Not-Retrieved/Not-Relevent: \overline{A} \cap \overline{B}$

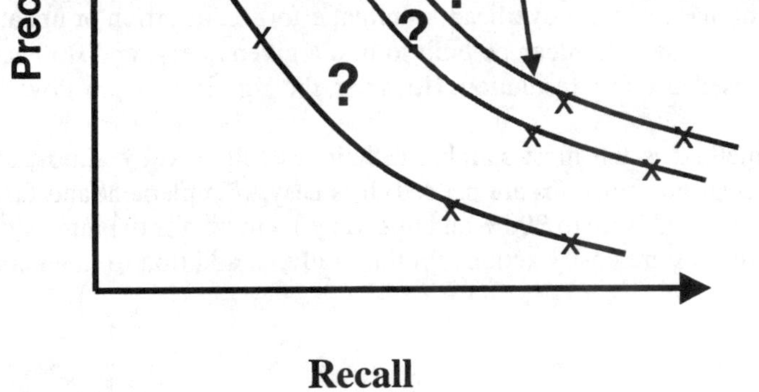

Figure 1. Inverse relationship between Precision and Recall.

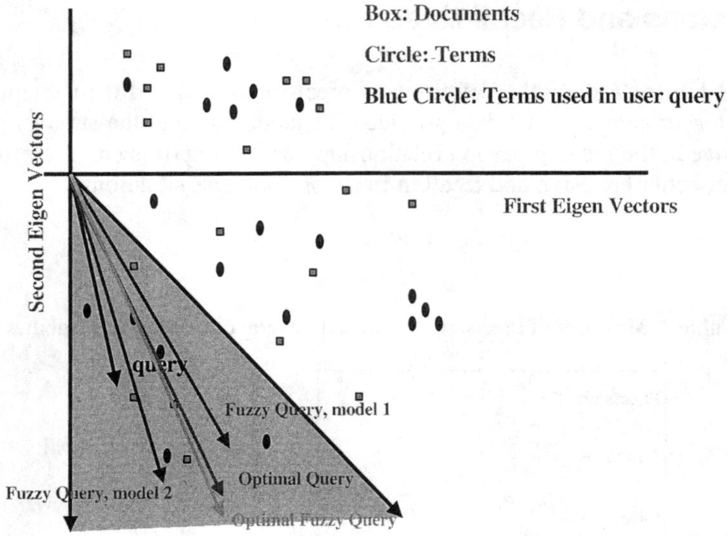

Figure 2. Schematic diagram of the performance of the Fuzzy-Latent Semantic Indexing method.

2.3 Search Strategy

There are several ways to search and query in databases such as *Latent Semantic Indexing (LSI),* full text scanning, inversion, and the use of signature files. While *LSI* has limitations, it is highly rewarding, since it is easy to implement and update; it is fast; it works in a reduced domain; it is scaleable; and it can be used for parallel processing. One solution to its Boolean model is to use an extended Boolean model or fuzzy logic. In this case, one can add a fuzzy quantifier to each term or concept. In addition, one can interpret the AND as a fuzzy-MIN function and the OR as a fuzzy-MAX function respectively.

The most straightforward way to search is ***full text scanning***. The technique is simple to implement; has no space overhead; minimal effort on insertion or update is needed; a finite state automaton can be built to find a given query; and Boolean expressions can be used as query resolution. However, the algorithm is too slow.

The ***inversion*** method is the most suitable techniques followed by almost all commercial systems (if no semantics are needed).It is easy to implement and fast. However, storage overhead is up to 300% and updating the index for dynamic systems and merging of lists are costly actions. In this study, in addition to inversion techniques, ***Fuzzy-Latent Semantic Indexing (FLSI)*** originally developed for text retrieval has been used (Nikravesh, 2001a and 2001b). ***Figure 2*** shows a schematic diagram of the performance of ***FLSI***. ***Figure 3*** and ***Figure 4*** show the performance of ***FLSI*** for text retrieval purposes. The following briefly describes the **FLSI** technique (Nikravesh, 2001a and 2001b):

Fuzzy-based decompositions are used to approximate the matrix of document vectors.

Terms in the document matrix may be presented using linguistic terms (or fuzzy terms such as most likely, likely, etc) rather than frequency terms or crisp values.

Decompositions are obtained by placing a fuzzy approximation onto the eigensubspace spanned by all the fuzzy vectors.

Empirically, we establish our technique such that the approximation errors of the fuzzy decompositions are close to the best possible; namely, to truncated singular value decompositions.

The followings are the potential applications of the FLSI:

1. *Search Engines:* The recent explosion of online information on the World Wide Web has given rise to a number of query-base search engines. However, this information is useless unless it can be effectively and efficiently searched.

2. *Fuzzy Queries in Multimedia Database Systems:* Even though techniques exist for locating exact matches for traditional database, finding relevant partial matches for Multimedia database systems might be a problem. It may not be also easy to specify query requests precisely and completely - resulting in a situation known as a fuzzy-querying.

3. *Query Based on User Profile:* It employs as combinations of technologies that take the result of the queries and organize them into categories for presentation to the user. The system can then save such document organizations in user profiles, which can then be used to help classify future query results by the same user.

4. *Information Retrievals:* The goal in information retrieval is to find documents that are relevant to a given user query.

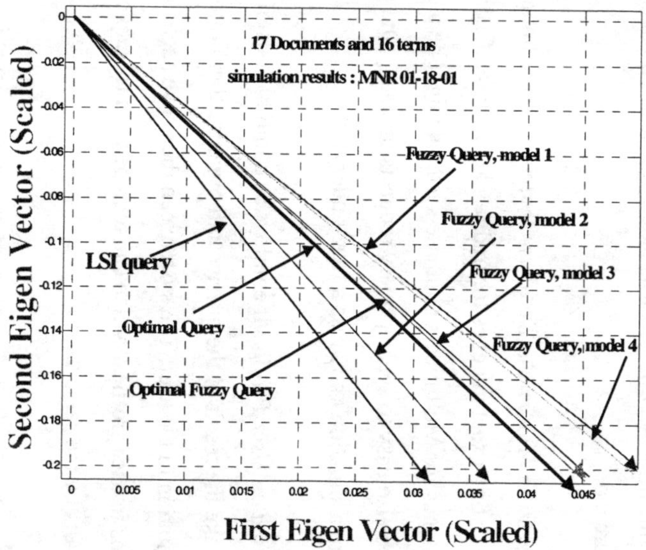

Figure 3. Example 1 of FLSI for text retrieval.

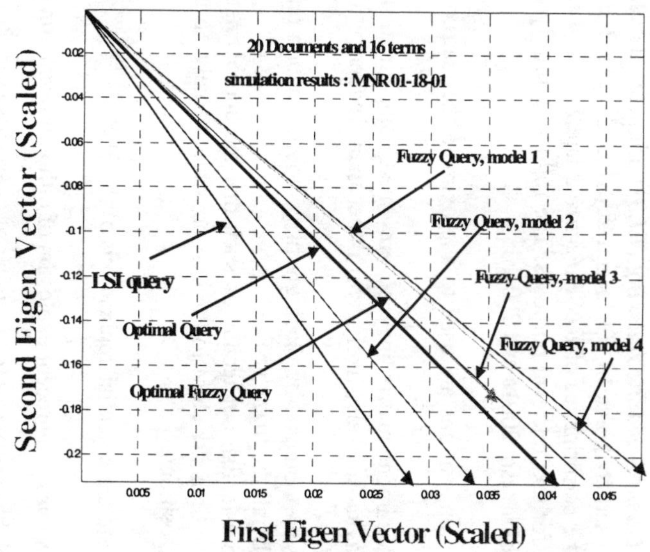

Figure 4. Example 2 of FLSI for text retrieval.

5. *Summary of Documents:* Human-quality text summarization systems are difficult to design, and even more difficult to evaluate, in part because documents can differ along several dimensions, such as length, writing style and lexical usage.

   - Text Summarization-Single Document
   - Text Summarization-Multi Documents

   Multi-document summarization differs from single in that the issues of compression, speed, redundancy and passage selection are critical in the formation of useful summaries.

6. *Information Fusion Such as Medical Records, Research Papers, News, etc:* Two groups of database or News are generated independently of each other, quantified the same n terms in the same m documents. The documents or NEWS form the two groups are similar but not necessarily identical. We are interested in merging documents or NEWS.

7. *File and Folder Organiser:* Organizers operate on data matrix (e.g., terms X file or folder; names or date X file or folder; etc.) to derive similarities, degree of match, clusters, and derive rules.

8. *Matching People*: Matching People operate on data matrix (e.g., Interests X People; Articles X people; etc.) to derive similarities and degree of match.

9. *Association Rule Mining for Terms-Documents:* Association Rule Mining algorithm operates on data matrix (e.g., Terms X Documents) to derive rules.

   i) Documents Similarity; Search Personalization-User Profiling. *Often time it is hard to find the "right" term and even in some cases the term does not exist.* The User Profile is automatically constructed from text document collection and can be used for Query Refinement and provide suggestions and for ranking the information based on pre-existence user profile.

ii) Terms Similarity; Automated Ontology Generation and Automated Indexing The ontology is automatically constructed from text document collection and can be used for Query Refinement.

10. *E-mail Notification:* E-mail notification whenever new matching documents arrive in the database, with link directly to documents or sort incoming messages in right mailboxes

11. *Modelling Human Memory:* The Technique can be used in some degree to model some of the associative relationships observed in human memory abased on term-term similarities.

12. *Calendar Manager*: automatically schedule meeting times.

13. *Others:* Telephony, Call Center, Workgroup Messages, E-Mail, Web-Mail, Personal Info, Home-Device Automation, etc.

2.4 Intelligent Data Mining: Fuzzy- Evolutionary Computing (Nikravesh 2002, 2003a, and 2003b and Loia et al. 2003)

2.4.1. Pattern Recognition

In the 1960s and 1970s, pattern recognition techniques were used only by statisticians and were based on statistical theories. Due to recent advances in computer systems and technology, artificial neural networks and fuzzy logic models have been used in many pattern recognition applications ranging from simple character recognition, interpolation, and extrapolation between specific patterns to the most sophisticated robotic applications. To recognize a pattern, one can use the standard multi-layer perceptron with a back-propagation learning algorithm or simpler models such as self-organizing networks (Kohonen, 1997) or fuzzy c-means techniques (Bezdek, 1981; Jang and Gulley, 1995). Self-organizing networks and fuzzy c-means techniques can easily learn to recognize the topology, patterns, and distribution in a specific set of information.

2.4.2 Clustering

Cluster analysis encompasses a number of different classification algorithms that can be used to organize observed data into meaningful structures. For example, k-means is an algorithm to assign a specific number of centers, k, to represent the clustering of N points (k<N). These points are iteratively adjusted so that each point is assigned to one cluster, and the centroid of each cluster is the mean of its assigned points.

In general, the k-means technique will produce exactly k different clusters of the greatest possible distinction. Alternatively, fuzzy techniques can be used as a method for clustering. Fuzzy clustering partitions a data set into fuzzy clusters such that each data point can belong to multiple clusters. Fuzzy c-means (FCM) is a well-known fuzzy clustering technique that generalizes the classical (hard) c-means algorithm and can be used where it is unclear how many clusters there should be for a given set of data. Subtractive clustering is a fast, one-pass algorithm for estimating the number of clusters and the cluster centers in a set of data. The cluster estimates obtained from subtractive clustering can be used to initialize iterative optimization-based clustering methods and model identification methods.

In addition, the self-organizing map technique known as Kohonen's self-organizing feature map (Kohonen, 1997) can be used as an alternative for clustering purposes. This technique converts patterns of arbitrary dimensionality (the pattern space) into the response of one- or two-dimensional arrays of neurons (the feature space). This unsupervised learning model can discover any relationship of interest such as patterns, features, correlations, or regularities in the input data, and translate the discovered relationship into outputs.

2.4.3 Mining and Fusion of Data

In the past, classical data processing tools and physical models solved many real-world complex problems. However, this should not obscure the fact that the world of information processing is changing rapidly. Increasingly we are faced on the one hand with more unpredictable and complex real-world, imprecise, chaotic, multi-dimensional and multi-domain problems with many interacting parameters in situations where small variability in parameters can change the solution completely. On the other hand, we are faced with profusion and complexity of computer-generated data. Unfortunately, making sense of these complex, imprecise and chaotic data which are very common in Engineering and science applications, is beyond the scope of human ability and understanding. What this implies is that the classical data processing tools and physical models that have addressed many complex problems in the past may not be sufficient to deal effectively with present and future needs.

Tables 9 and **10** show the list of the Data Fusion (dominated by Integration process) and Data Mining techniques (Dominated by Interpretation process)

Table 9. Data Mining Techniques (Interpretation)

Deductive Database Client
Inductive Learning
Clustering
Case-based Reasoning
Visualization
Statistical Package

Table 10. Data Fusion Techniques (Integration)

Deterministic

-Transformation based (projection, ...)
-Functional evaluation based (vector quantization, ...)
-Correlation based (pattern match, if/then productions)
-Optimization based (gradient-based, feedback, LDP, ...)
Non-deterministic

-Hypothesis testing (classification, ...)
-Statistical estimation (maximum likelihood, ...)
-Discrimination function (linear aggregation, ...)
-Neural network (supervised learning, clustering, ...)
-Fuzzy Logic (fuzzy c-mean clustering, ...)
-Hybrid (genetic algorithm, Bayesian network, ...)

2.4.4 Intelligent Information Processing

In conventional information processing technique, once all the pertinent data is properly fused, one has to extract the relevant information from the data and draw the necessary conclusions. This can be done either true reliance on human expert or an intelligent system that has the capability to learn and modify its knowledge base as new information become available. In intelligent information processing techniques, the process of information fusion is an integrated part of the information mining. **Table 11** shows the comparison between Conventional and intelligent techniques for information processing.

Table 11. Conventional Vs. Intelligent

Conventional ------------- -Data assumption: a certain probability distribution -Model: weight functions come from varigram trend and probability constraints -Simulation: Stochastic, not optimized Intelligent ------------ -Data automatic clustering and expert-guided segmentation -Classification of relationship between data and targets -Model: weight functions come from supervised training based on initial known information Simulation: optimized by GA, SA, ANN, and BN

2.4.5 Data Mining

Data Mining or "classification to explore a dataset" is a trend in clustering techniques in which the user has no or little prior assumptions about the data, but wants to explore if data or subset of data falls into "meaningful group" (a term for which the user may not even have a specific definition). Many clustering and data mining algorithm assume a certain type of input such as numerical (in case of k-means) or categorical input. In addition, most techniques either use a prior knowledge to define distance or similarity measure or use probabilistic techniques which break down as the dimensionality of the corresponding feature space increases. It is also require a prior knowledge about the problem domain to fix the number and starting points in which it is clearly not accessible easily where the number of input pararemeters are very large in hyperspace. Finally the clustering problem is an optimization problem and is known to be NP-hard problem.

When data is imprecise and has mix nature (numerical and categorical) and several objectives to be matched at the same time, the optimization problem may be more complex and will fall into Multi-Objective and Multi-Criteria with conflicting objectives which in this case, the conventional techniques could not be applied.

A unified approach based on soft computing will help fill the existing technology gap and is bound to play a key role in solving the above problems. Soft computing is consortium of computing methodologies (Fuzzy Logic (GL), Neuro Computing (NC), Genetic Computing (GC), and Probabilistic Reasoning (PR) including ; Genetic Algorithms (GA), Chaotic Systems (CS), Belief Networks (BN), Learning Theory (LT)) which collectively provide a foundation for the Conception, Design and Deployment of Intelligent Systems. Among main components of soft computing are the artificial neural computing, fuzzy logic computation, and the evolutionary computing.

The intelligent computing techniques will establish a unified framework to solve the above challenges using Soft Computing Techniques (SCT) to utilize the specific strength of each method to address different aspects of the problem. Fuzzy Logic ideal for handling subjective and imprecise information, uncertainty management and knowledge integration. Neural network powerful tool for self-learning and data integration and does not require specification of structural relationships between the input and output data. Evolutionary Computing is effective for handling scale problems, dynamic updating, for pattern extraction, reduce the complexity of the neuro-fuzzy model, and robust optimization along the multidimensional, highly nonlinear and non-convex search hyper-surfaces.

Motivated by current advances in DNA computing which has been showed promises toward solving complex problem including "NP-complete" problems such as Hamiltonian path problem and Satisfiability Problem with ability to pursue an unbounded number of independent computational searches in parallel, we will use Artificial DNA computing to solve the optimization problem.

2.4.6. Genetic Algorithm

Genetic algorithm (GA) is one of the stochastic optimization methods which is simulating the process of natural evolution. GA follows the same principles as those in nature (survival of the fittest, Charles Darwin).

GA first was presented by John Holland as an academic research. However, today GA turns out to be one of the most promising approaches for dealing with complex systems which at first nobody could imagine that from a relative modest technique. GA is applicable to multi-objectives optimization and can handle conflicts among objectives. Therefore, it is robust where multiple solutions exist. In addition, it is highly efficient and it is easy to use.

Another important feature of GA is its capability of extraction of knowledge or fuzzy rules. GA is now widely used and applied to discovery of fuzzy rules. However, when the data sets are very large, it is not easy to extract the rules.

2.4.7 DNA Computing: Intelligent Data Mining Techniques

To overcome such a limitation, a new coding technique is needed. Motivated by current advances in DNA computing which has been showed promises toward solving complex problem including "NP-complete" problems such as Hamiltonian path problem and Satisfiability Problem with ability to pursue an unbounded number of independent computational searches in parallel, we will use a new coding method based on biological DNA and Artificial DNA computing to solve the optimization problem.

The DNA can have many redundant parts which is important for extraction of knowledge. In addition, this technique allows overlapped representation of genes and it has no constraint on crossover points. Also, the same type of mutation can be applied to every locus. In this technique, the length of chromosome is variable and it is easy to insert and/or delete any part of DNA chromosomes. Since the length of the chromosome in artificial DNA coding is variable, it will be very easy to include genetic operations such as virus and enzyme operations. This process and the overlap and redundancy of genes will give the genes the ability to adapt, which increases the chance of survival of genes far beyond the lifetime of individuals.

Artificial DNA algorithm can be used in a hierarchical fuzzy model for pattern extraction and to reduce the complexity of the neuro-fuzzy models. In addition, artificial DNA can be use to extract the number of the membership functions required for each parameter and input variables.

The DNA coding method and the mechanism of development from artificial DNA are suitable for knowledge extraction including fuzzy IF ...THEN from large data set for Data Mining purposes. The rules are extracted from DNA chromosomes as follows. Each artificial amino acid has several meaning. The meaning of genes is determined by the combination of the amino acids. Each amino acid can be translated into an input variable and its membership function. A sequence of amino acids (one genes) corresponds to one fuzzy rule. The Artificial DNA chromosomes having several genes make up a set of fuzzy rules. Each rule represent a subset of data. Therefore, not only data will be mined and clustered but also will be translated into factual knowledge given the linguistic nature of the IF ... THEN rules. This will give a new ability to the user such that the rules based on factual knowledge (data) and knowledge drawn from human experts (inference) will be combined, ranked, and clustered based on the confidence level of human and factual support. This will effectively provide validation of an interpretation, a model, a hypothesis, or alternatively indicate a need for rejection or reevaluation. This will also provide the ability to answer "What if?" questions in order to decrease uncertainty during the process of data Mining and knowledge extraction.

We claim that Fuzzy- artificial DNA can be used for robust optimization along the multidimensional, highly nonlinear and non-convex search hyper-surfaces, generalize its estimation through evolution and manage the uncertainty through fuzzy based technique, even though the environment may partially observable.

The main features of the new methodologies are:

- It uses minimal prior knowledge with respect to the input structure of data and its probability distribution
- Minimal a prior knowledge require about the problem domain to fix the number and starting points
- Can be used to solve optimization problems known as NP-hard problem.
- Can be used when data is imprecise and has mix nature (numerical and categorical)
- Can be used when several objectives to be matched at the same time
- Can be used for Multi-Objective and Multi-Criteria optimization with conflicting objectives
- Scalability/parallel processing
- Can be used for high dimensionality in the feature space with respect to data/problem-space (sparse-data)
- Can extract both the cluster and association rules given certain objective

3 Implementation - Fuzzy Query and Ranking

In this section, we introduce fuzzy query and fuzzy aggregation for credit scoring, credit card ranking, and university admissions.

3.1 Application to Credit Scoring

Credit scoring was first developed in the 1950's and has been used extensively in the last two decades. In the early 1980's, the three major credit bureaus, Equitax, Experian, and TransUnion worked with the Fair Isaac Company to develop generic scoring models that allow each bureau to offer an individual score based on the contents of the credit bureau's data. FICO is used to make billions of financing decisions each year serving a 100 billion dollar industry. Credit scoring is a statistical method to assess an individual's credit worthiness and the likelihood that the individual will repay his/her loans based on their credit history and current credit accounts. The credit report is a snapshot of the credit history and the credit score is a snapshot of the risk at a particular point in time. Since 1995, this scoring system has made its biggest contribution in the world of mortgage lending. Mortgage investors such as Freddie Mac and Fannie Mae, the two main government-chartered companies that purchase billion of dollars of newly originated home loans annually, endorsed the Fair Isaac credit bureau risk, ignored subjective considerations, but agreed that lenders should also focus on other outside factors when making a decision.

When you apply for financing, whether it's a new credit card, car or student loan, or a mortgage, about 40 pieces of information from your credit card report are fed into a model (***Table 1***). This information is categorized into the following five categories with different level of importance (% of the score):

- Past payment history (35%)
- Amount of credit owed (30%)
- Length of time credit established (15%)
- Search for and acquisition of new credit (10%)
- Types of credit established (10%)

When a lender receives your Fair Isaac credit bureau risk score, up to four "score reason codes" are also delivered. These explain the reasons why your score was not higher. Followings are the most common given score reasons (Fair Isaac);

- Serious delinquency
- Serious delinquency, and public record or collection filed
- Derogatory public record or collection filed
- Time since delinquency is too recent or unknown
- Level of delinquency on accounts
- Number of accounts with delinquency
- Amount owed on accounts
- Proportion of balances to credit limits on revolving accounts is too high
- Length of time accounts have been established
- Too many accounts with balances

By analyzing a large sample of credit file information on people who recently obtained new credit, and given the above information and that contained in Table 1, a statistical model has been built. The model provides a numerical score designed to predict your risk as a borrower. Credit scores used for mortgage lending range from 0 to 900 (usually above 300). The higher your score, the less risk you represent to lenders. Most lenders will be happy if your score is 700 or higher. You may still qualify for a loan with a lower score given all other factors, but it will cost you more. For example, given a score of around 620 and a $25,000 car loan for 60 months, you will pay approximately $4,500 more than with a score of 700. You will pay approximately $6,500 more than if your score is 720. Thus, a $25,000 car loan for 60 months with bad credit will cost you over $10,000 more for the life of the loan than if you have an excellent credit score.

Given the factors presented earlier and the information provided in ***Table 1***, a simulated model has been developed. A series of excellent, very good, good, not good, not bad, bad, and very bad credit scores have been recognized (without including history). Then, fuzzy similarity and ranking have been used to rank the new user and define his/her credit score. ***Figure 5*** shows the simplified flow diagram and flow of information for PNL-Based Fuzzy Query. In the inference engine, the rules based on factual knowledge (data) and knowledge drawn from human experts (inference) are combined, ranked, and clustered based on the confidence level of human and factual support. This information is then used to build the fuzzy query model with associated weights. In the query level, an intelligent knowledge-based search engine provides a means for specific queries. Initially we blend traditional computation with fuzzy reasoning. This effectively provides validation of an interpretation, model, hypothesis, or alternatively, indicates the need to reject or reevaluate. Information must be clustered, ranked, and translated to a format amenable to user interpretation.

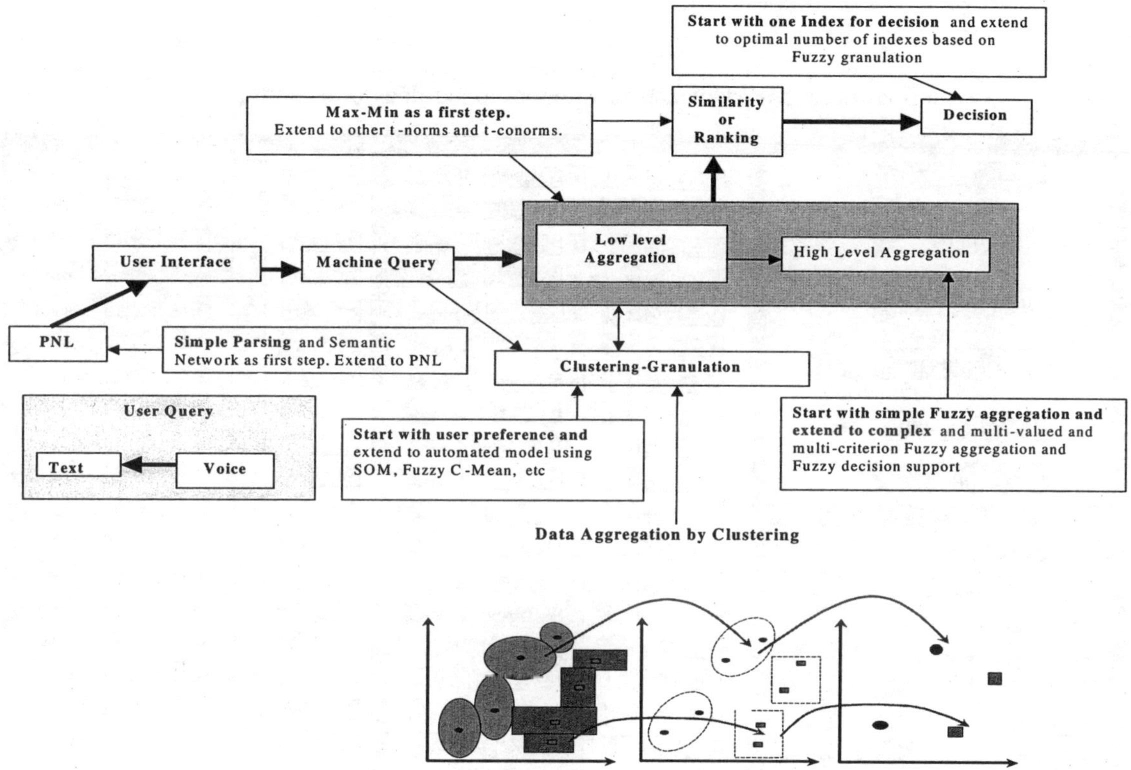

Figure 5. Simplified flow diagram and flow of information for PNL-Based Fuzzy Query.

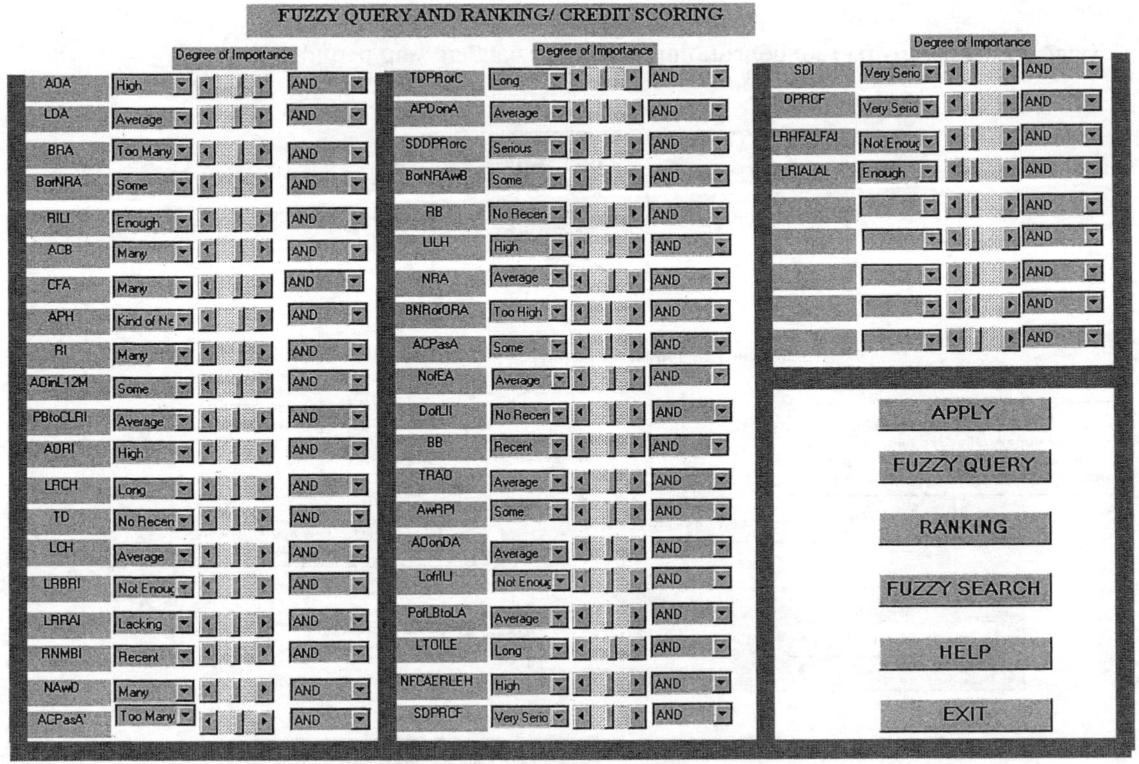

Figure 6. A snapshot of the software developed for credit scoring.

Figure 6 shows a snapshot of the software developed for credit scoring. ***Table 1*** shows the granulation of the variables that has been used for credit scoring/ranking. To test the performance of the model, a demo version of the software is available at: http://zadeh.cs.berkeley.edu/ (Nikravesh, 2001a). Using this model, it is possible to have dynamic interaction between model and user. This provides the ability to answer "What if?" questions in order to decrease uncertainty, to reduce risk, and to increase the chance to increase a score.

3.2 Application to Credit Card Ranking

Credit ratings that are compiled by the consumer credit organization such as the U.S. Citizens for Fair Credit Card Terms (CFCCT) (U.S Citizens for Fair Credit Card Terms) could simply save you hundreds of dollars in credit card interest or help you receive valuable credit card rebates and rewards including frequent flyer miles (free airline tickets), free gas, and even hundreds of dollars in cash back bonuses.

CFCCT has developed an objective-based method for ranking credit cards in US. In this model, interest rate has the highest weighting in the ranking formula. FCC rates credit cards based on the following criteria (U.S Citizens for Fair Credit Card Terms):

- Purchase APR
- Cash Advance APR
- Annual Fees
- Penalty for cards that begin their grace periods at the time of purchase/posting instead of at the time of billing
- Bonuses for cards that don't have cash advance fees
- Bonuses for cards that limit their total cash advance fees to $10.00
- Bonuses for introductory interest rate offers for purchases and/or balance transfers
- Bonuses for cards that have rebate/perk programs
- Bonuses for cards that have fixed interest rates.

Table 12. Credit cards ranked by the CFCCT.

Classic Cards	Type	Gold Cards	Type	Platinum Cards	Type
Pulaski B& T	V	Pulaski	MC	Capital One	VP
Ark. Natl	MC/V	Capital One	VP	NextCard	VP
Capital One	V	SFNB	V	BofA	VP
NextCard	V	NextCard	V	Simmons	VP
Wachovia	V	BofA	V	G&L Bank	MCP/VP
MCP/VPBlue	AMEX	Wachovia	V	Aria	VP
Helena Natl	MC/V	Blue	AMEX	Ever	VP
Simmons	V	Helena	MC/V	Blue	AMEX
Metro. Natl.	V	Simmons	V	AF	VP
Umbrella	V	Metro.	V	Banco	VP

V=Visa; MC=MasterCard; AMEX=American Express

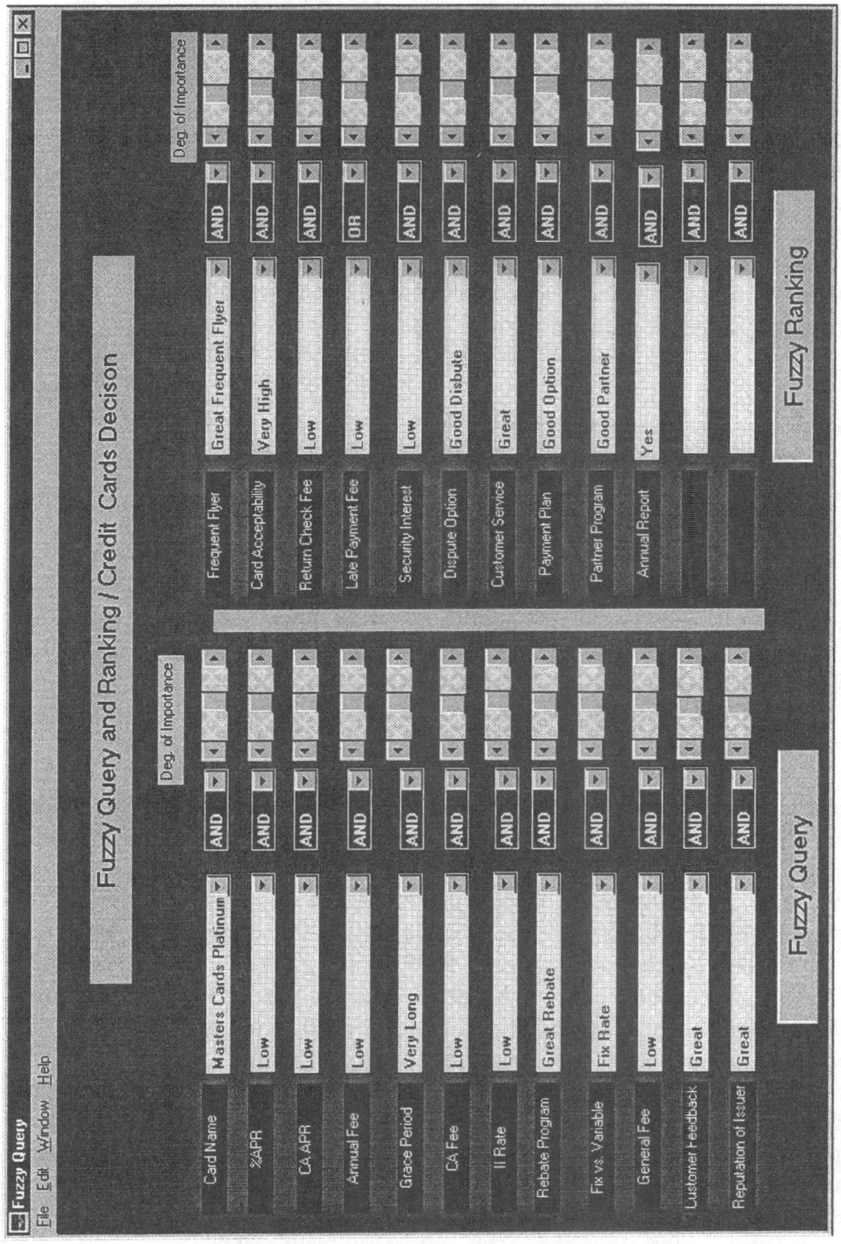

Figure 7. A snapshot of the software developed to rank credit cards.

Table 12 shows the top 10 classic cards, the top 10 gold cards, and the top 10 platinum cards which have been ranked by the CFCCT method (U.S Citizens for Fair Credit Card Terms) as of March 2001. Given the above factors and the information provided in ***Table 8***, a simulated model has been developed. A series of excellent, very good, good, not good, not bad, bad, and very bad credit cards have been recognized for the credit cards listed in ***Table 9***. Then, fuzzy similarity and ranking has been used to rank the cards and define a credit score. ***Figure 7*** shows a snapshot of the software developed to rank credit cards. ***Table 2*** shows the granulation of the variables that has been used for the rankings. To test the performance of the model, a demo version of the software is available at: http://zadeh.cs.berkeley.edu/ (Nikravesh, 2001a).

3.3 University Admissions

Hundreds of millions of applications were processed by U.S. universities resulting in more than 15 million enrollments in the year 2000 for a total revenue of over $250 billion. College admissions are expected to reach over 17 million by the year 2010, for total revenue of over $280 billion. In Fall 2000, UC Berkeley was able to admit about 26% of the 33,244 applicants for freshman admission (University of California-Berkeley). In Fall 2000, Stanford University was only able to offer admission to 1168 men from 9571 applications (768 admitted) and 1257 women from 8792 applications (830 admitted), a general admit rate of 13% (Stanford University Admission).

The UC Berkeley campus admits its freshman class on the basis of an assessment of the applicants' high school academic performance (approximately 50%) and through a comprehensive review of the application including personal achievements of the applicant (approximately 50%) (University of California-Berkeley). For Fall 1999, the average weighted GPA of an admitted freshman was 4.16, with a SAT I verbal score range of 580-710 and a SAT I math score range of 620-730 for the middle 50% of admitted students (University of California-Berkeley). While there is no specific GPA for UC Berkeley applicants that will guarantee admission, a GPA of 2.8 or above is required for California residents and a test score total indicated in the University's Freshman Eligibility Index must be achieved. A minimum 3.4 GPA in A-F courses is required for non-residents. At Stanford University, most of the candidates have an un-weighted GPA between 3.6 and 4.0 and verbal SAT I and math SAT I scores of at least 650 (Stanford University Admission) At UC Berkeley, the academic assessment includes student's academic performance and several measured factors such as:

- College preparatory courses
- Advanced Placement (AP)
- International Baccalaureate Higher Level (IBHL)

- Honors and college courses beyond the UC minimum and degree of achievement in those courses
- Uncapped UC GPA
- Pattern of grades over time
- Scores on the three required SAT II tests and the SAT I (or ACT)
- Scores on AP or IBHL exams
- Honors and awards which reflect extraordinary, sustained intellectual or creative achievement
- Participation in rigorous academic enrichment
- Outreach programs
- Planned twelfth grade courses
- Qualification for UC Eligibility in the Local Context

All freshman applicants must complete courses in the University of California's A-F subject pattern and present scores from SAT I (or ACT) and SAT II tests with the following required subjects:

a. History/Social Science - 2 years required
b. English - 4 years required
c. Mathematics - 3 years required, 4 recommended
d. Laboratory Science - 2 years required, 3 recommended
e. Language Other than English - 2 years required, 3 recommended
f. College Preparatory Electives - 2 years required

At Stanford University, in addition to the academic transcript, close attention is paid to other factors such as student's written application, teacher references, the short responses and one-page essay (carefully read for quality, content, and creativity), and personal qualities.

The information provided in this study is a hypothetical situation and does not reflect the current UC system or Stanford University admissions criteria. However, we use this information to build a model to represent a real admissions problem. For more detailed information regarding University admissions, please refer to the University of California-Berkeley and Stanford University, Office of Undergraduate Admission (University of California-Berkeley; Stanford University Admission).

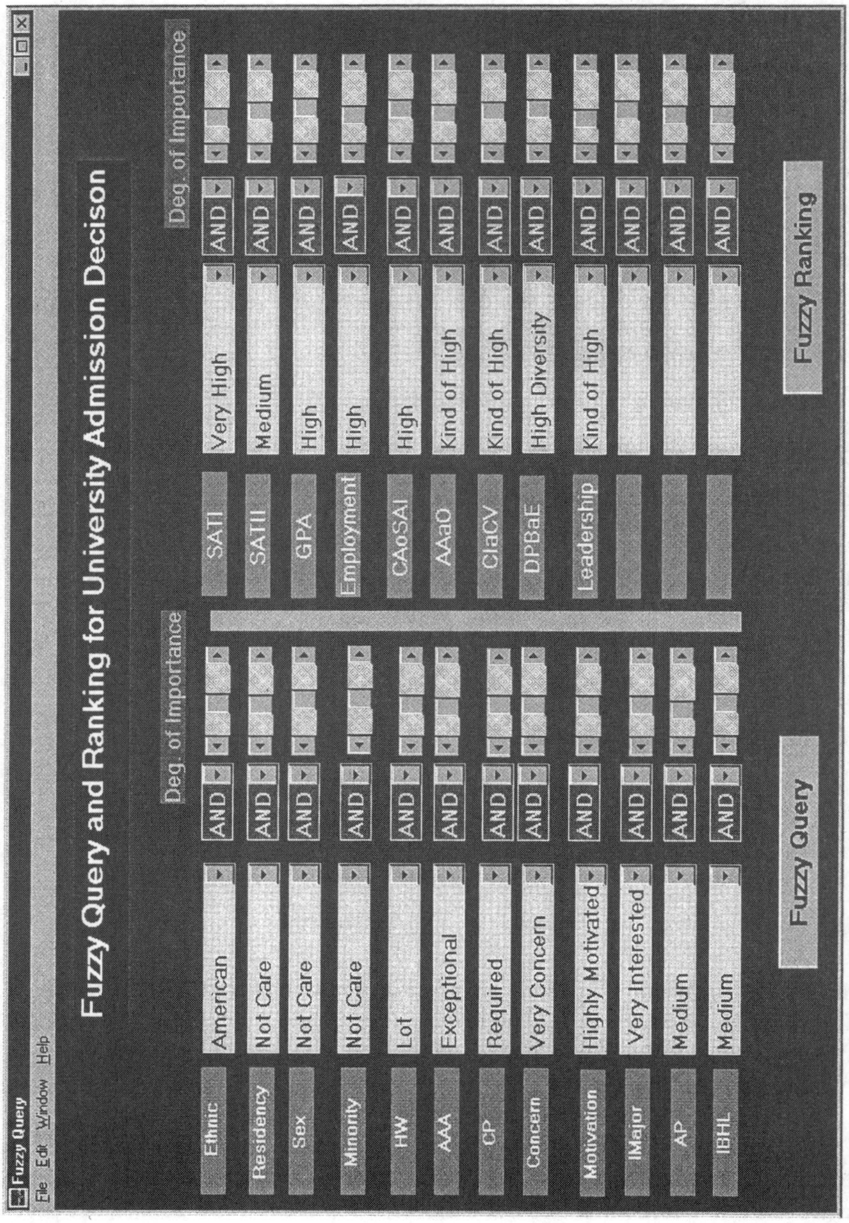

Figure 8. A snapshot of the software for University Admission Decision Making.

Given the factors above and the information contained in ***Table 3***, a simulated-hypothetical model (a Virtual Model) was developed. A series of excellent, very good, good, not good, not bad, bad, and very bad student given the criteria for admission has been recognized. These criteria over time can be modified based on the success rate of students admitted to the university and their performances during the first, second, third and fourth years of their education with different weights and degrees of importance given for each year. Then, fuzzy similarity and ranking can evaluate a new student rating and find it's similarity to a given set of criteria.

Figure 8 shows a snapshot of the software developed for university admissions and the evaluation of student applications. Table 3 shows the granulation of the variables that was used in the model. To test the performance of the model, a demo version of the software is available at: http://zadeh.cs.berkeley.edu/ (Nikravesh, 2001a). Incorporating an electronic intelligent knowledge-based search engine, the results will eventually be in a format to permit a user to interact dynamically with the contained database and to customize and add information to the database. For instance, it will be possible to test an intuitive concept by dynamic interaction between software and the human mind.

This will provide the ability to answer "What if?" questions in order to decrease uncertainty and provide a better risk analysis to improve the chance for "increased success" on student selection or it can be used to select students on the basis of "diversity" criteria. The model can be used as for decision support and for a more uniform, consistent and less subjective and biased way. Finally, the model could learn and provide the mean to include the feedback into the system through time and will be adapted to the new situation for defining better criteria for student selection.

In this study, it has been found that ranking and scoring is a very subjective problem and depends on user perception (***Figure 9 and Figure 10***) and preferences in addition to the techniques used for the aggregation process which will effect the process of the data mining in reduced domain (***Figure 11***). Therefore, user feedback and an interactive model are recommended tools to fine-tune the preferences based on user constraints. This will allow the representation of a multi-objective optimization with a large number of constraints for complex problems such as credit scoring or admissions. To solve such subjective and multi-criteria optimization problems, GA-fuzzy logic and DNA-fuzzy logic models [2] are good candidates.

In the case of the GA-Fuzzy logic model, the fitness function will be defined based on user constraints. For example, in the admissions problem, assume that we would like to select students not only on the basis of their achievements and criteria defined in Table 3, but also on the basis of diversity which includes gender distribution, ethnic background distribution, geophysical location distribution, etc.

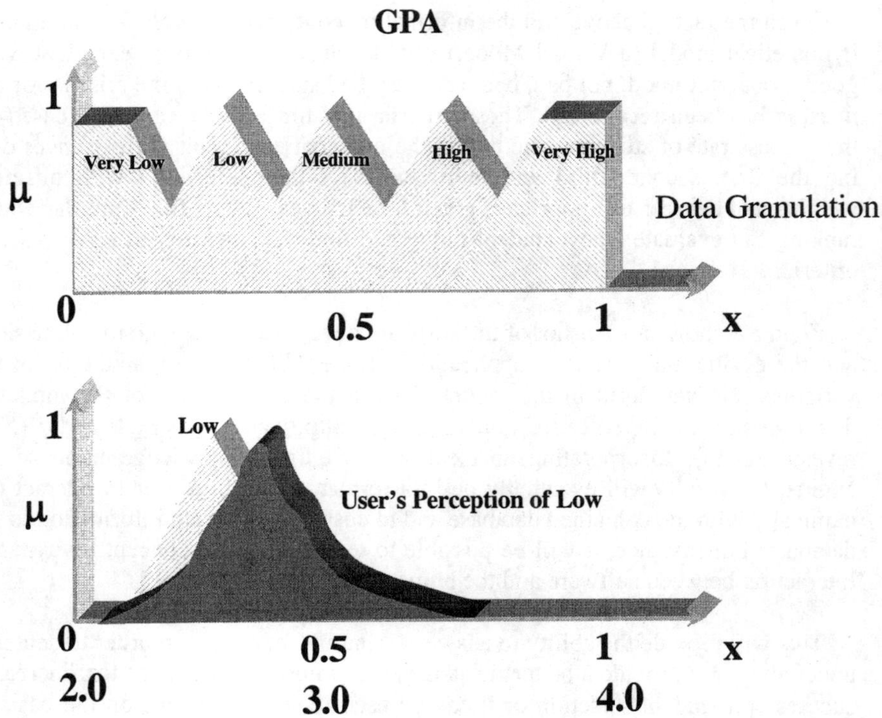

Figure 9. User's perception of "GPA Low"

The question will be "what are the values for the preferences and which criteria should be used to achieve such a goal?" In this case, we will define the genes as the values for the preferences and the fitness function will be defined as the degree by which the distribution of each candidate in each generation match the desired distribution. fuzzy similarity can be used to define the degree of match which can be used for better decision analysis.

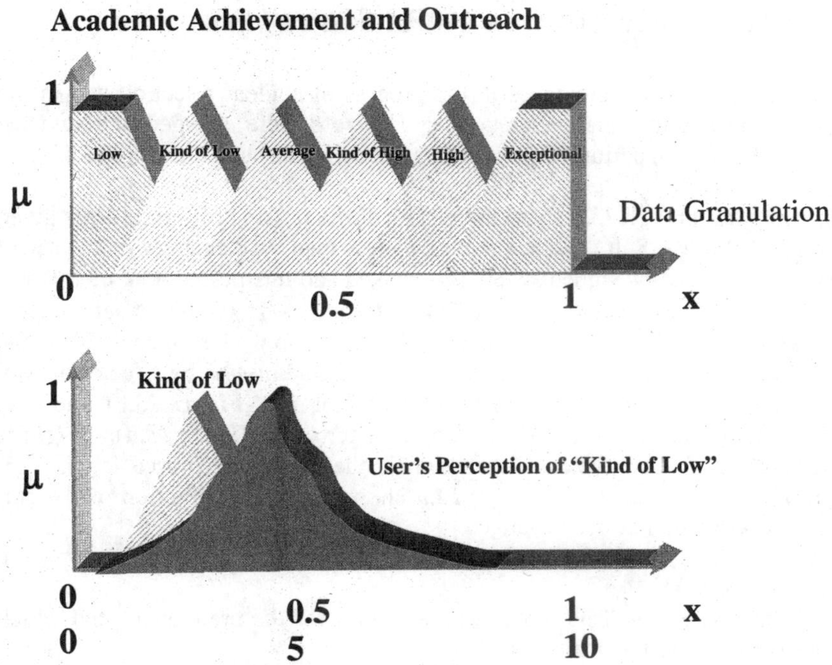

Figure 10. User's perception of Academic

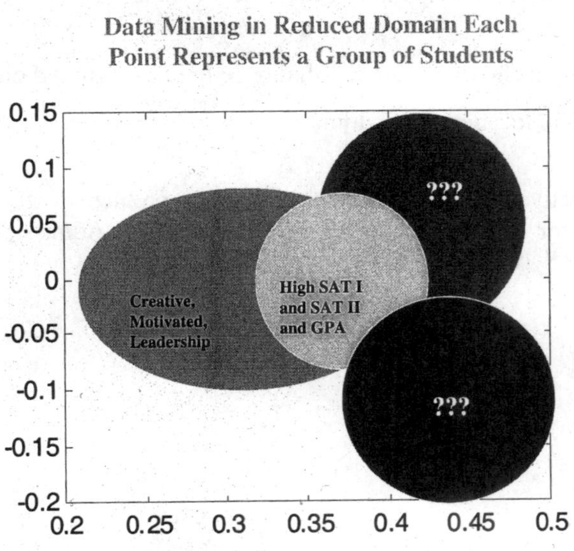

Figure 11. Typical Text and Rule Data Mining based on Techniques described in "Search Strategy and Figure 5.

3.3.1 Effect of Preferences on Ranking of Students

To study the effect of preferences in the process of student selection and in the process of the ranking, the preferences in ***Figure 8*** were changed and students were ranked based on perturbed preferences, models 1 through 5 in ***Figure 12.***

Figures 13.a through ***13.d*** show the results of the ranking of the students given the models 1 through 5. It is shown that given less than %10 changes on the actual preferences, most of the students were mis-ranked and mis-placed. Out of 100 students, less than %50 students or as an average only %41 of the actual students were selected (***Figure 13.a***). ***Figure 13.b*** shows that only less than %70 of the students will be correctly selected if we increase the admission by a factor of two, around %85 if we increase the admission by a factor of 3 (***Figure 13.c***), and less than %90 if we increase the admission by a factor of 4 (***Figure 13.d***). ***Figures 14.a*** through ***14.d*** show typical distribution of the 21 variables used for the Admission model. ***Figures 14.a*** through ***14.d*** show that the distribution of the students also drastically has been changed.

Now, the question will be "what are the values for the preferences and which criteria should be used to achieve such a goal?"

- Given a set of successful students, we would like to adjust the preferences such that the model could reflect this set of students.
- Diversity which includes gender distribution, ethnic background distribution, geophysical location distribution, etc.

To solve such subjective and multi-criteria optimization problems with a large number of constraints for complex problems such as University Admissions, the BISC Decision Support System is an excellent candidate.

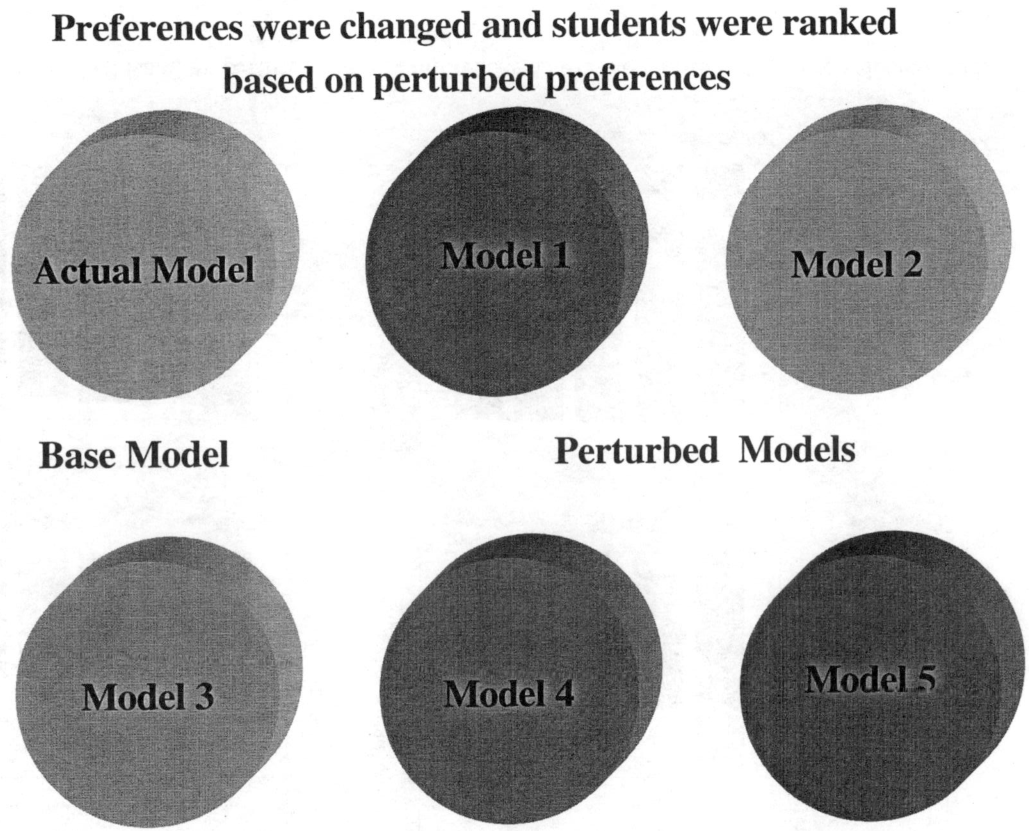

Figure 12. Models 1 through 5 are models based on preferences were perturbed around the actual value.

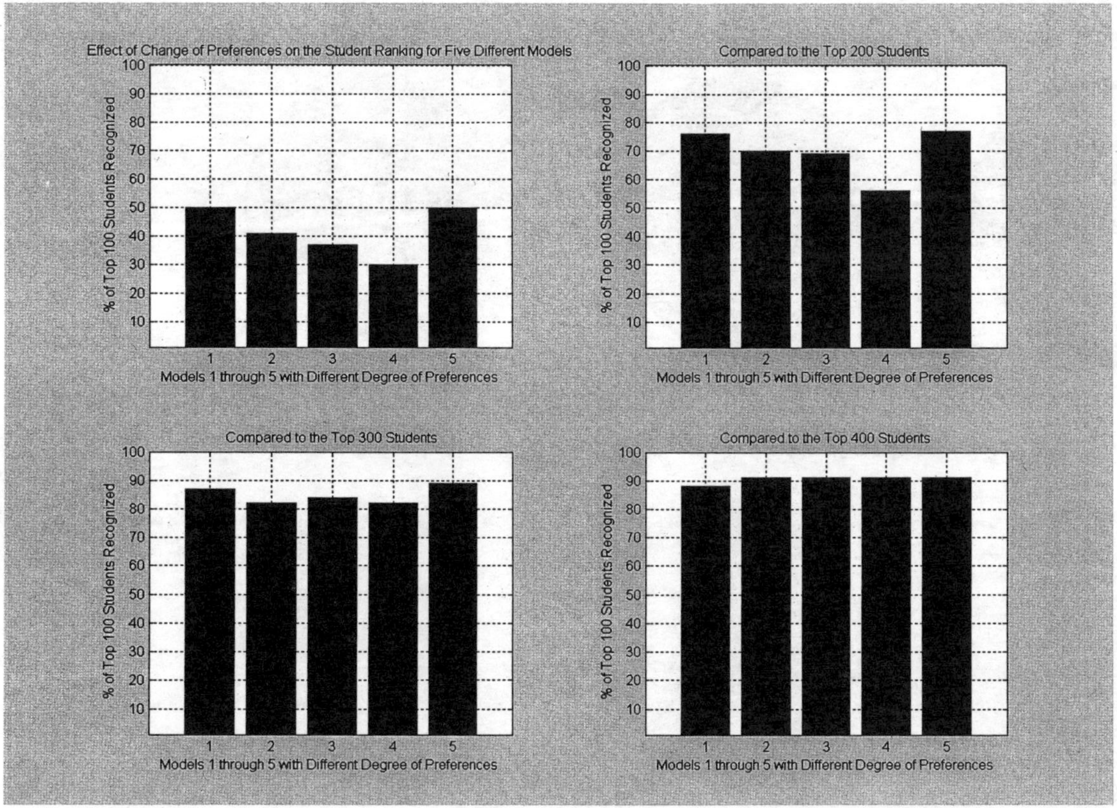

Figure 13. Effect of less than +-%10 Random perturbation on Preferences on the recognition of the pre-selected students given actual model.

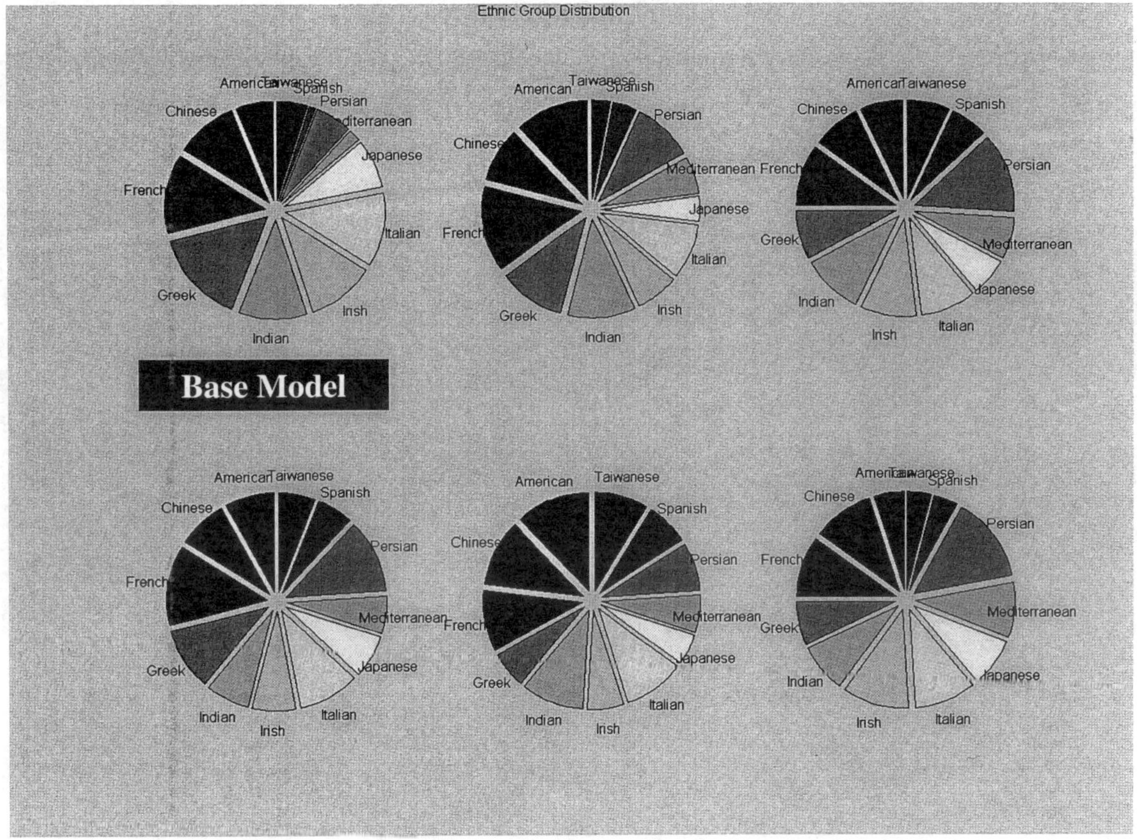

Figure 14.a. Ethnic Group Distribution

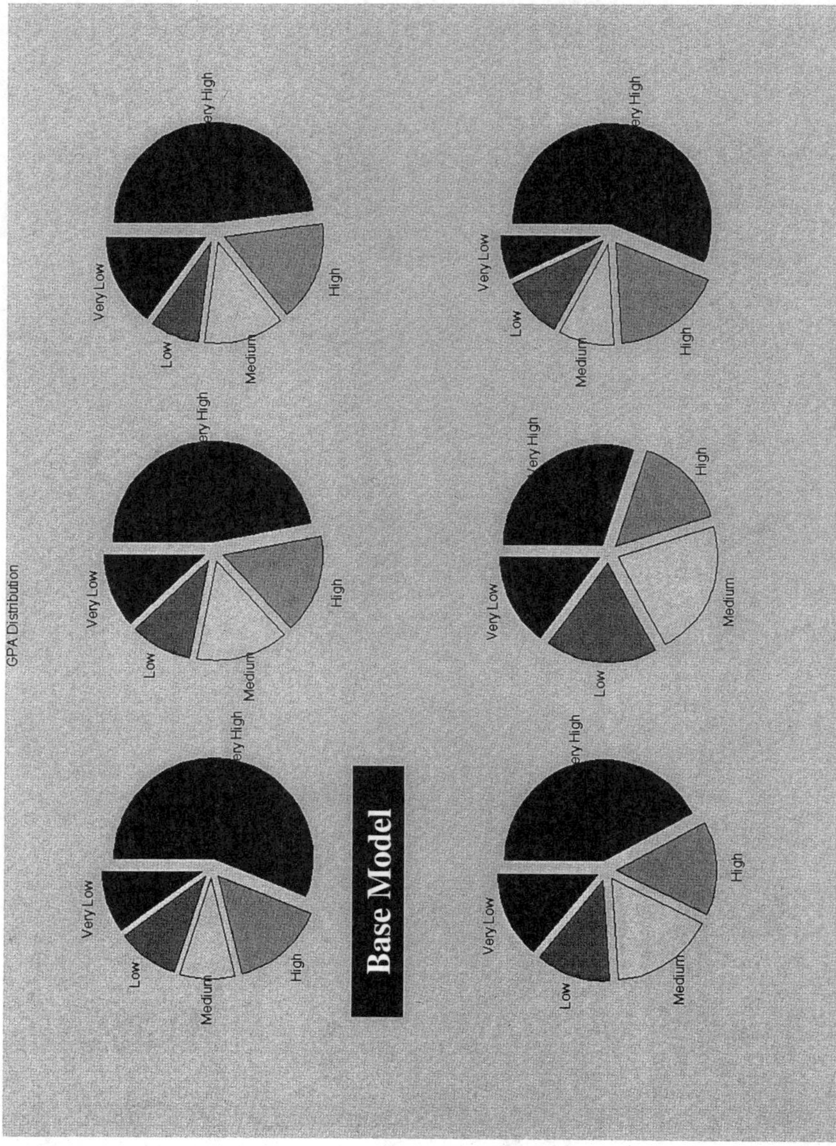

Figure 14.b. GPA Distribution

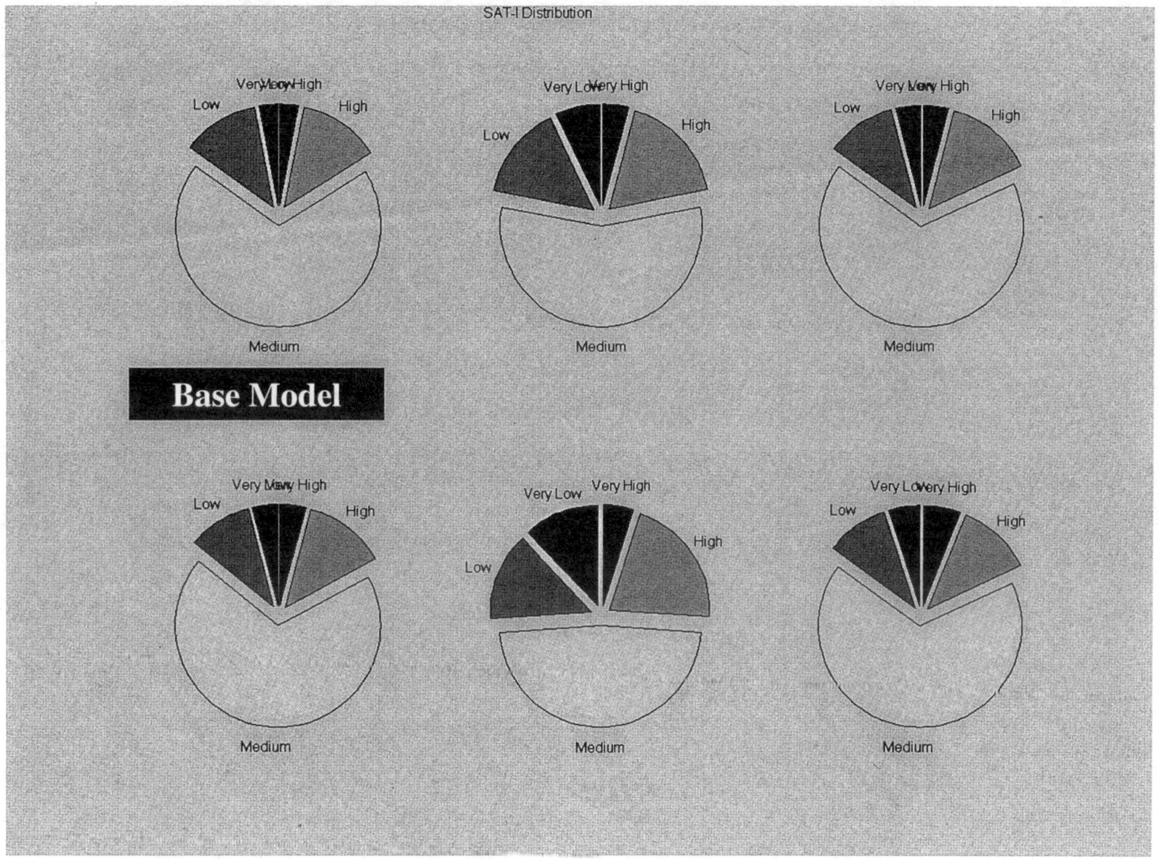

Figure 14.c. SAT-I Distribution

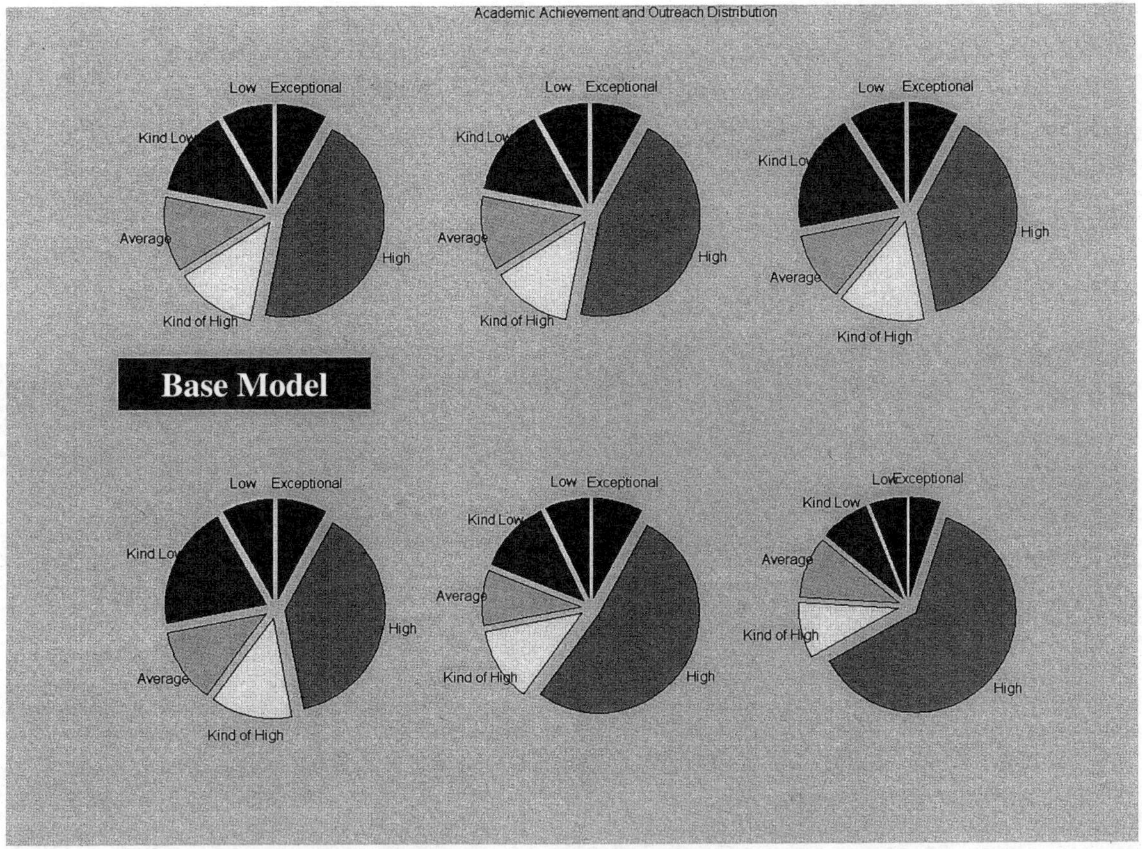

Figure 14.d. Academic Achievement Distribution

4 BISC Decision Support System

Decision Support systems may represented in either of the following forms 1) physical replica of a system, 2) analog or physical model, 3) mathematical (qualitative) model, and 4) mental models. Decision support system is an approach or a philosophy rather than a precise methodology that can be used mainly for

- strategic planning such as resource allocation
- management control such as efficient resources utilization
- operational control for efficient and effective execution of specific tasks

Decision support system is an approach or a strategy rather than a precise methodology, which can be used for 1) use intelligently the vast amounts of important data in organizations in an optimum way as a decision support system and 2) share intelligently and securely company's data internally and with business partners and customers that can be process quickly by end users and more specifically for :

- strategic planning such as resource allocation
- management control such as efficient resources utilization
- operational control for efficient and effective execution of specific tasks

The main key features of the Decision Support System for the internet applications are 1) to use intelligently the vast amounts of important data in organizations in an optimum way as a decision support system and 2) To share intelligently and securely company's data internally and with business partners and customers that can be process quickly by end users. In this section, we describe the use of the BISC Decision Support System as an intelligent real-time decision-making and management model based on two main motivations:

- In recent years, needs for more cost effective strategy and multicriteria and multiattribute optimization in an imprecise and uncertain environment have emphasized the need for risk and uncertainty management in the complex dynamic systems. There exists an ever-increasing need to improve technology that provides a global solution to modeling, understanding, analyzing and managing imprecision and risk in real-time automated decision-making for complex dynamic systems.

- As a result intelligent dynamic systems with growing complexity and technological challenges are currently being developed. This requires new technology in terms of development, engineering design and virtual simulation models. Each of these components adds to the global sum of uncertainty about risk of during decision-making process. While the technological expertise of each component becomes increasingly complex, there is a need for better integration of each component into a global model adequately capturing the uncertainty on key strategic parameters. The uncertainty quantification on such key parameters is required in any type of decision analysis.

The BISC (Berkeley Initiative in Soft Computing) Decision Support System Components include (***Figure 15***):

- Data Management: database(s) which contains relevant data for the decision process

- User Interface
    - users and decision support systems (DSS) communication

- Model Management and Data Mining
    - includes software with quantitative and fuzzy models including aggregation process, query, ranking, and fitness evaluation

- Knowledge Management and Expert System: model representation including
    - linguistic formulation,
    - functional requirements
    - constraints
    - goal and objectives

  - o linguistic variables requirements

- Evolutionary Kernel and Learning Process
  - o Includes software with quantitative and fuzzy models including, Fuzzy-GA, fuzzy aggregation process, ranking, and fitness evaluation

- Data Visualization: Allows end-users or decision makers can intervene in the decision-making process and see the results of the intervention

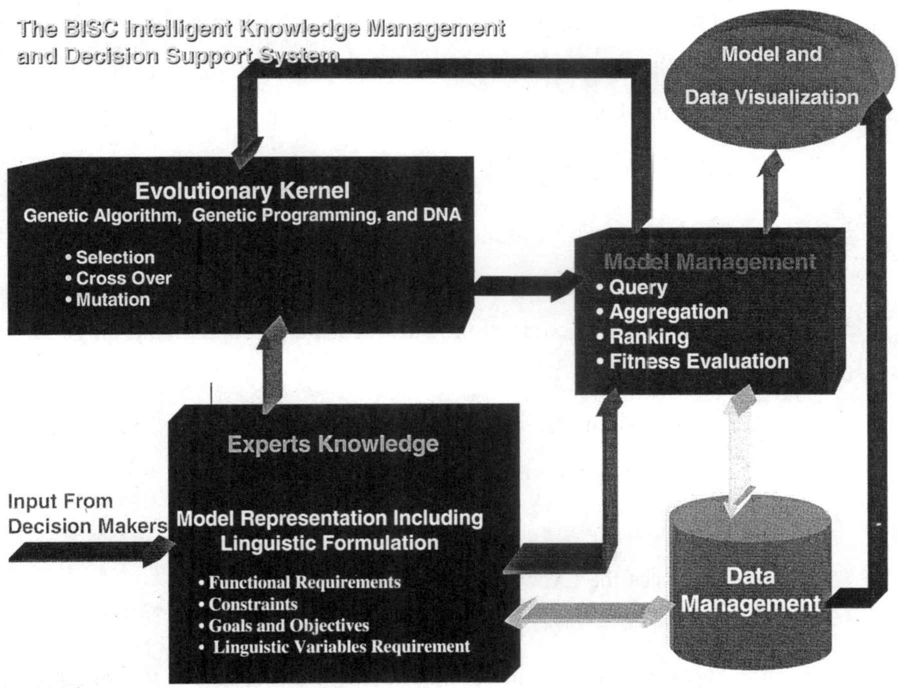

Figure 15. The BISC Decision Support System

Data Visualization and Visual Interactive Decision Making allows end-user or decision makers to recognize trends, patterns, and anomalies that can not be predicted or recognized by standard analysis methods and include the following components:

- Visual interactive modeling (VIM): user can intervene in the decision-making process and see the results of the intervention

- Visual interactive simulation (VIS): users may interact with the simulation and try different decision strategies

The Expert System uses both Fuzzy Logic and Case-Based Reasoning (CBR) for the following reasons:

- Case-Based Reasoning (CBR)
  - solve new problems based on history of given solved old problems
  - Provide a framework for knowledge acquisition and information system development
  - enhance learning capability
  - generate explanations and recommendation to users

- Fuzzy Logic
  - simulating the process of human reasoning
  - framework to computing with word and perception, and linguistics variables.
  - deals with uncertainties
  - creative decision-making process

The components of the Expert System include (***Figure 16***)

- the knowledge base contains engineering knowledge for model representation which provide problem solving environment
- the inference engine provide reasoning, conclusions, and recommendation
- the user interface and knowledge based editor provide dialog environment for questions and answers

- the advisor and translator can translate the machine inference to a human understandable advice, recommendation, and logical explanation

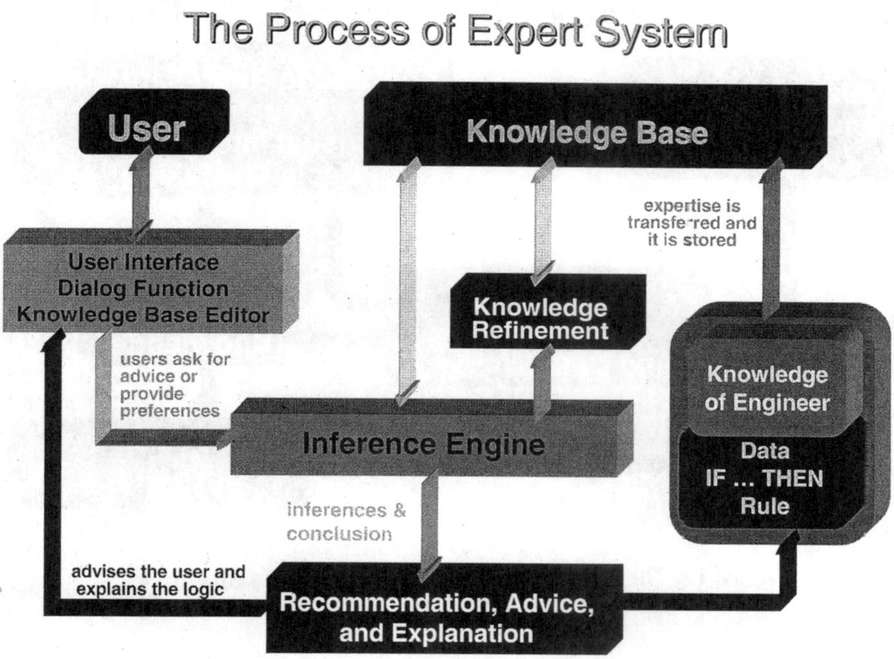

Figure 16. The components of the Expert System

The Data and Knowledge Management model include the following components (***Figure 17***)

- knowledge discovery and data mining- using search engines, databases, data mining, and online analytical processing, the proper knowledge must be found, analyzed, and put into proper context
- organize knowledge bases - it stores organizational knowledge and best practices
- knowledge acquisition - determines what knowledge (information) is critical to decision making

- knowledge representation - target audiences are defined and technologies are put into place to enable knowledge delivery when needed

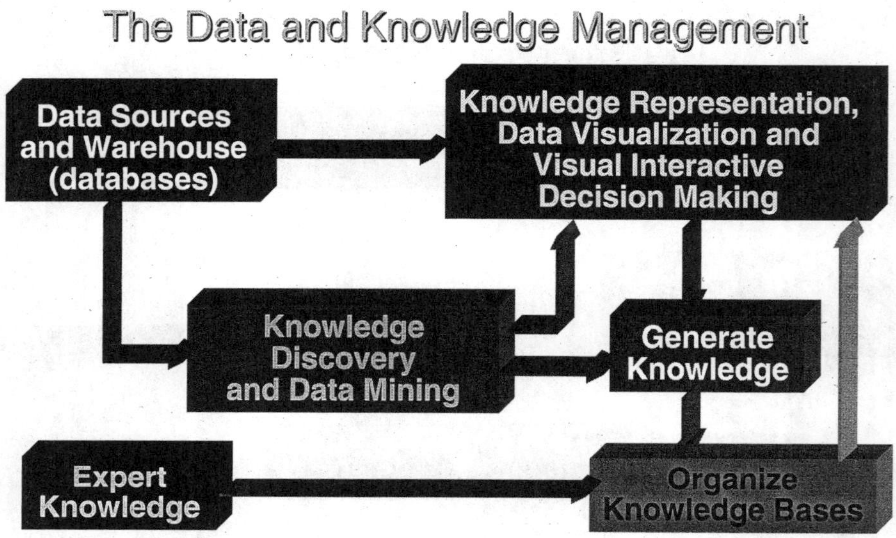

Figure 17. The Data and Knowledge Management Model

4.1 Implementation- BISC Decision Support System

In this section, we will introduce the BISC-DSS system for university admissions. In the case study, we used the GA-Fuzzy logic model for optimization purposes. The fitness function will be defined based on user constraints. For example, in the admissions problem, assume that we would like to select students not only on the basis of their achievements and criteria defined in Table 3 as a successful student, but also on the basis of diversity which includes gender distribution, ethnic background distribution, geophysical location distribution, etc. The question will be "what are the values for the preferences and which criteria should be used to achieve such a goal?" In this case, we will define the genes as the values for the preferences and the fitness function will be defined as the degree by which the distribution of each candidate in each generation match the desired distribution. Fuzzy similarity can be used to define the degree of match, which can be used for better decision analysis.

Figure 18 shows the performance of the conventional GA. The program has been run for 5000 generations and ***Figure 18*** shows the last 500 GA generations. As it is shown, the GA technique has been approached to a fitness of 80% and no further improvement was expected. Given what has been learned in each generation with respect to trends in the good genes, a series of genes were selected in each generation and has been used to introduce a new initial population to be used for GA. This process has been repeated until it was expected no improvement be achieved. ***Figure 19*** shows the performance of this interaction. The new model has reached a new fitness value, which is over 95%. ***Figure 20*** show the results of the ranking of the students given the actual model, predicted model (Model number 1) and models 2 through 4 which has been used to generate the initial population for training the fuzzy-GA model. It is shown that the predicted model ranked and selected most of the predefined students (***Figures 20.a-20.d***) and predefined distributions (***Figures 21.a-21.f***) and properly represented the actual model even though the initial models to generate the initial population for training were far from the actual solution (***Figures 20.a-20.d and 21.a-21.f***). Out of 100 students, more than 90% students of the actual students were selected (***Figure 20.a***). ***Figure 20.b*** shows that %100 of the students will be correctly selected if we increase the admission by a factor of less than two. In has been concluded for this case study that %100 of students were selected if we increase the student admission by a factor of less than 1.15. ***Figures 20.a-20.d*** and ***21.a-21.f*** show that the initial models, model 2 through 5, were far from the actual model. Out of 100 students, less than 3% of the actual students were selected (***Figure 20.a***), around 5% if we increase the admission by a factor of 2 (***Figure 20.b***), around 10% if we increase the admission by a factor of 3 (***Figure 20.c***), and less than 15% if we increase the admission by a factor of 4 (***Fiure. 20.d***). ***Figures 21.a-21.f*** show typical distribution of the 21 variables used for the admission model. ***Figures 21.a*** through ***21.f*** show that the distribution of the student are properly presented by the predicted model and there is an excellent match between the actual model and the predicted model, even though the distributions of the initial populations are far from the actual model.

To show if the new technique is robust, we tested the methodology with different initial populations and different constraints. In addition, we have used the methodology for different problems. It has been concluded that in all cases, we were able to design a model, which represents the actual model given that all the constraints have been defined. ***Figure 22*** shows the results from data mining in reduced domain using part of a selected dataset as shown on ***Figure 11*** as a typical representation and techniques and strategy represented in ***Figure 5.***

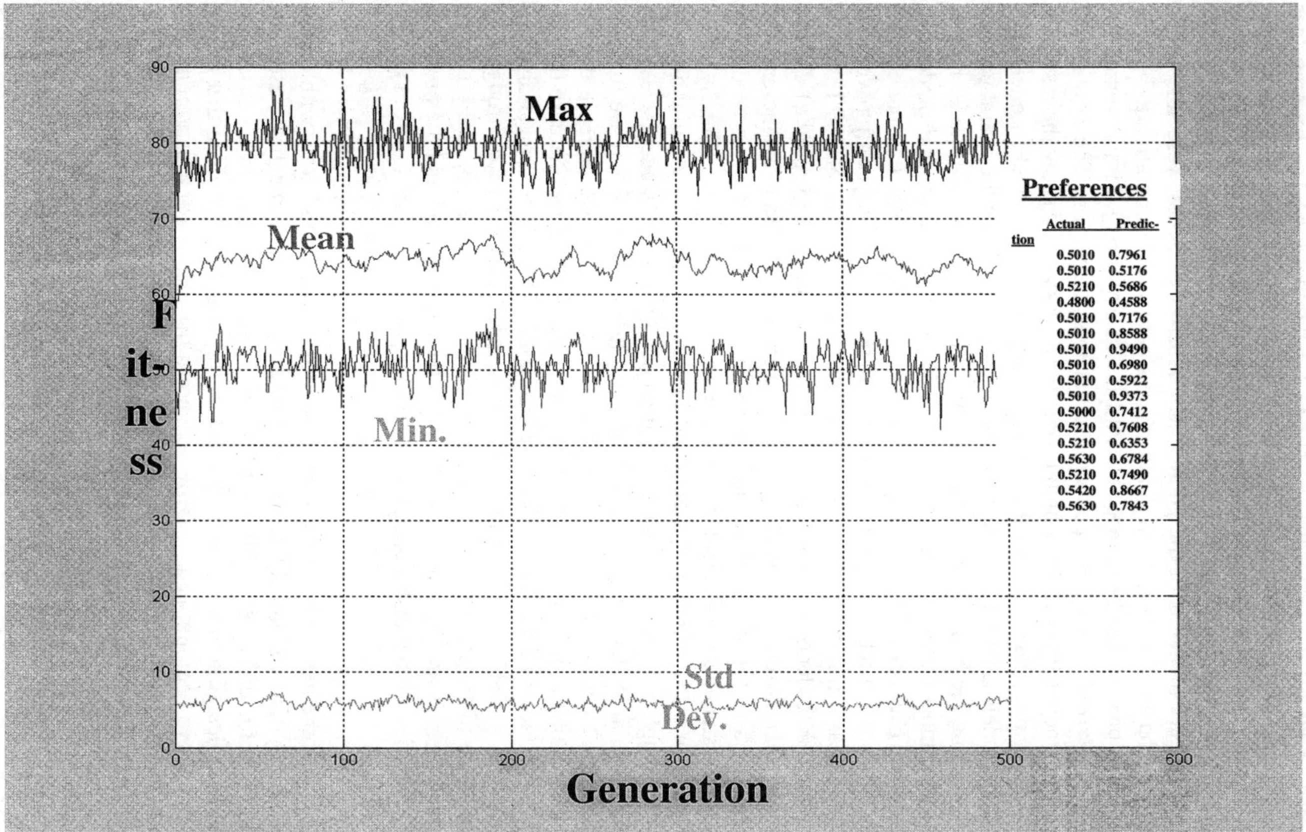

Figure 18. Conventional GA: Multi-Objective Multi-Criteria Optimization for the University Admission

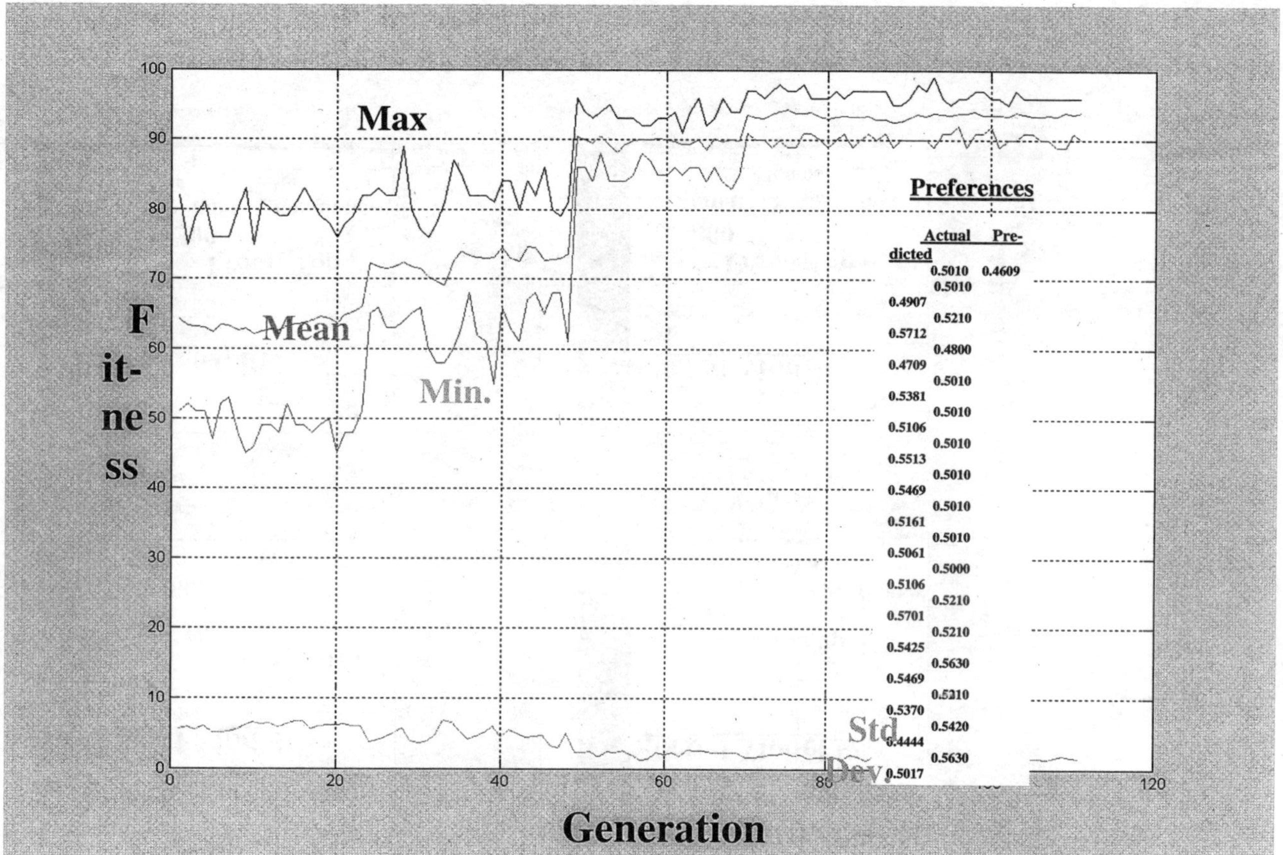

Figure 19. Interactive-GA Multi-Objective Multi-Criteria Optimization for the University Admission

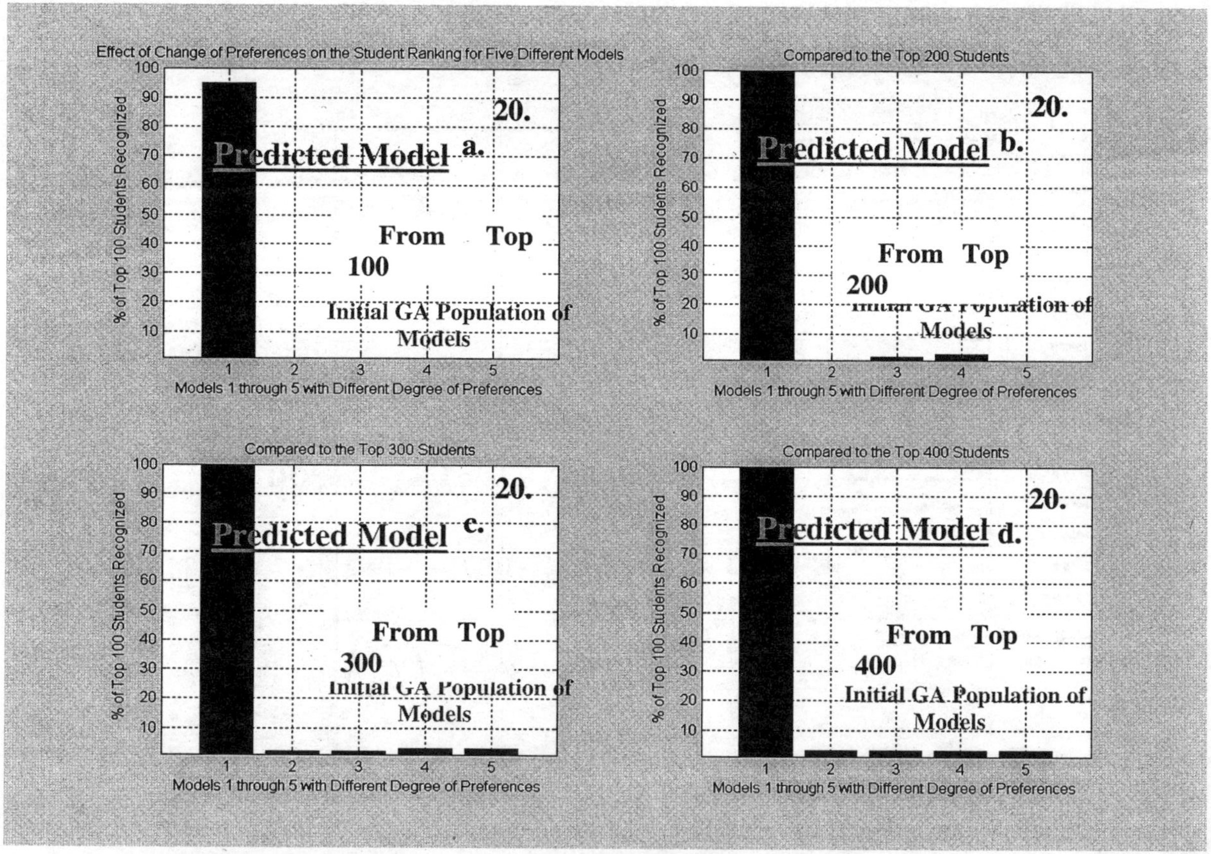

Figure 20. Results of the Ranking of the Students given Predicted Model and initial population for Fuzzy-GA Model

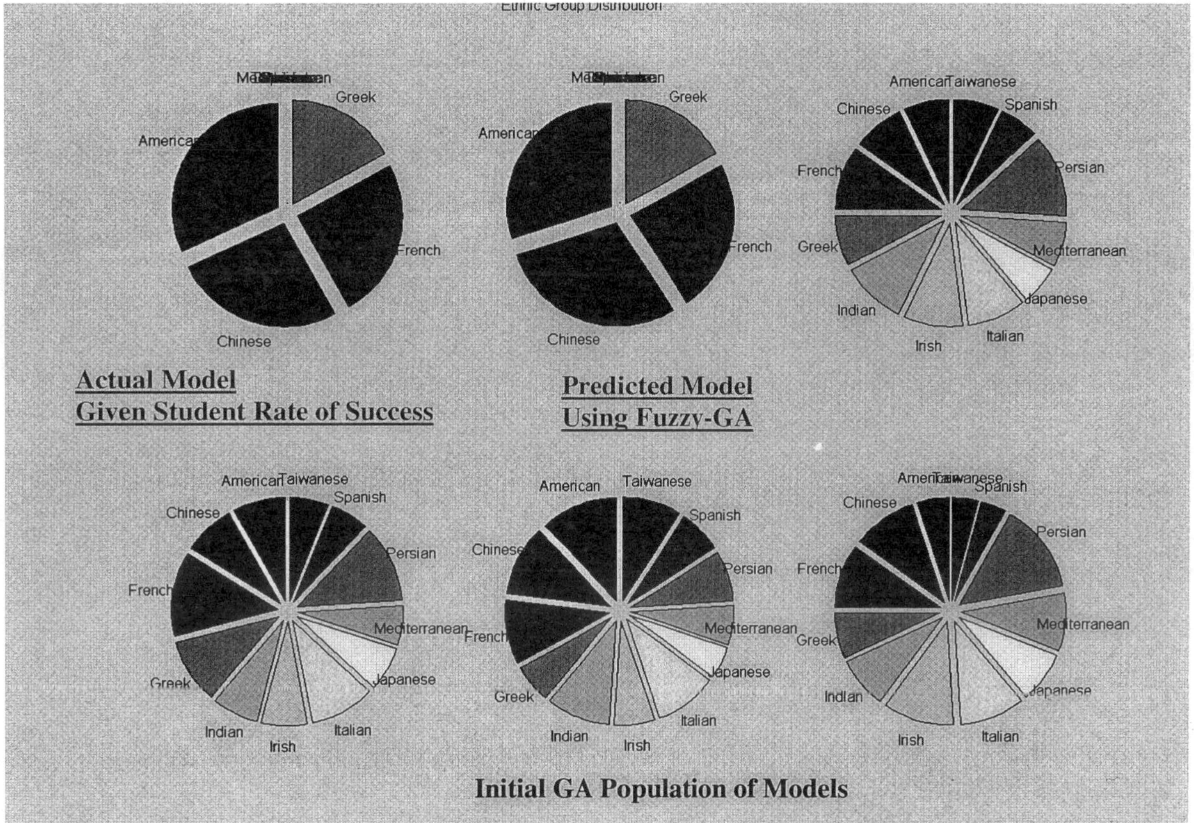

Typical Distribution of the Variables used for the Admission Model; Actual, Predicted and Initial Models for Fuzzy-GA

Figure 21.a. Ethnic Group Distribution

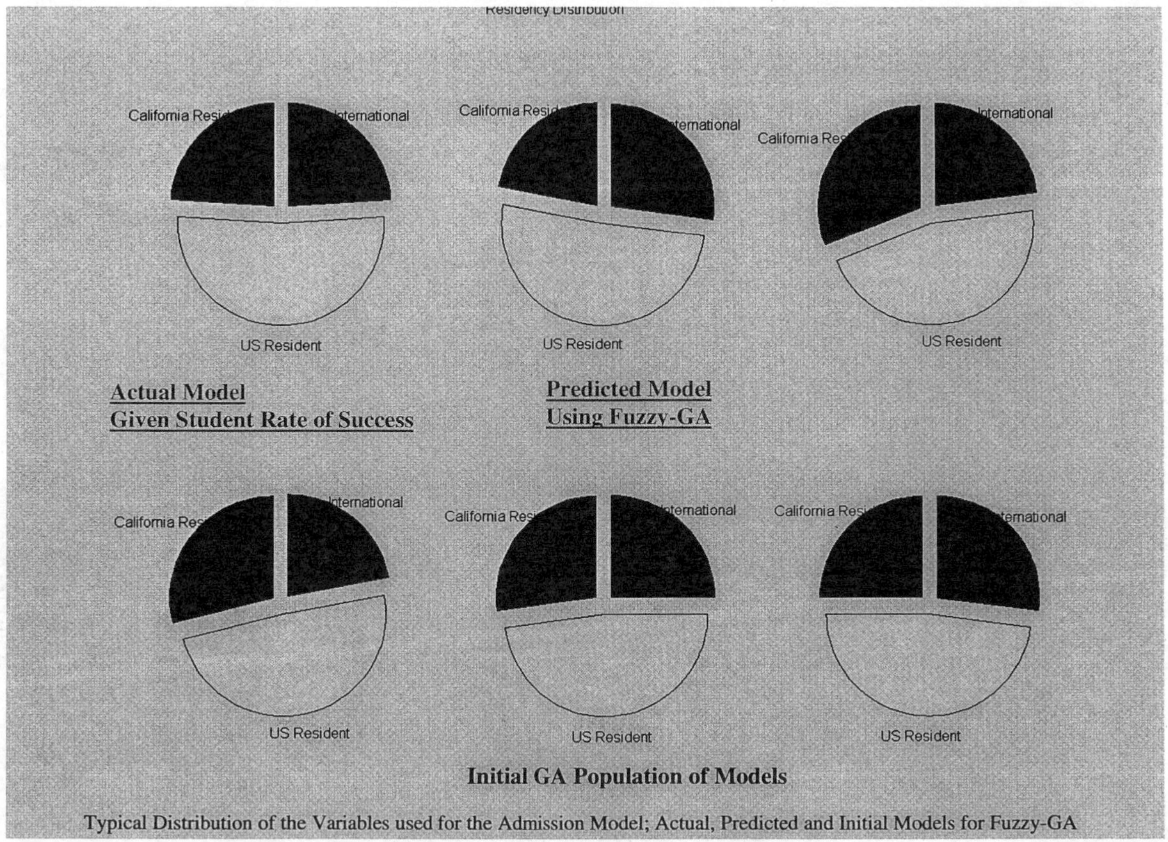

Figure 21.b. Residency Distribution

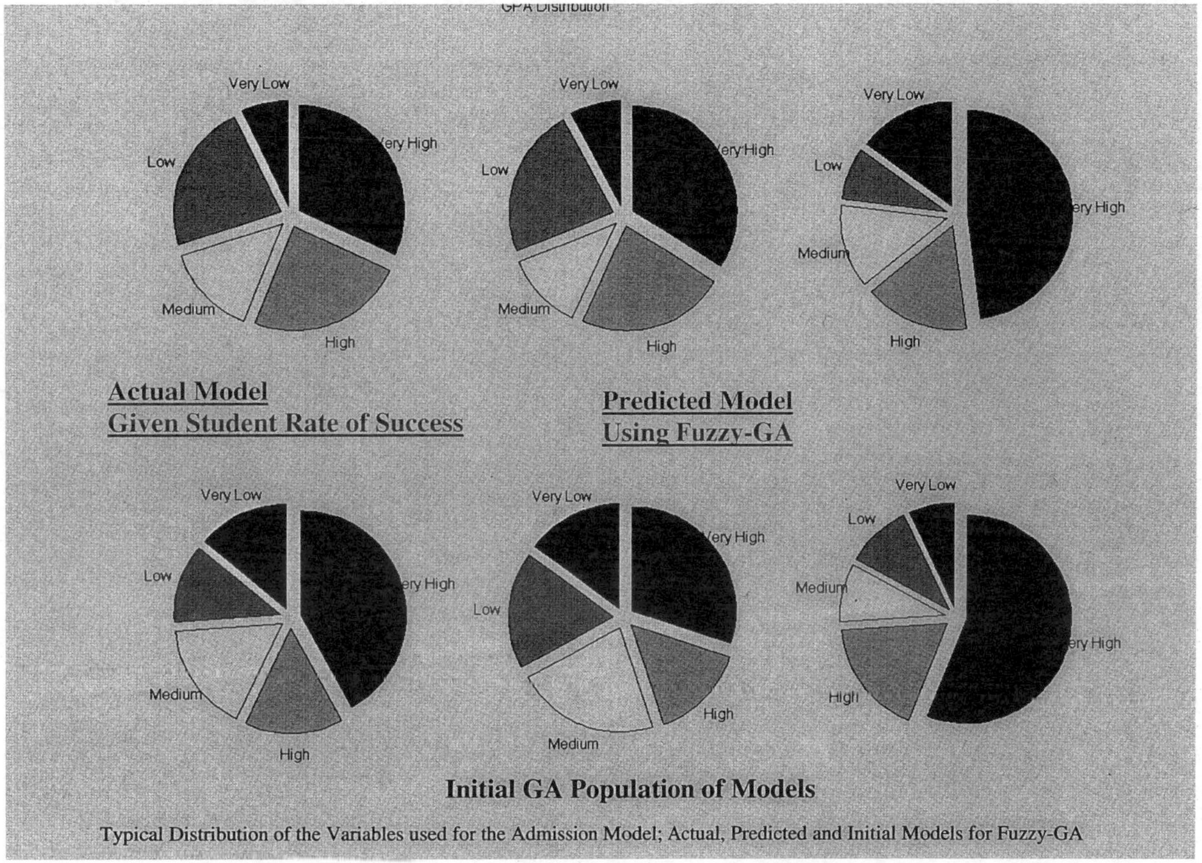

Typical Distribution of the Variables used for the Admission Model; Actual, Predicted and Initial Models for Fuzzy-GA

Figure 21.c. GPA Distribution

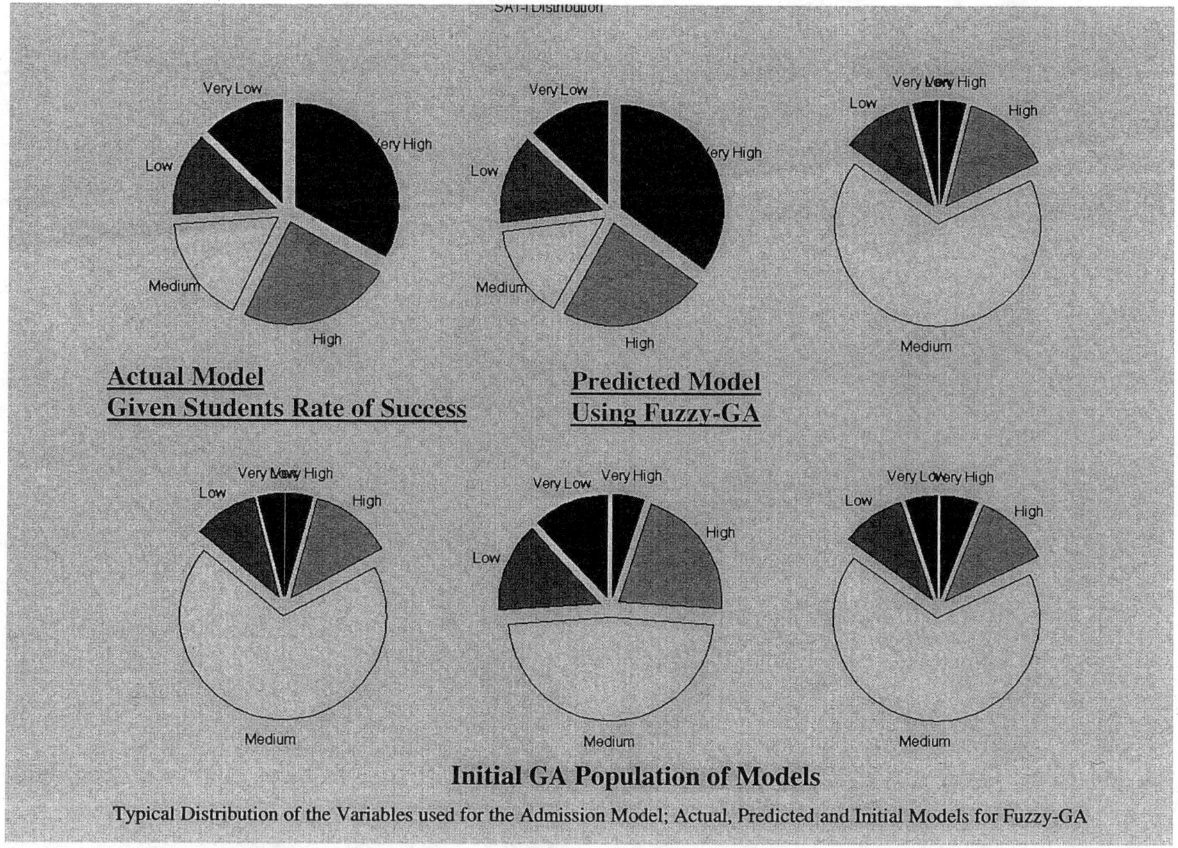

Figure 21.d. SAT-I Distribution

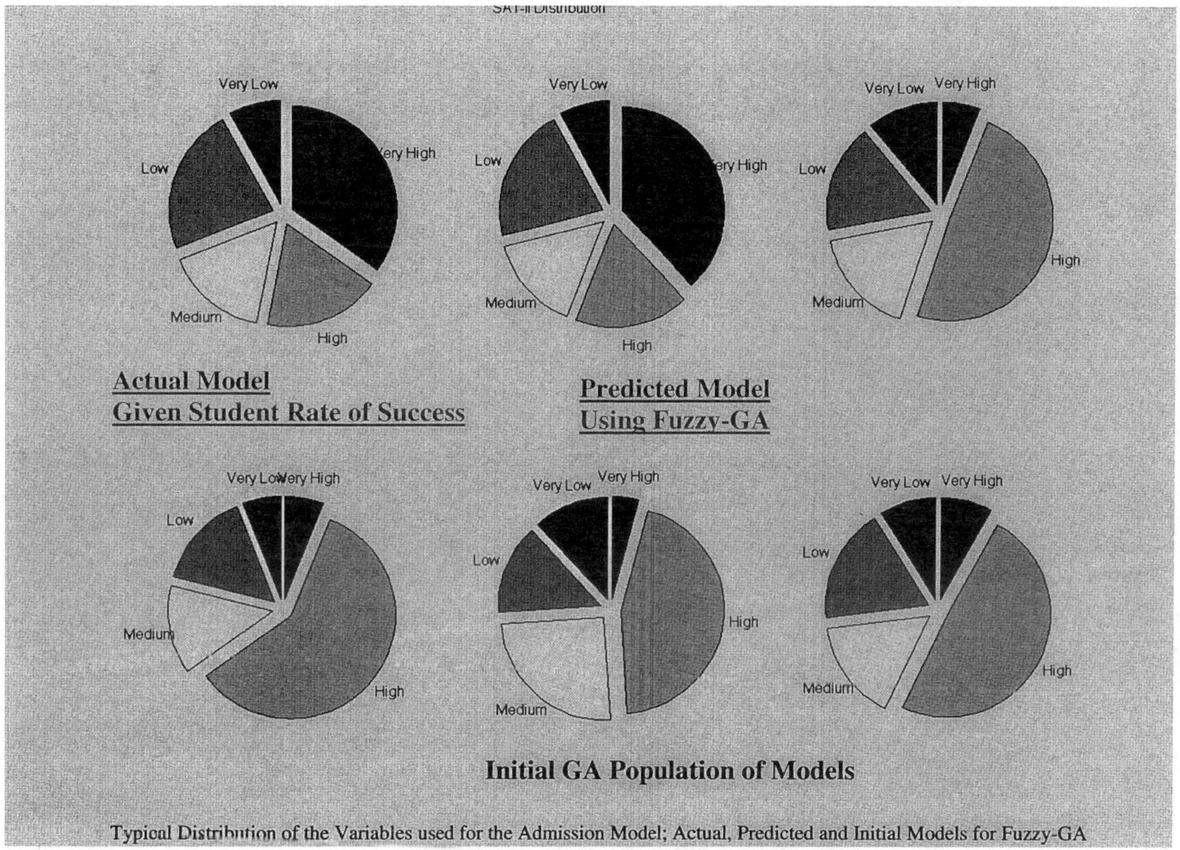

Figure 21.e. SAT-II Distribution

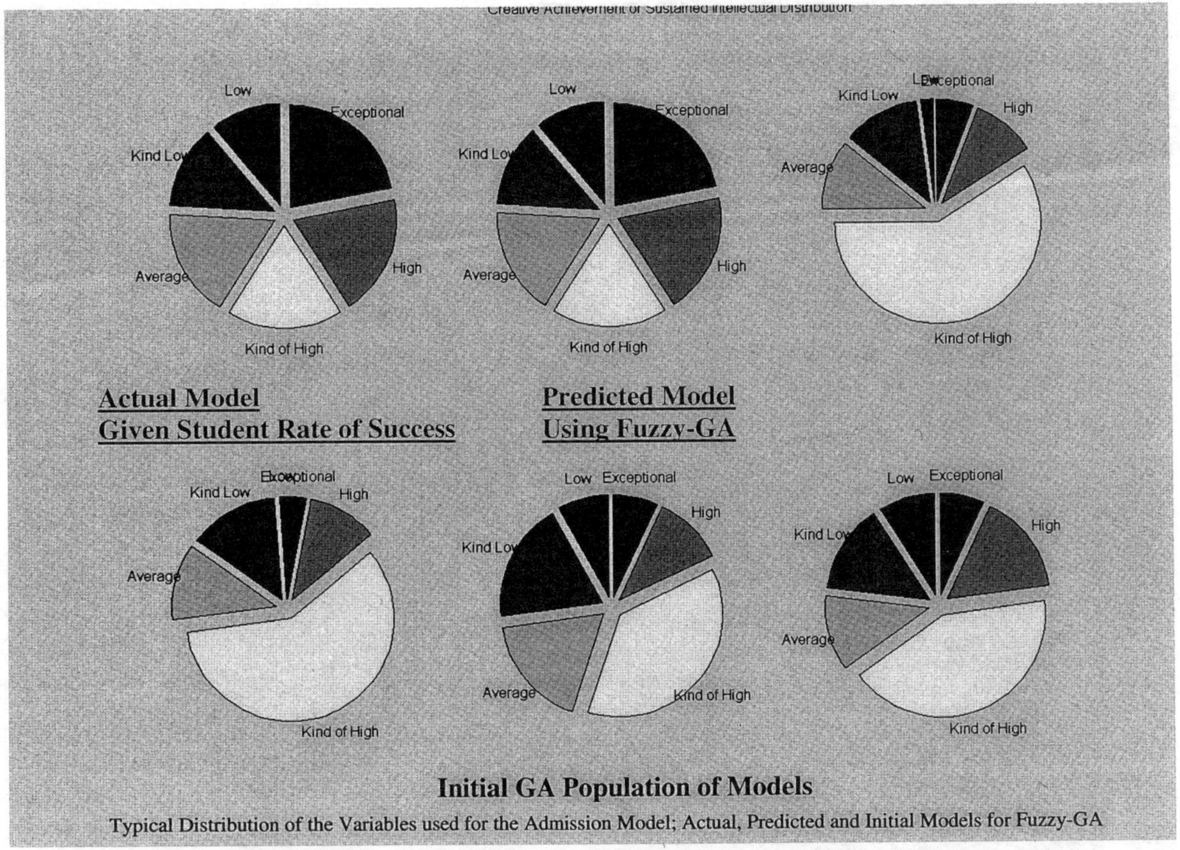

Figure 21.f. Creative Achievement or Sustained Intellectual Distribution

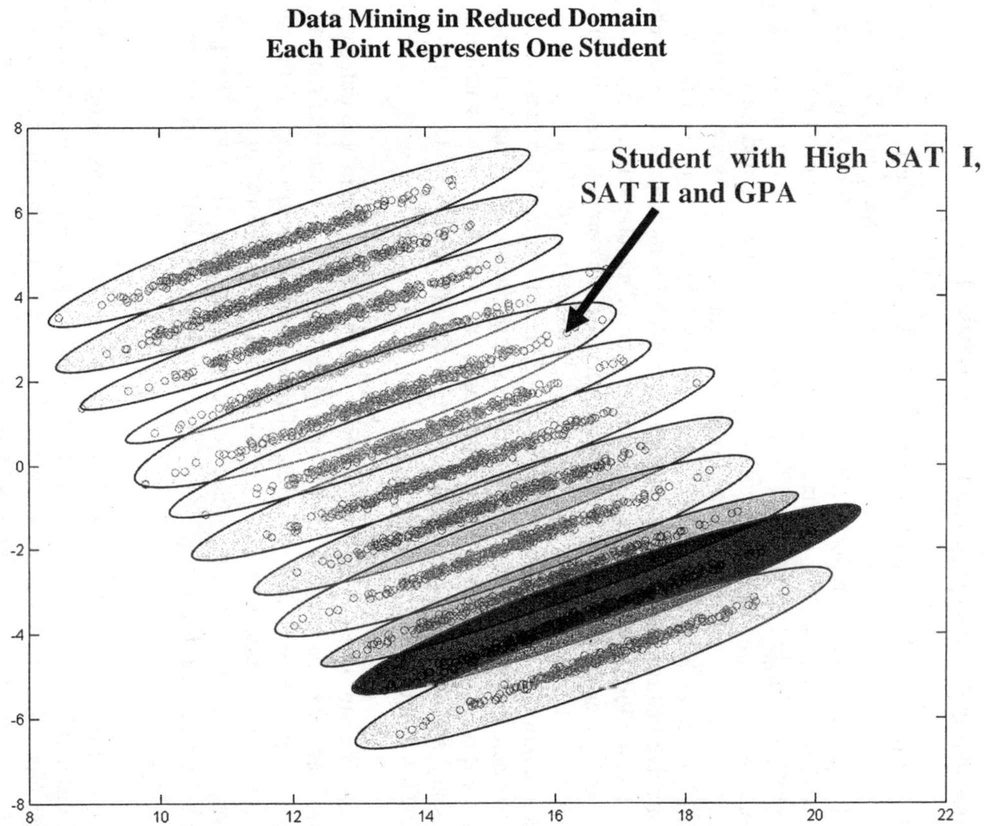

Figure 22. Data Mining based on Techniques described in "Search Strategy" and Fig. 5. on selected dataset

4.2 Date Matching

The main objective of this project was to find the best possible match in the huge space of possible outputs in the databases using the imprecise matching such as fuzzy logic concept, by storing the query attributes and continuously refining the query to update the user's preferences. We have also built a Fuzzy Query system, which is a java application that sits on top of a database.

With traditional SQL queries (relational DBMS), one can select records that match the selection criteria from a database. However, a record will not be selected if any one of the conditions fails. This makes searching for a range of potential candidates difficult. For example, if a company wants to find an employee who is proficient in skill A, B, C and D, they may not get any matching records, only because some candidates are proficient in 3 out of 4 skills and only semi-proficient in the other one. Since traditional SQL queries only perform Boolean matching, some qualities of real life, like "far" or "expensive" or "proficient", which involve matters of degree, are difficult to search for in relational databases. Unlike Boolean logic, fuzzy logic allows the degree of membership for each element to range over an interval. So in a fuzzy query, we can compute how similar a record in the database is to the desired record. This degree of similarity can be used as a ranking for each record in the database. Thus, the aim of the fuzzy query project for date matching is to add the capability of imprecise querying (retrieving similar records) to traditional DBMS. This makes some complex SQL statements unnecessary and also eliminates some repetitious SQL queries (due to empty-matching result sets).

In this program, one can basically retrieve all the records from the database, compare them with the desired record, aggregate the data, compute the ranking, and then output the records in the order of their rankings. Retrieving all the records from the database is a naïve approach because with some preprocessing, some very different records are not needed from the database. However, the main task is to compute the fuzzy rankings of the records so efficiency is not the main concern here.

The major difference between this application and other date matching system is that a user can input his hobbies in a fuzzy sense using a slider instead of choosing crisp terms like "Kind of" or "Love it". These values are stored in the database according to the slider value (**Figures 23 and 24**) .

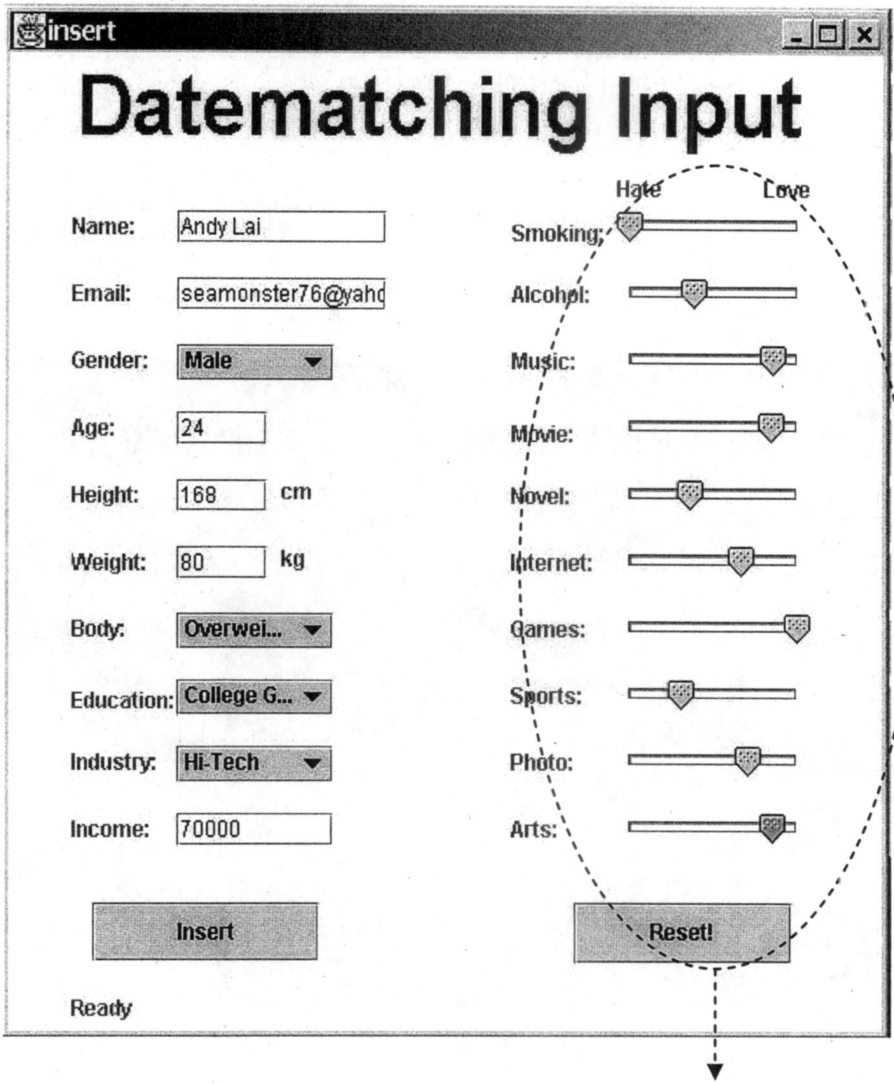

Figure 23. Date matching input form

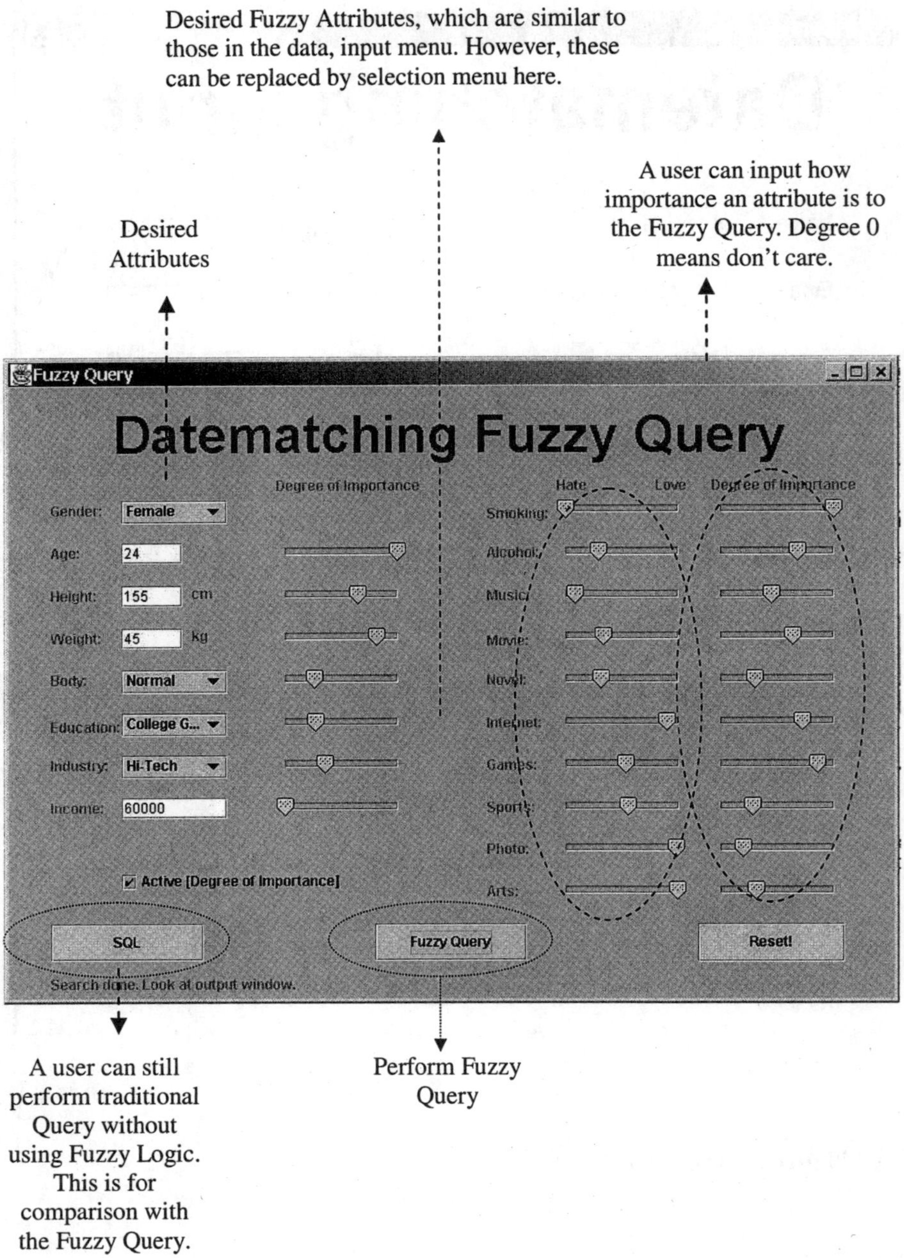

Figure 24. Snapshot of the Date Matching Software

Figure 25 shows the results are obtained from fuzzy query using the search criteria in the previous page. The first record is the one with the highest ranking – 80%. Note that it matches the age field of the search criteria but it's off a bit from the height and weight fields. So one can do imprecise querying.

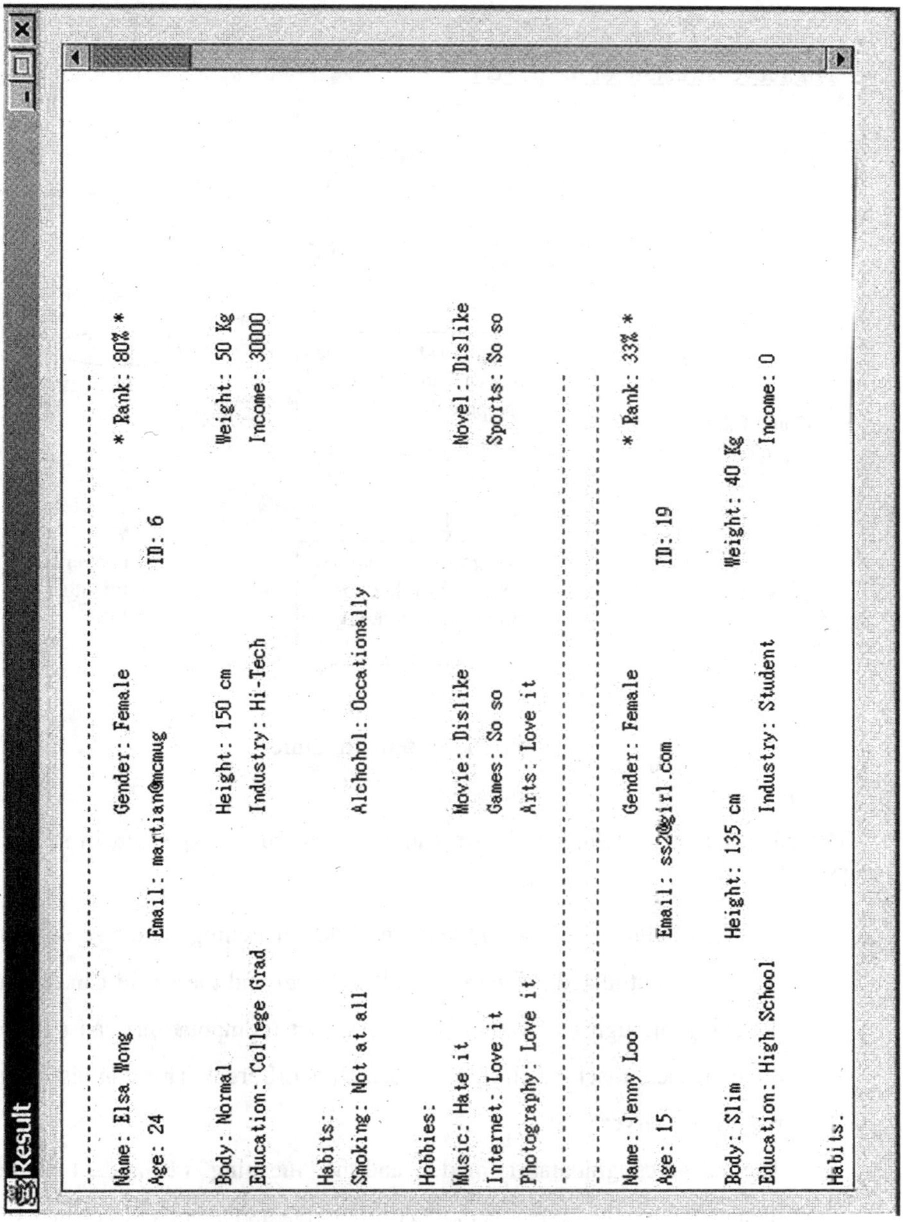

Figure 25. Sample of the output from Date Matching software

The system is modulated into three main modules **(Figure 26).** The core module is the fuzzy engine which accepts input from a GUI module and outputs result to another GUI module. The GUIs can be replaced by other processing modules such that the input can be obtained from other system and the result can be used for further analysis.

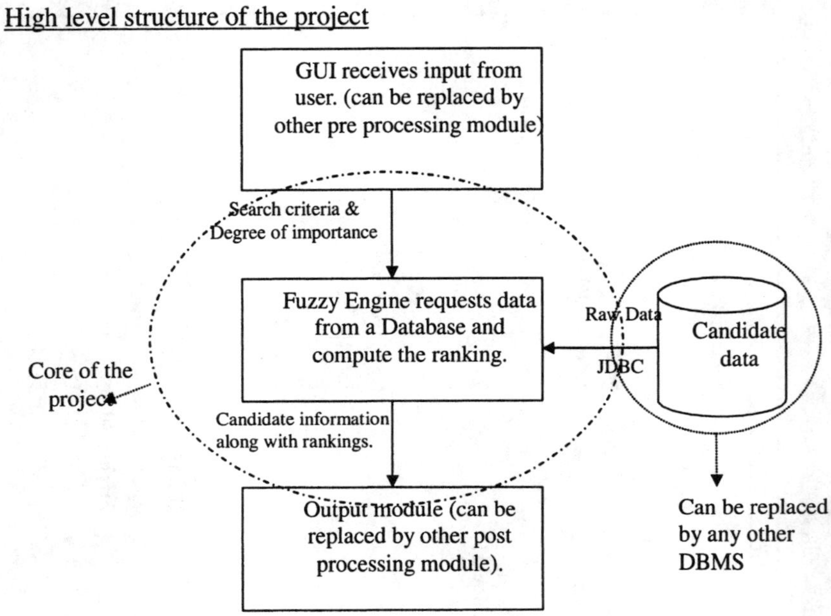

Figure 26. System Structure

The current date matching software can be modified or expanded in several ways:

1. One can build a server/client version of date-matching engine so that we can use a centralized database and all users around the world can do the matching through the web. The ranking part (computation) can still be done on local machine since every search is different. This can also help reduce the server load.
2. The attributes, granulation models and the "meaning" of the data can be tunable so that the system is more configurable and adaptive to changes.

3. User preference capability can be added to the system. (The notion of "overweight" and "tall" can be different to different people.)
4. The GUI needs to be changed to meet real user needs.
5. One can build a library of fuzzy operators and aggregation functions such that one can choose the operator and function that matches the application.
6. One can instead build a generic fuzzy engine framework which is tunable in every way to match clients' needs.
7. The attributes used in the system are not very complete compared to other data matching systems online. However, the attributes can be added or modified with some modification to the program without too much trouble.

Recently, we have added a web interface to the existing software and built the database framework for further analysis in user profiling so that users could find the best match in the huge space of possible outputs. We saved user profiles and used them as basic queries for that particular user. Then, we stored the queries of each user in order to "learn" about this user's preference. In addition, we rewrote the fuzzy search engine to be more generic so that it would fit any system with minimal changes. Administrator can also change the membership function to be used to do searches. Currently, we are working on a new generic software to be developed for a much more diverse applications and to be delivered as stand alone software to both academia and businesses.

4.3 BISC-DSS Potentials

The followings are the potential applications of the BISC Decision Support System:

1. *Physical Stores or E-Store:* A computer system that could instantly track sales and inventory at all of its stores and recognize the customer buying trends and provide suggestion regarding any item that may interest the customer

    - to arrange the products

- on pricing, promotions, coupons, etc
- for advertising strategy

2. *Profitable Customers:* A computer system that uses customer data that allows the company to recognize good and bad customer by the cost of doing business with them and the profits they return

    - keep the good customers
    - improve the bad customers or decide to drop them
    - identify customers who spend money
    - identify customers who are profitable
    - compare the complex mix of marketing and servicing costs to access to new customers

3. *Internet-Based Advising: :* A computer system that uses the expert knowledge and the customer data (Internet brokers and full-service investment firms) to recognize the good and bad traders and provide intelligent recommendation to which stocks buy or sell

    - reduce the expert needs at service centers
    - increase customer confidence
    - ease-of-use
    - Intelligent coaching on investing through the Internet
    - allow customers access to information more intelligently

4. *Managing Global Business:* A computer system responding to new customers and markets through integrated decision support activities globally using global enterprise data warehouse
    - information delivery in minutes
    - lower inventories

- intelligent and faster inventory decisions in remote locations

5. *Resource Allocator:* A computer system that intelligently allocate resources given the degree of match between objectives and resources available

- resource allocation in factories floor
- for human resource management
- find resumes of applicants posted on the Web and sort them to match needed skill and can facilitate training and to manage fringe benefits programs
- evaluate candidates predict employee performance

6. *Intelligent Systems to Support Sales:* A computer system that matching products and services to customers needs and interest based on case-based reasoning and decision support system to improve

- sale
- advertising

7. *Enterprise Decision Support:* An interactive computer-based system that facilitates the solution of complex problems by a group of decision makers either by speeding up the process of the decision-making process and improving the quality of the resulting decisions through expert and user (company-customer) collaboration and sharing the information, goals, and objectives.

8. *Fraud Detection:* An Intelligent Computer that can learn the user's behavior through in mining customer databases and predicting customer behaviours (normal and irregularities) to be used to uncover, reduce or prevent fraud.

- in credit cards

- stocks
- financial markets
- telecommunication
- insurance

9. *Supply-Chain Management (SCM):* Global optimization of design, manufacturing, supplier , distribution, planning decisions in a distributed environment

10. *BISC-DSS and Autonomous Multi-Agent System:* A key component of any autonomous multi-agent system –especially in an adversarial setting -- is decision module, which should be capable of functioning in an environment of imprecision, uncertainty and imperfect reliability. BISC-DSS will be focused on the development of such system and can be used as a decision-support system for ranking of decision alternatives. BISC-DSS can be used :

- As global optimizer for planning decisions in a distributed environment
- To facilitates the solution of complex problems by a group of autonomous agents by speeding up the process of decision-making, collaboration and sharing the information, goals, and objectives
- To intelligently allocate resources given the degree of match between objectives and resources available
- Assisting autonomous multi-agent system in assessing the consequences of decision made in an environment of imprecision, uncertainty, and partial truth and providing a systematic risk analysis
- Assisting multi-agent system answer "What if Questions", examine numerous alternatives very quickly, ranking of decision alternatives, and find the value of the inputs to achieve a desired level of output

11. *BISC-DSS can be integrated into TraS toolbox to develop:* Intelligent Tracking System (ITraS): Given the information about suspicious activities such as phone calls, emails, meetings, credit card information, hotel and airline reservations that are stored in a database containing the originator, recipient, locations, times, etc. we can use BISC-DSS and visual data mining to find suspicious pattern in data using geographical maps. The technology developed can detect unusual patterns, raise alarms based on classification of activities and offer explanations based on automatic learning techniques for why a certain activity is placed in a particular class such as "Safe", "Suspicious", "Dangerous" etc. The underlying techniques can combine expert knowledge and data driven rules to continually improve its classification and adapt to dynamic changes in data and expert knowledge.

12. *BISC-DSS can be integrated into fuzzy conceptual set toolbox to develop TIKManD:* A new Tool for Intelligent Knowledge Management and Discovery (TIKManD). The model can be used to recognize terrorism activities through data fusion & mining and pattern recognition technology given online textual information through Email or homepages and voice information given the wire tapping and/or chat lines or huge number of "tips" received immediately after the attack.

The followings are the potential applications areas of the BISC Decision Support System:

- *Finance:* stock prices and characteristics, credit scoring, credit card ranking
- *Military:* battlefield simulation and decision making
- *Medicine:* diagnosis

- *Marketing:* store and product display and electronic shopping
- *Internet:* provide knowledge and advice to large numbers of user
- *Education:* university admission

5 Web Intelligence: Web-Based BISC Decision Support system

Most of the existing search systems 'software' are modeled using crisp logic and queries. In this chapter we introduce fuzzy querying and ranking as a flexible tool allowing approximation where the selected objects do not need to match exactly the decision criteria resembling natural human behavior. The model consists of five major modules: the Fuzzy Search Engine, the Application Templates, the User Interface, the Database and the Evolutionary Computing. The system is designed in a generic form to accommodate more diverse applications and to be delivered as stand-alone software to academia and businesses.

5.1 Web Intelligence: Introduction

Searching database records and ranking the results based on multi-criteria queries is central for many database applications used within organizations in finance, business, industrial and other fields. Most of the available systems 'software' are modeled using crisp logic and queries, which results in rigid systems with imprecise and subjective process and results. In this chapter we introduce fuzzy querying and ranking as a flexible tool allowing approximation where the selected objects do not need to match exactly the decision criteria resembling natural human behavior.

The model consists of five major modules: the Fuzzy Search Engine (FSE), the Application Templates (AT), the User Interface (UI), the Database (DB) and the Evolutionary Computing (EC). We developed the software with many essential features. It is built as a web-based software system that users can access and use over the Internet. The system is designed to be generic so that it can run different application domains. To this end, the Application Template module provides information of a specific application as attributes and properties, and serves as a guideline structure for building a new application.

The Fuzzy Search Engine (FSE) is the core module of the system. It has been developed to be generic so that it would fit any application. The main FSE component is the query structure, which utilizes membership functions, similarity functions and aggregators.

Through the user interface a user can enter and save his profile, input criteria for a new query, run different queries and display results. The user can manually eliminate the results he disapproves or change the ranking according to his preferences.

The Evolutionary Computing (EC) module monitors ranking preferences of the users' queries. It learns to adjust to the intended meaning of the users' preferences.

5.2 Model framework

The DSS system starts by loading the application template, which consists of various configuration files for a specific application (see section 5.4) and initializing the database for the application (see section 5.6), before handling user's requests, see **Figure 27.**

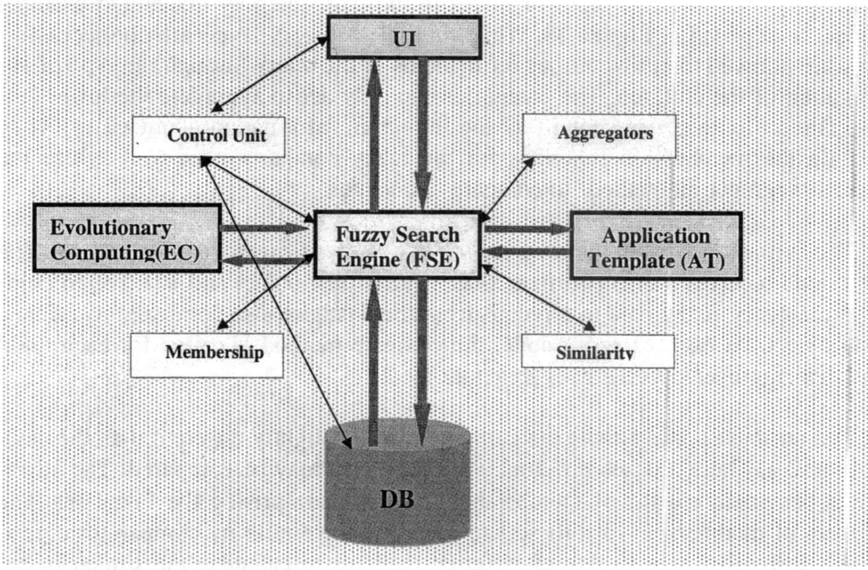

Figure 27. The BISC-DSS general framework

Once the DSS system is initialized, users can enter in the user interface their own profiles or make a search with their preferences. These requests are handled by the control unit of the system. The control unit converts user input into data objects that are recognized by the DSS system then, based on the request types, it forwards them to the appropriate modules.

If the user wants to create a profile, the control unit will send the profile data directly to the database module, which stores the data in the database for the application. If the user wants to query the system, the control unit will direct the user's preferences to the Fuzzy Search Engine, which queries the database (see section 5.3). The query results will be sent back to the control unit and displayed to the users.

5.3 Fuzzy Engine

5.3.1 Fuzzy Query, search and Ranking

To support generic queries, the fuzzy engine has been designed to have a tree structure. There are two types of nodes in the tree, category nodes and attribute nodes, as depicted in **Figure 28.** While multiple category levels are not necessary, they are allowed to allow various refinements of the query through the type of aggregation of the children. The categories can only act to aggregate the lower levels. The attribute nodes contain all the important information about query. They contain the membership functions for the fuzzy comparison as well as use the various aggregation methods to compare two values.

The flow of control in the program when a query is executed is as follows. The root node receives a query formatted as a fuzzy data object and is asked to compare the query fuzzy data to a record from the database also formatted as a fuzzy data object. At each category node, the compare method is called for each child and then aggregated using an aggregator object.

The attribute nodes handle the compare method slightly different than the category nodes. There are two different ways attributes may be compared. The attribute nodes contain a list of membership functions comprising the fuzzy set. The degrees of membership for this set are passed to the similarity comparator object, which currently has a variety of different methods to calculate the similarity between the two membership vectors. In the other method, the membership vector created by having full membership to a single membership function specified in the fuzzy data object, but no membership value for the other functions.

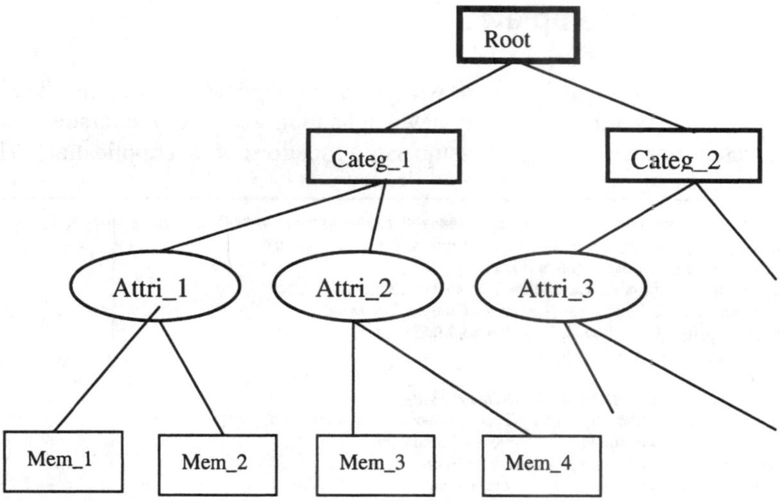

Figure 28. The Fuzzy search engine tree structure.

The resulting comparison value returned from the root node is assigned to the record. The search request is then added to a sorted list ordered by this ranking in descending value. Each of the records from the database is compared to the query and the results are returned. For certain search criteria, it may be desirable to have exact values in the query. For such criteria, the database is used to filter the records for comparison.

5.3.2 Membership function

Currently there are three membership functions implemented for the Fuzzy Engine. A generic interface has been created to allow several different types of membership functions to be added to the system. The three types of membership functions in the system are: Gaussian, Triangular and Trapezoidal. These functions have three main points, for the lower bound, upper bound and the point of maximum membership. For other functions, optional extra points may be used to define the shape (an extra point is required for the trapezoidal form).

5.4 Application Template

The DSS system is designed to work with different application domains. The application template is a format for any new application we build, it contains data of different categories, attributes and membership functions of that application. The

```
##########################################################################
#This is a properties file for membership definition. We should specify
#the following properties for an attribute:
# -  A unique identifier for each defined membership function.
# -  A type from the following: {Gaussian, Triangle, Trapezoid}
# -  Three points: Lowerbound, Upperbound, Maximum
# -  Optional point: Auxillary Maximum
# Format:
# <MF_Name>.membershipFunctionName = <MF_Name>
# <MF_Name>.membershipFunctionType = {Gaussian/Triangle/Trapezoid}
# <MF_Name>.lowerBound          = lowerBoundValue
# <MF_Name>.upperBound          = upperBoundValue
# <MF_Name>.maxValue            = maxValue
# <MF_Name>.optionPoint         = pt1, pt2, pt3 ...
#
##########################################################################
```

```
##########################################################################
#
# Gender Membership Functions
#
male.membershipFunctionName = male
male.membershipFunctionType = Triangle
male.lowerbound        = 1
male.upperbound        = 1
male.maxValue          = 1

female.membershipFunctionName = female
female.membershipFunctionType = Triangle
female.lowerbound        = 0
female.upperbound        = 0
female.maxValue          = 0
#
# Age Membership Functions
#
young.membershipFunctionName = young
young.membershipFunctionType = Triangle
young.lowerbound        = 0
young.upperbound        = 35
young.maxValue          = 20

middle.membershipFunctionName = middle
middle.membershipFunctionType = Triangle
middle.lowerbound        = 20
middle.upperbound        = 50
middle.maxValue          = 35

old.membershipFunctionName = old
old.membershipFunctionType = Triangle
old.lowerbound        = 35
old.upperbound        = 100
old.maxValue          = 50
```

Figure 29. Template of the date matching application

application template module consists of two parts the application template data file, and the application template logic. The application template data file specifies all the membership functions, attributes and categories of an application. We can consider it as a configuration data file for an application. It contains the definition of membership functions, attributes and the relationship between them.

The application template logic parses and caches data from the data file so that other modules in the system can have faster access to definitions of membership functions, attributes and categories. It also creates a tree data structure for the fuzzy search engine to transverse. **Figure 29** shows part of the sample configuration file from the Date Matching application.

5.5 User interface

It is difficult to design a generic user interface that suits different kind of applications for all the fields. For example, we may want to have different layouts for user interfaces for different applications. To make the DSS system generic while preserving the user friendliness of the interfaces for different applications, we developed the user interfaces into two parts.

First, we designed a specific HTML interface for each application we developed. Users can input their own profiles, make queries by specifying preferences for different attributes. Details for the DSS system are encapsulated from the HTML interface so that the HTML interface design would not be constrained by the DSS system.

The second part of our user interface module is a mapping between the parameters in the HTML files and the attributes in the application template module for the application. The input mapping specifies the attribute names each parameter in the HTML interface corresponds to. With this input mapping, user interface designer can use any input method and parameter names freely (**Figure 30**).

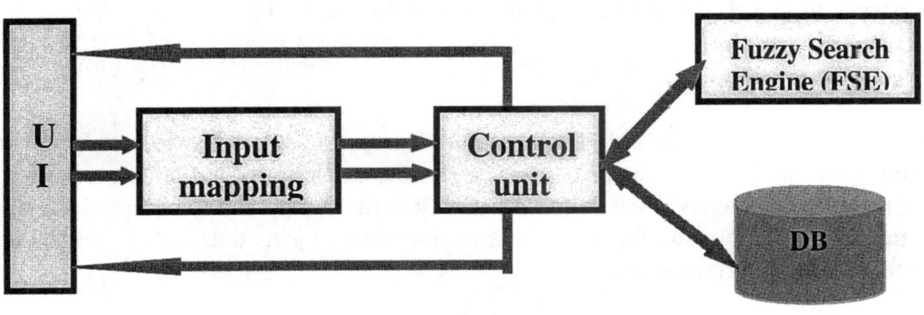

Figure 30. User interface data flow

5.6 Database (DB)

The database module is responsible for all the transactions between the DSS system and the database. This module handles all queries or user profile creations from the Fuzzy Engine and the Control Unit respectively. For queries from the Fuzzy Search Engine, it retrieves data from the database and returns it in a data object form. Usually queries are sets of attribute values and their associated weights. The database module (**Figure 31**) returns the matching records in a format that can be manipulated by the user, as eliminating one or more record or changing their order. For creating user profile, on the other hand, it takes data objects from the Control Unit and stores it in the database. There are three components in the DB module: the DB Manager (DBMgr), the DB Accessor (DBA) and DB Accessor Factory (DBA Factory).

5.6.1 DB Manager

The DB Manager is accountable for two things: setting up database connections and allocating database connections to DB Accessor objects when needed. During the initialization of the DSS system, DB Manager loads the right driver, which is used for the communications between the database and the system. It also supplies information to the database for authentication purposes (e.g. username, password, path to the database etc).

5.6.2 DB Accessor Factory

The DB Accessor Factory creates DB Accessor objects for a specific application. For example, if the system is running the date matching application, DB Accessor Factory will create DB Accessor objects for the date matching application. The existence of this class serves the purpose of using a generic Fuzzy Search Engine.

5.6.3 DB Accessor

DB Accessor is responsible for storing and getting user profiles to and from the database. It also saves queries from users to the database so that other modules in the system can analyze user's preferences. It is the component that queries the database and wrap result from the database into data objects that are recognized by our application framework.

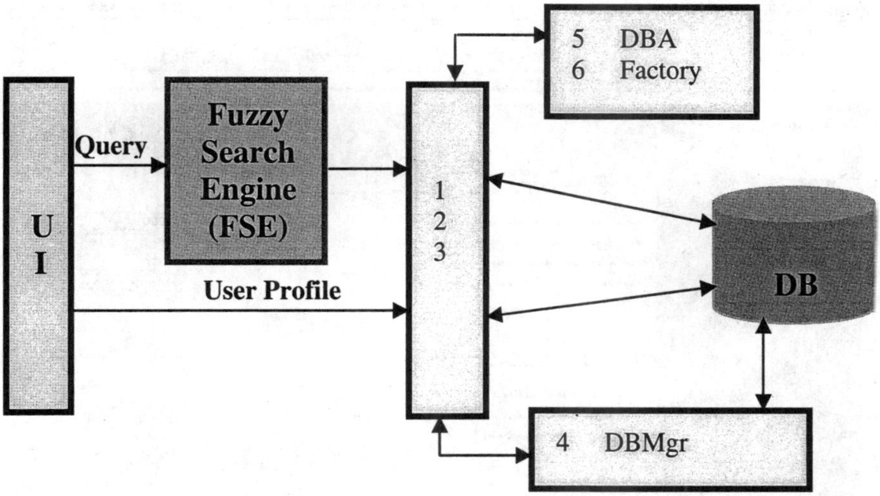

Figure 31. Database module components

5.7 Applications

In this work, we implemented our approach on four important applications: Credit ranking (scoring) (**Figure 32.a. 32.b**), which has been used to make financing decisions concerning credit cards, cars and mortgage loans; the process of college admissions where hundreds of thousands of applications are processed yearly by U.S. Universities (**Figure 33**); and date matching (**Figures 34.a and 34.b**) as one of the most popular internet programs. Even though we implemented three applications, the system is designed in a generic form to accommodate more diverse applications and to be delivered as stand-alone software to academia and businesses.

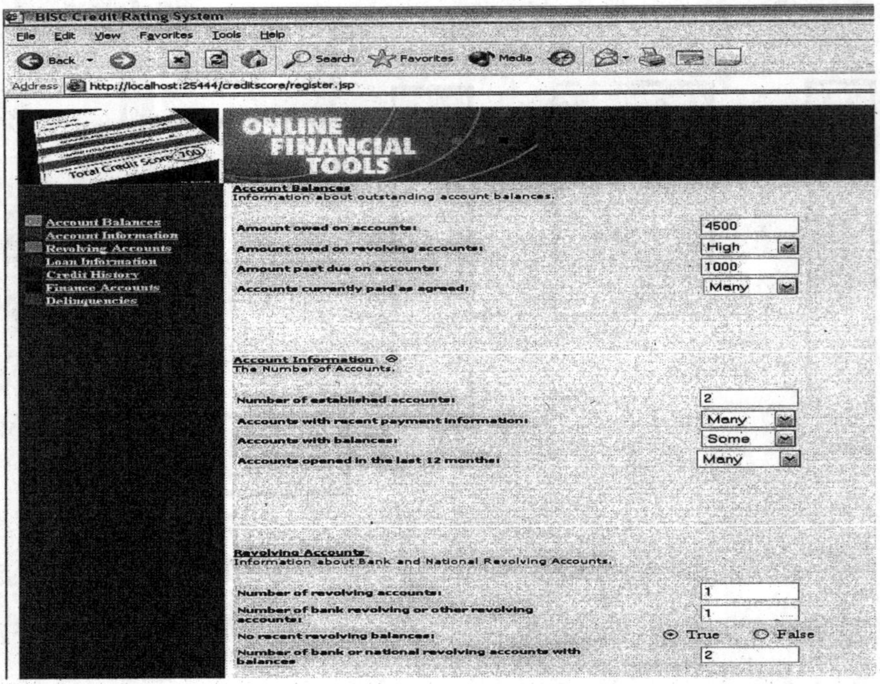

Figure 32.a. A snapshot of the variable input for credit scoring software.

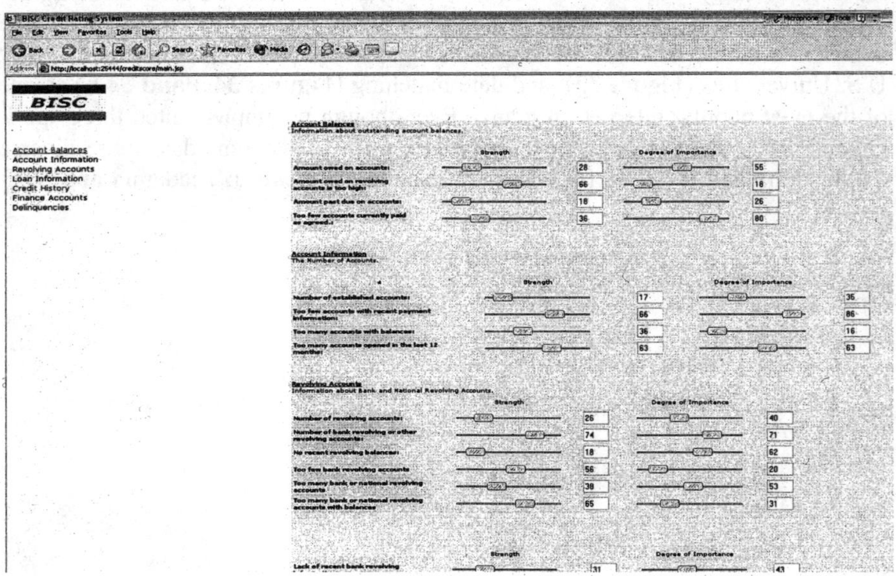

Figure 32.b. A snapshot of the software developed for credit scoring.

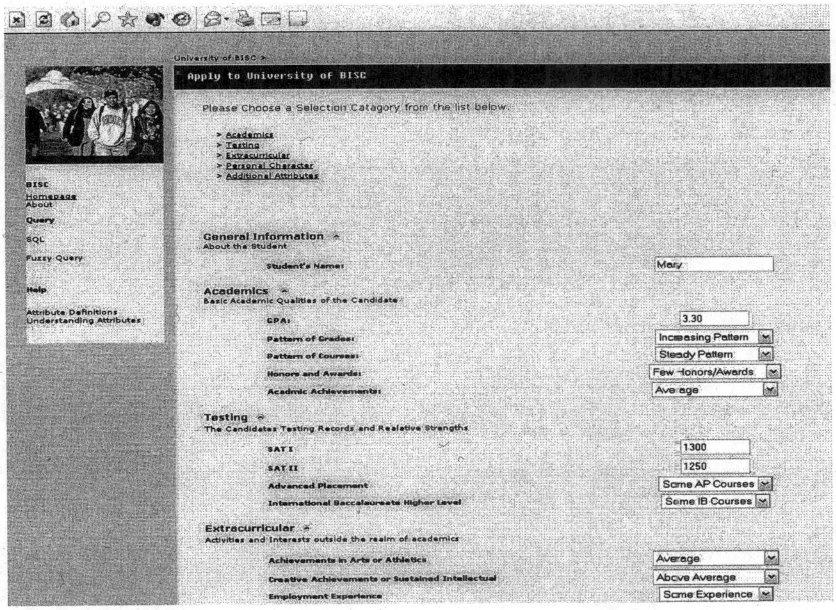

Figure 33. A snapshot of the software for University Admission Decision Making.

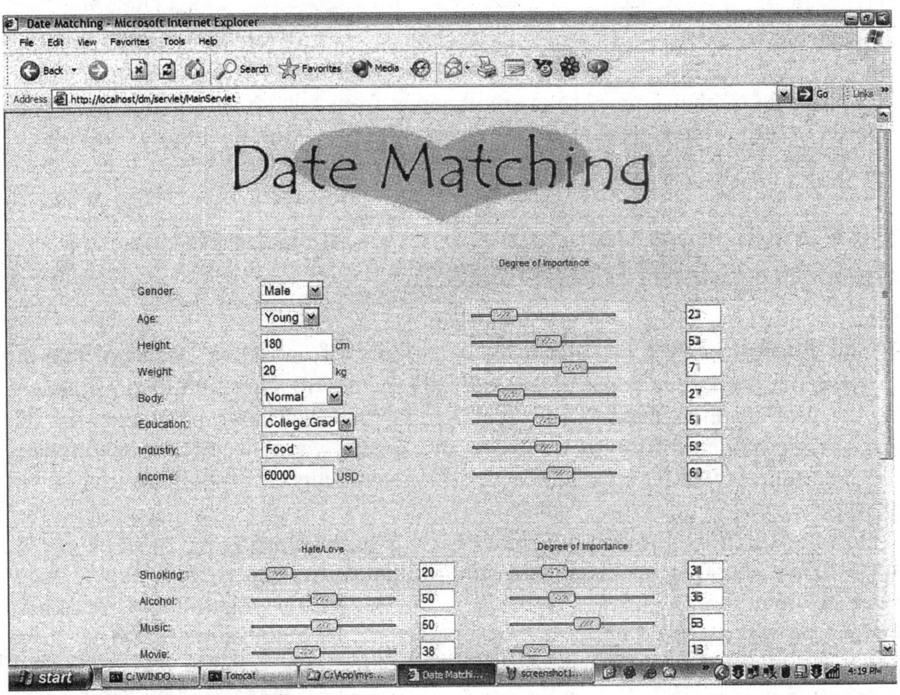

Figure 34.a. Date matching input form

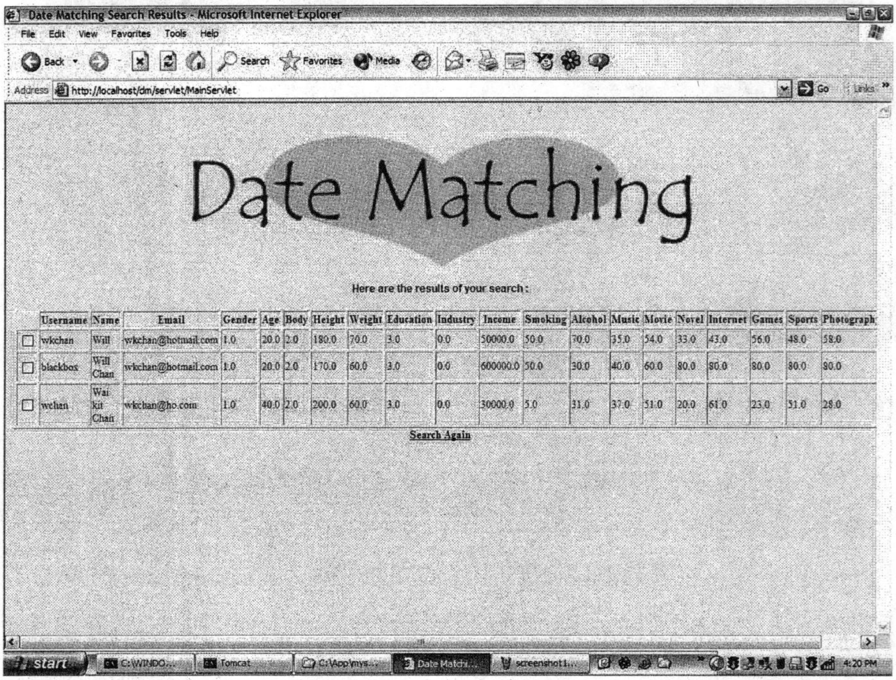

Figure 34.b. shows the results are obtained from fuzzy query using the search criteria in the previous page. The first record is the one with the highest ranking.

5.8 Evolutionary Computing for the BISC Decision Support system (EC-BISC-DSS)

In the Evolutionary Computing (EC) module of the BISC Decision Support System, our purpose is to use an evolutionary method to allow automatic adjusting of the user's preferences. These preferences can be seen as parameters of the fuzzy logic model in form of weighting of the used variables. These preferences are then represented by a weight vector and genetic algorithms will be used to fix them.

In the fuzzy logic model, the variables are combined using aggregation operators. These operators are fixed based on the application expert knowledge. However, we may have to answer to the question: how to aggregate these variables? Indeed, to make decision regarding the choice of the aggregators that have to be used in addition to the preferences the application expert might need help. We propose to automatically select the appropriate aggregators for a given application according to some corresponding training data. Moreover, we propose to combine these selected aggregators in a decision tree. In the Evolutionary Computation approach, Genetic Programming, which is an extension of Genetic Algorithms, is the closest technique to our purpose. It allows us to learn a tree structure which represents the combination of aggregators. The selection of these aggregators is included to the learning process using the Genetic Programming.

Genetic algorithms and genetic programming will be first introduced in the next section. Then, their adaptation to our decision system will be described.

5.8.1 Genetic algorithms and genetic programming

Introduced by J. Holland (1992), Genetic Algorithms (GAs) constitute a class of stochastic searching methods based on the mechanism of natural selection and genetics. They have recently received much attention in a number of practical problems notably in optimization problems as machine learning processes (Banzhaf et al., 1982).

5.8.1.1 Basic description

To solve an optimization problem, usually we need to define the search method looking for the best solution and to specify a measure of quality that allows to compare possible solutions and to find the best one. In GAs, the search space corresponds to a set of individuals represented by their DNA. These individuals are evaluated by a measure of their quality called fitness function which has to be defined according to the problem itself. The search method consists in a evolutionary process inspired by the Darwinian principle of reproduction and survival of the fittest individual.

This evolutionary process begins with a set of individuals called population. Individuals from one population are selected according to their fitness and used to form a new population with the hope to produce better individuals (offspring). The population is evolved through successive generations using genetic operations until some criterion is satisfied.

The evolution algorithm is resumed in **Figure 35**. It starts by creating randomly a population of individuals which constitute an initial generation. Each individual is evaluated by calculating its fitness. Then, a selection process is performed based on their fitness to choose individuals that participate to the evolution. Genetic operators are applied to these individuals to produce new ones. A new generation is then created by replacing existing individuals in the previous generation by the new ones. The population is evolved by repeating individuals' selection and new generations creation until the end criterion is reached in which case the evolution is stopped.

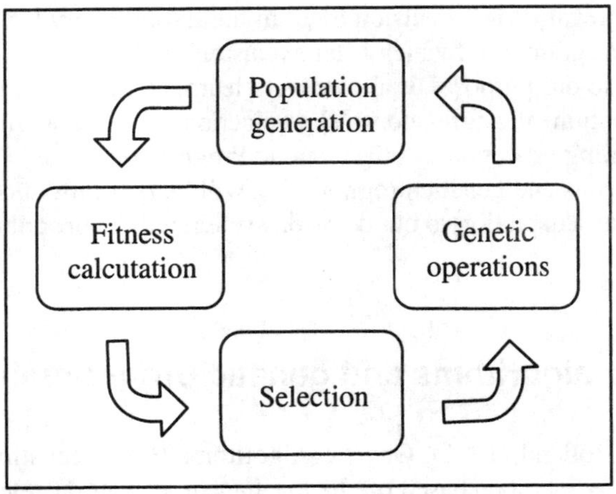

Figure 35. Genetic Algorithm Cycle

The construction of a GA for any problem can be separated into five tasks: choice of the representation of the individuals, design of the genetic operators, determination of the fitness function and the selection process, determination of parameters and variables for controlling the evolution algorithm, and definition of the termination criterion.

In the conventional GAs, individuals' DNA are usually represented by fixed-length character strings. Thus, the DNA encoding requires a selection of the string length and the alphabet size. Binary strings are the most common encoding because its relative simplicity. However, this encoding might be not natural for many problems and sometimes corrections must be made on the strings provided by genetic operations. Direct value encoding can be used in problems where use of binary encoding would be difficult. In the value encoding, an individual's DNA is represented by a sequence of some values. Values can be anything connected to the problem, such as (real) numbers.

5.8.1.2 Genetic operators

The evolution algorithm is based on the reproduction of selected individuals in the current generation breeding a new generation composed of their offspring. New individuals are created using either sexual or asexual reproduction. In sexual reproduction, known as crossover, two parents are selected and DNA from both parents is inherited by the new individual. In asexual reproduction, known as mutation, the selected individual (parent) is simply copied, possibly with random changes.

Crossover operates on selected genes from parent DNA and creates new offspring. This is done by copying sequences alternately from each parent and the points where the copying crosses is chosen at random. For example, the new indivudal can be bred by copying everything before the crossover point from the first parent and then copy everything after the crossover point from the other parent. This kind of crossover is illustrated in **Figure 36** for the case of binary string encoding. There are other ways to make crossover, for example by choosing more crossover points. Crossover can be quite complicated and depends mainly on the encoding of DNA. Specific crossover made for a specific problem can improve performance of the GA.

Mutation is intended to prevent falling of all solutions in the population into a local optimum of the solved problem. Mutation operation randomly changes the offspring resulted from crossover. In case of binary encoding we can switch a few randomly chosen bits from 1 to 0 or from 0 to 1 (see **Figure 37**.). The technique of mutation (as well as crossover) depends mainly on the encoding of chromosomes. For example when permutations problem encoding, mutation could be performed as an exchange of two genes.

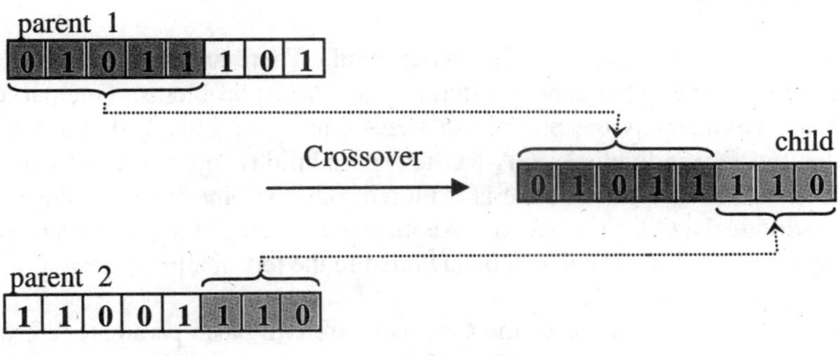

Figure 36. Genetic Algorithm - Crossover

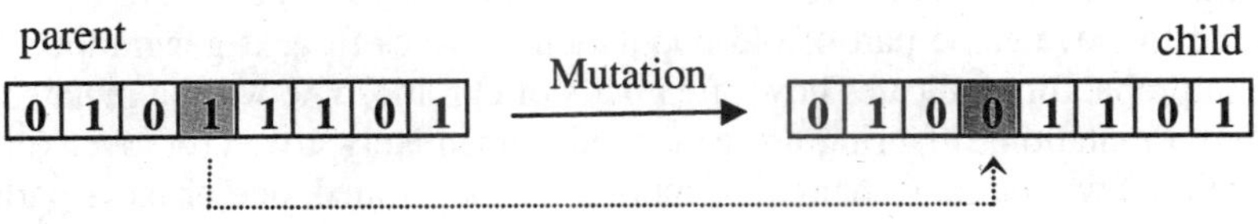

Figure 37. Genetic Algorithm - Mutation

5.8.1.3 Selection process

Individuals that participate to genetic operations are selected according to their fitness. Even that the main idea is to select the better parents in the hope that they will produce better offspring, the problem of how to do this selection remains. This can be done in many ways. We will describe briefly some of them. The (μ,λ) selection, consists in breeding λ offspring from μ parents and then μ offspring will be selected for the next generation. In the Steady-State Selection, in every generation a few good (with higher fitness) individuals are selected for creating new offspring. Then some bad (with lower fitness) individuals are removed and replaced by the new offspring. The rest of population survives to new generation. In the tournament selection, a group of individuals is chosen randomly and the best individual of the group is selected for reproduction. This kind of selection allows to give a chance to some weak individual in the population which could contain good genetic material (genes) to participate to reproduction if it is the best one in its group. Elitism selection aims at preserving the best individuals. So it first copies the best individuals to the new population. The rest of the population is constructed in ways described above. Elitism can rapidly increase the performance of GA, because it prevents a loss of the best found solution.

5.8.1.4 Parameters of GA

The outline of the Basic GA is very general. There are many parameters and settings that can be implemented differently in various problems. One particularly important parameter is the population size. On the one hand, if the population contains too few individuals, GA has few possibilities to perform crossover and only a small part of search space is explored. On the other hand, if there are too many individuals, GA slows down. Another parameter to take into account is the number of generations which can be included in the termination criterion.

For the evolution process of the GA, there are two basic parameters: crossover probability and mutation probability. The crossover probability indicates how often crossover will be performed. If there is no crossover, offspring are exact copies of parents. If there is crossover, offspring are made from parts of both parent's DNA. Crossover is made in hope that new chromosomes will contain good parts of old chromosomes and therefore the new chromosomes will be better. However, it is good to leave some part of old population survives to next generation. The mutation probability indicates how often parts of chromosome will be mutated. If there is no mutation, offspring are generated immediately after crossover (or directly copied) without any change. If mutation is performed, one or more parts of a chromosome are changed.

5.8.1.5 Genetic programming

Genetic programming (GP) is a technique pioneered by J. Koza (1992) which enables computers to solve problems without being explicitly programmed. It is an extension of the conventional GA in which each individual in the population is a computer program. It works by using GAs to automatically generate computer programs that can be represented as linear structures, trees or graphs. Tree encoding is the most used form to represent the programs. Tree structures are composed of primitive functions and terminals appropriate to the problem domain. The functions may be arithmetic operations, programming commands, and mathematical logical or domain-specific functions. To apply GP to a problem, we have to specify the set functions and terminals for the tree construction. Also, besides the parameters of the conventional GA, other parameters which are specific to the individual representation can be considered such as tree size for example.

Genetic operations are defined specifically for the type of encoding used to represent the individuals. In the case of tree encoding, new individuals are produced by removing branches from one tree and inserting them into another. This simple process ensures that the new individual is also a tree and so is also syntactically valid. The crossover and mutation operations are illustrated in **Figures 38 and 39**. The mutation consists in randomly choosing a node in the selected tree, creating a new individual and replacing the sub-tree rooted at the selected node by the created individual. The crossover operation is performed by randomly choosing nodes in the selected individuals (parents) and exchanging the sub-trees rooted at these nodes which produce two new individuals (offspring).

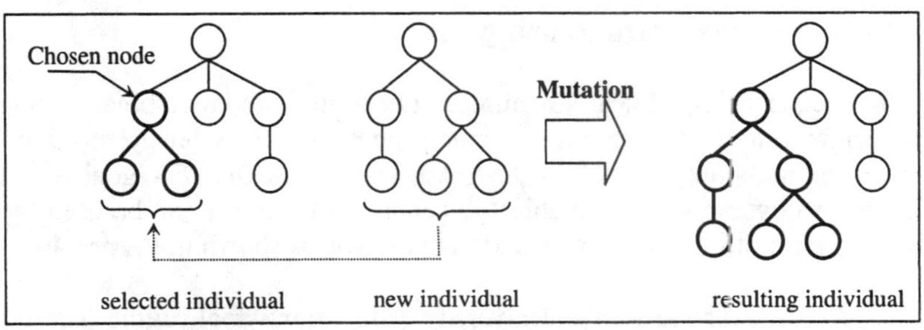

Figure 38. Genetic programming - Tree-encoding individual mutation

5.8.2 Implementation

After having introduced the GA and GP background, now we are going to describe their application to our problem. Our aim is at learning fuzzy-DSS parameters which are the weight vector representing the user preferences associated to the variables that have to be aggregated on the one hand, and the adequate decision tree representing the combination of the aggregation operators that have to be used, on the other hand.

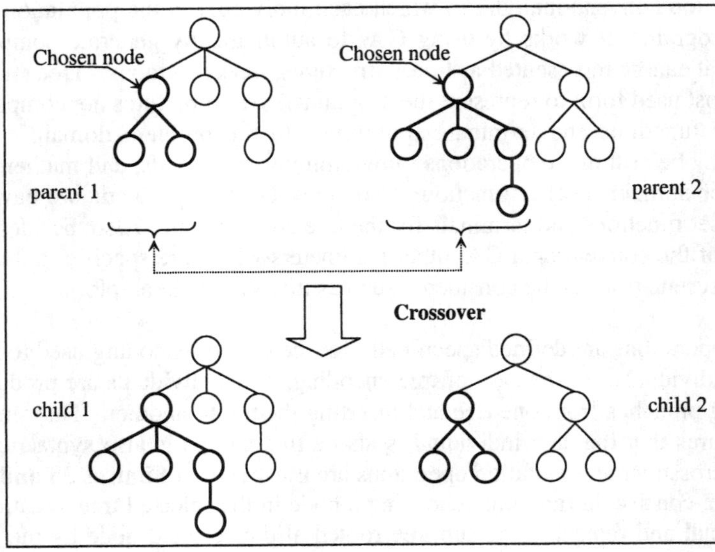

Figure 39. Genetic programming - Tree-encoding individual crossover.

5.8.2.1 Preferences learning using GA

Weight vector being a linear structure, can be represented by a binary string, in which weight values are converted to binary numbers. This binary string corresponds to the individual's DNA in the GA learning process. The goal is to find the optimal weighting of the variables. A general GA module can be used by defining a specific fitness function for each application as shown in **Figure 40.**

Let's see the example of the University Admissions application. The corresponding fitness function is shown **Figure 41.** The fitness is computed based on a training data set composed of vectors $\vec{X}_1,\cdots,\vec{X}_N$ of fuzzy values $(X_{i1},\cdots X_{ik})$ for each $\vec{X}_i$. Each value of a fuzzy variable is constituted of a crisp value between 0 and 1 and a set of membership functions. During the evolution process, for each weighting vector $(W_1,W_2,\cdots,W_k)$, the corresponding fitness function is computed. Using these weights, a score is calculated for each vector. Afterward, these scores are ranked and compared with the actual ranking using similarity measure.

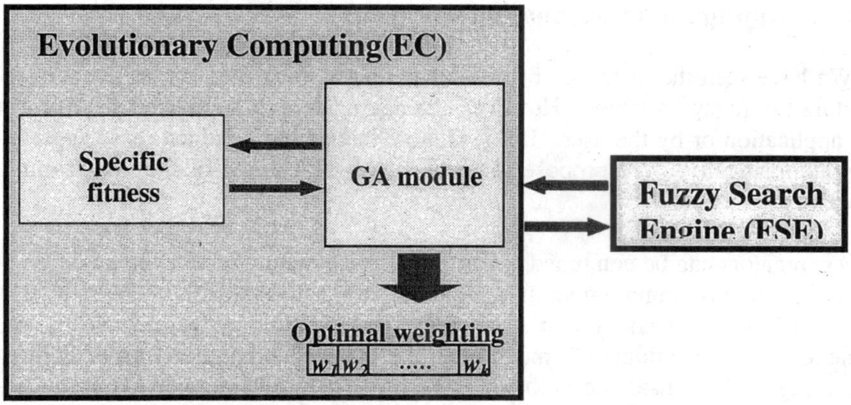

Figure 40. Evolutionary Computing Module: preferences learning.

Let's assume that we have N students and the goal is to select among them n students that will be admitted. Each student is then represented by value vector in the training data set. The similarity measure could the common vectors in the n top ones between the computed and the actual ranking. This intersection has then to be maximized. We can also consider the intersection on a larger number $n_1 > n$ of top vectors. This measure can be combined to the first one with different degrees of importance. The Fitness value will be a weighted sum of these two similarity measures.

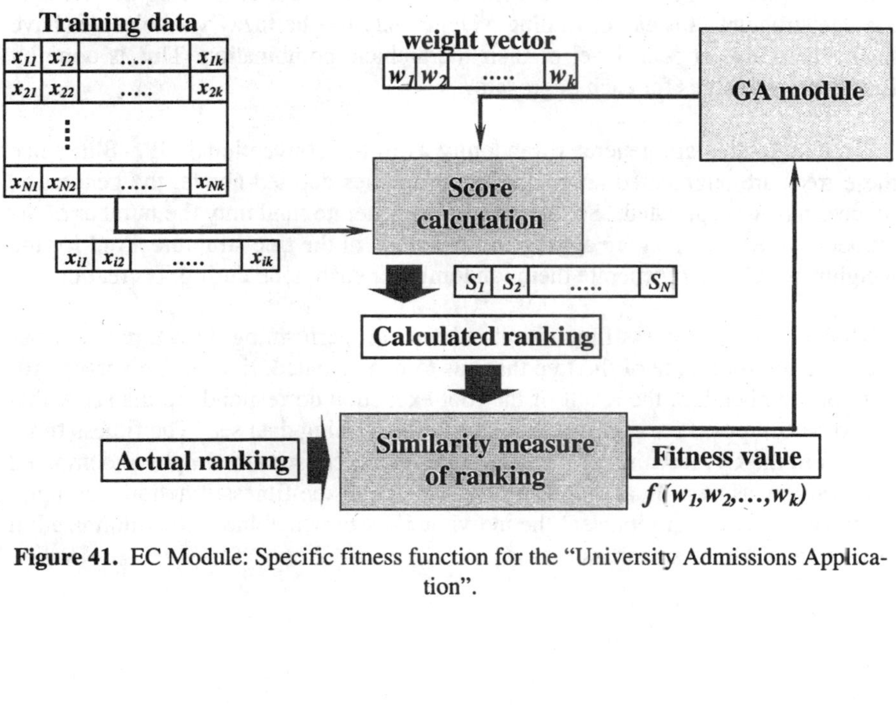

Figure 41. EC Module: Specific fitness function for the "University Admissions Application".

5.8.2.2 Aggregation tree learning using GP

We have seen the learning of the weights representing the user preferences regarding the fuzzy variables. However, the aggregators that are used are fixed in the application or by the user. But it is more interesting to adjust these aggregators automatically. We propose to include this adjustment in the GA learning process.

Aggregators can be combined in form of a tree structure which can be built using a Genetic Programming learning module. It consists in evolving a population individuals represented by tree structures. The evolution principle remains the same as in a conventional GP module but the DNA encoding needs to be defined according to the considered problem. We propose to define an encoding for aggregation trees which is more complex than for classical trees and which is common to all considered applications. As shown in **Figure 42**, we need, in addition to the fitness function specification, to define a specific encoding.

We need to specify the functions (tree nodes) and terminals that are used to build aggregation trees. The functions correspond to aggregation operators and terminals (leaves) are the fuzzy variables that have to be aggregated. Usually, in GP the used functions have a fixed number of arguments. In our case, we prefer not to fix the number of arguments for the aggregators. We might however define some restrictions such as specifying minimal and maximal number of arguments. These numbers can be considered as parameters of the learning process. This encoding property allows a largest search space to solve our problem. Another property which is indispensable specificity is the introduction of weights values in the tree structure. Instead of finding weights only for the fuzzy variables, we have to fix them also at each level of their hierarchical combination. This is done by fixing weight values for each aggregator.

Tree structures are generated randomly as in the conventional GP. But, since these trees are augmented according the properties defined above, the generation process has to be updated. So, we decided to generate randomly the number of arguments when choosing an aggregator as a node in the tree structure. And for the weights, we chose to generate them randomly for each node during its creation.

Concerning the fitness function, it is based on performing the aggregation operation a the root node of the tree that has to be evaluated. For the University Admissions application, the result of the root execution corresponds to the score that has to be computed for each value vector in the training data set. The fitness function, as in the GA learning of the user preferences, consists in simple or combined similarity measures. In addition, we can include to the fitness function a complementary measure that represent the individual's size which has to be minimized in order to avoid huge size trees.

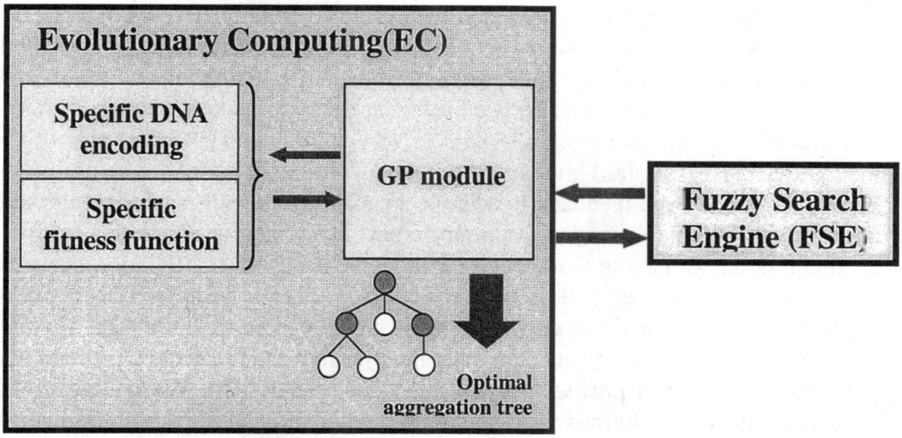

Figure 42. Evolutionary Computing Module: aggregation tree learning.

6 Conclusions

Most of the existing search systems 'software' is modeled using crisp logic and queries. In this paper, we introduced fuzzy querying and ranking as a flexible tool allowing approximation where the selected objects do not need to match exactly the decision criteria resembling natural human behavior. Searching database records and ranking the results based on multi-criteria queries is central for many database applications used within organizations in finance, business, industrial and other fields. The model consists of five major modules: the Fuzzy Search Engine (FSE), the Application Templates (AT), the User Interface (UI), the Database (DB) and the Evolutionary Computing (EC). We developed the software with many essential features. It is built as a web-based software system that users can access and use over the Internet. The system is designed to be generic so that it can run different application domains. To this end, the Application Template module provides information of a specific application as attributes and properties, and serves as a guideline structure for building a new application.

The Fuzzy Search Engine (FSE) is the core module of the system. It has been developed to be generic so that it would fit any application. The main FSE component is the query structure, which utilizes membership functions, similarity functions and aggregators.

Through the user interface a user can enter and save his profile, input criteria for a new query, run different queries and display results. The user can manually eliminate the results he disapproves or change the ranking according to his preferences.

The Evolutionary Computing (EC) module monitors ranking preferences of the users' queries. It learns to adjust to the intended meaning of the users' preferences.

The BISC decision support system key features are 1) intelligent tools to assist decision-makers in assessing the consequences of decision made in an environment of imprecision, uncertainty, and partial truth and providing a systematic risk analysis, 2) intelligent tools to be used to assist decision-makers answer "What if Questions", examine numerous alternatives very quickly and find the value of the inputs to achieve a desired level of output, and 3) intelligent tools to be used with human interaction and feedback to achieve a capability to learn and adapt through time In addition, the following important points have been found in this study 1) no single ranking function works well for all contexts, 2) most similarity measures work about the same regardless of the model, 3) there is little overlap between successful ranking functions, and 4) the same model can be used for other applications such as the design of a more intelligent search engine which includes the user's preferences and profile (Nikravesh, 2001a and 2001b). We have also described the use of evolutionary computation methods for optimization problem in the BISC decision support system. It is an original idea in combining fussy logic, machine learning and evolutionary computation. We gave some implementation precisions for the University Admissions application. We plan also to apply our system to many other applications.

7 Acknowledgement

Funding for this research was provided by the British Telecommunication (BT) and the BISC Program of UC Berkeley. The authors would like to acknowledge the effort by has been done by CS199 (BISC search engine group) and special thanks to Jonathan K. Lee, Wai-Kit Chan, Shuangyu Chang, Harmon Singh, Neema Raphael, Arthur Bobel, Naveen Sridhar, Wai-Lam Chan (Suke), Thanh Thanh Le, Trai Le, Nelly Tanizar, Kaveh Moghbeli, and Kit Hoi Lai (Andy).

8 References

1. Banzhaf, W., P. Nordin, R.E. Keller, F.D. Francone, Genetic Programming : An Introduction On the Automatic Evolution of Computer Programs and Its Applications, dpunkt.verlag and Morgan Kaufmann Publishers, San Francisco, CA, USA, 1998, 470 pages.

2. Bezdek, J.C., 1981, Pattern Recognition with Fuzzy Objective Function Algorithm, Plenum Press, New York.
3. Bonissone P.P., Decker K.S. (1986) Selecting Uncertainty Calculi and Granularity: An Experiment in Trading; Precision and Complexity, in Uncertainty in Artificial Intelligence (L. N. Kanal and J. F. Lemmer, Eds.), Amsterdam.
4. Detyniecki M (2000) Mathematical Aggregation Operators and their Application to Video Querying, Ph.D. thesis, University of Paris VI.
5. Fagin R. (1998) Fuzzy Queries in Multimedia Database Systems, Proc. ACM Symposium on Principles of Database Systems, pp. 1-10.
6. Fagin R. (1999) Combining fuzzy information from multiple systems. J. Computer and System Sciences 58, pp 83-99.
7. Fair, Isaac and Co.: http://www.fairisaac.com/.
8. Grabisch M (1996) K-order additive fuzzy measures. In Proc of 6th intl Conf on Information Processing and Management of Uncertainty in Knowledge-based Systems, Spain, pp 1345-50
9. Grabisch M, Murofushi T, Sugeno M (2000) Fuzzy Measures and Integrals:Theory and Applications, Physica-Verlag, NY
10. Holland, J. H.. Adaptation in Natural and Artificial Systems: An Introductory Analysis with Applications to Biology, Control and Artificial Intelligence. MIT Press, 1992. First Published by University of Michigan Press 1975.
11. Jang, J.S.R., and N. Gulley, 1995, Fuzzy Logic Toolbox, The Math Works Inc., Natick, MA.
12. Kohonen, T., 1997, Self-Organizing Maps, Second Edition, Springer.Berlin.
13. Kohonen, T., 1987, Self-Organization and Associate Memory, 2nd Edition, Springer Verlag., Berlin.
14. Koza, J. R., Genetic Programming: On the Programming of Computers by Means of Natural Selection, Cambridge, Mass. : MIT Press, USA 1992, 819 pages.
15. Mizumoto M. (1989) Pictorial Representations of Fuzzy Connectives, Part I: Cases of T-norms, T-conorms and Averaging Operators, Fuzzy Sets and Systems 31, pp. 217-242.
16. Murofushi T, Sugeno M (1989) An interpretation of fuzzy measure and the Choquet integral as an integral with respect to a fuzzy measure. Fuzzy Sets and Systems, (29): pp 202-27
17. Nikravesh M. (2001a) Perception-based information processing and retrieval: application to user profiling, 2001 research summary, EECS, ERL, University of California, Berkeley, BT-BISC Project. (http://zadeh.cs.berkeley.edu/ & http://www.cs.berkeley.edu/~nikraves/ & http://www-bisc.cs.berkeley.edu/).
18. Nikravesh M. (2001b) Credit Scoring for Billions of Financing Decisions, Joint 9th IFSA World Congress and 20th NAFIPS International Conference. IFSA/NAFIPS 2001 "Fuzziness and Soft Computing in the New Millenium", Vancouver, Canada, July 25-28, 2001.
19. Masoud Nikravesh, F. Aminzadeh, and Lotfi A. Zadeh, (2003a), Soft Computing and Intelligent Data Analysis in Oil Exploration, Development in Petroleum Science, # 51, Elesevier Science B. V., The Netherlands, 2003.
20. Masoud Nikravesh and Ben Azvine (2002), Fuzzy Queries, Search, and Decision Support System, Journal of Soft Computing, Volume 6 (5), August 2002.

21. Masoud Nikravesh, B. Azvine, R. Yagar, and Lotfi A. Zadeh (2003b) "New Directions in Enhancing the power of the Internet", to be published in the Series Studies in Fuzziness and Soft Computing, Physica-Verlag, Springer (August 2003)
22. Stanford University Admission, http://www.stanford.edu/home/stanford/facts/undergraduate.html
23. Sugeno M (1974) Theory of fuzzy integrals and its applications. Ph.D. Dissertation, Tokyo Institute of Technology.
24. U.S. Citizens for Fair Credit Card Terms; http://www.cardratings.org/cardrepfr.html.
25. University of California-Berkeley, Office of Undergraduate Admission, http://advising.berkeley.edu/ouars/.
26. Vincenzo Loia, Masoud Nikravesh and Lotfi A. Zadeh (2003), Fuzzy Logic and the Internet", to be published in the Series Studies in Fuzziness and Soft Computing, Physica-Verlag, Springer (August 2003)
27. Yager R (1988), On ordered weighted averaging aggregation operators in multicriteria decision making, IEEE transactions on Systems, Man and Cybernetics (18), 183-190.

Evaluating Ontology Based Search Strategies

Chris Loer[1], Harman Singh[2], Allen Cheung[3], Sergio Guadarrama[4], and Masoud Nikravesh[5]

[1] email: cloer@cal.berkeley.edu
[2] email: hjsingh@berkeley.edu
[3] email: allenmhc@cal.berkeley.edu
[4] Dept. of Artificial Intelligence and Computer Science, Universidad Politécnica de Madrid Madrid, Spain email: sguada@dia.fi.upm.es
[5] Berkeley Initiative in Soft Computing (BISC) Computer Science Division, Dept. of EECS University of California Berkeley CA 94704, USA email: nikravesh@cs.berkeley.edu

Abstract. We present a framework system for evaluating the effectiveness of various types of "ontologies" to improve information retrieval. We use the system to demonstrate the effectiveness of simple natural language-based ontologies in improving search results and have made provisions for using this framework to test more advanced ontological systems, with the eventual goal of implementing these systems to produce better search results, either in restricted search domains or in a more generalized domain such as the World Wide Web.

1 Introduction and Motivation

Soft computing ideas have many possible applications to document retrieval and internet search, but the complexity of search tools, as well as the prohibitive size of the Internet (for which improved search technology is especially important), makes it difficult to test the effectiveness of soft computing ideas quickly (specifically, the usage of conceptual fuzzy set ontologies) to improve search results. To lessen the difficulty and tediousness of testing these ideas, we have developed a framework for testing the application of soft computing ideas to the problem of information retrieval [1]. Our framework system is loosely based on the "General Text Parser (GTP)" developed at the University of Tennessee and guided by "Understanding Search Engines", written by some of the authors of GTP [1]. This framework allows a user to take a set of documents, form a "vector search space" out of these documents, and then run queries within that search space.

[1] This project was developed under the auspices of the Berkeley Initiative in Soft Computing from January to May of 2004

Throughout this paper, we will use the term “ontology” to refer to a data structure that encodes the relationships between a set of terms and concepts.

Our framework takes an ontology and uses its set of relationships to modify the term-frequency values of every document[2] in its search space, with the goal of creating a search space where documents are grouped by semantic similarity rather than by simple coincidence of terms. The interface for specifying an ontology to this framework is a “Conceptual Fuzzy Set Network,” which is essentially a graph with terms and concepts as nodes and relations (which include activation functions) as the edges respectively[8].

As well as allowing users to directly manipulate a number of factors that control how the system indexes documents (i.e. the ontology that the system uses), the framework is specifically designed to be easily extensible and modular. We believe that there are a wealth of strategies for improving search results that have yet to be tested, and hope that for many of them, simple modifications to this framework will allow researchers to quickly evaluate the utility of the strategy.

2 System Description

The framework exists as a set of packages for dealing with various search tasks: it is currently tied together by a user interface that coordinates the packages into a simple search tool. This section will give a brief overview of the interesting features of the framework – for more detailed documentation of both the features and the underlying code, please see `http://www-bisc.cs.berkeley.edu/ontologysearch`. While making design decisions, we have tried to make every part of the code as extensible and modular as possible, so that further modifications to individual parts of the document indexing process can be made as easily as possible, usually without modifying the existing code save a few additions to the user interface. Our current implementation includes the following features:

- A web page parser with a word stemmer attached
- Latent Semantic Indexing (LSI) of a vector search space
- Linear “Fuzzification” based on an ontology specified in XML
- The ability to run queries on a defined search space created from a set of documents
- A visualization tool that projects an n-dimensional search space into two dimensions
- Fuzzy c-means clustering of documents
- Automatic generation of ontologies based on OMCSNet

[2] That is, the number of times a given term appears in a given document, for all terms. The vector of terms for every document is normalized for ease of calculation.

2.1 Search Spaces

After parsing, each document is represented as a vector mapping terms to frequencies, where the frequency "value" is measured with Term-Frequency Inverse Document Frequency (TF-IDF) indexing (although the system allows for alternative frequency measurements). These documents are represented as an n-dimensional vector space, where n is the number of unique terms in all of the documents, i.e. the union of all terms in all documents. From this initial vector space it is possible to construct an "LSI Space", which is a copy of the original vector space that has been modified using LSI; in our case, we use a Singular Value Decomposition (SVD) matrix decomposition method to optimally compress sparse matrices[3]. SVD compression is lossy, but the optimality of the compression ensures that semantically similar terms are the first to be conflated as the amount of compression increases [1]. Query matching is performed by calculating the cosine similarity between the query term vector and document term vectors within the search space.

2.2 Ontology Implementation

Our framework treats "ontologies" as a completely separate module, and its only requirement is that an ontology must be able to "fuzzify" a set of terms (i.e. relate terms to each other) according to its own rules. We have included an ontology parser which parses XML files of a certain format into a base ontology class. Figure 1 shows an example of the XML format of a simple ontology; this basic ontology class stores a set of words and for each word, a set of directed relations from that word to all other words in the ontology[4].

To reshape a search space using an ontology, the user must choose an activation function for increasing or decreasing the value of related terms as specified by the rules of the ontology. With our framework, we have included a linear propagation function for ontologies; it takes the frequency of every term in the document, looks for that term in the ontology, and increases the frequency of all related terms by the value specified in that ontology[5]. If sigmoid propagation is being used, then frequencies will actually be decreased if they fall below a certain threshold, so that only terms that have a high degree of "support" (that is, they occur along with other terms that are deemed to be related to them, and thus probably have to do with the central meaning of the document) end up becoming amplified.

---

[3] Our vector space is a sparse matrix, as every document has only a fraction of all the terms in the search domain

[4] Each relation contains a real value between 0 and 1, where 0 signifies a complete lack of relation and 1 signifies synonyms.

[5] For example, given that `farm` is related to `agriculture` by 0.45, `farm` has a frequency value of 2, and `agriculture` has a value of 0, linear propagation would give `agriculture` a new value of $0.45 * 2 = 0.9$.

```
  <?xml version="1.0" encoding="ISO-8859-1" ?>
  <!DOCTYPE Ontology (View Source for full doctype...)>
- <Topic>
  - <term word="right">
      <relation word="right wrong" weight="0.5" />
    </term>
  - <term word="bush">
      <relation word="on table" weight="0.4375" />
      <relation word="live underwater" weight="0.4375" />
      <relation word="tree" weight="0.7890625" />
      <relation word="compute" weight="0.25" />
      <relation word="at grocery store" weight="0.25" />
      <relation word="in horse mouth" weight="0.25" />
      <relation word="be yellow green and red" weight="0.25" />
      <relation word="stem" weight="0.68359375" />
      <relation word="hide in" weight="0.5" />
      <relation word="patriotism" weight="0.4375" />
```

Fig. 1. A simple example of the format of an ontology stored in XML. This particular ontology is automatically generated from a database of concepts, and thus has many relations that do not immediately seem useful.

2.3 Clustering

The framework includes a clustering unit that performs Fuzzy C-Means clustering on a search space[2]; the user interface allows users to specify whether to perform clustering, how many clusters to create, and what membership threshold to use. The clustering process assigns each document a degree of membership in each of the clusters, used in the visualization to illustrate document groupings (which should tend to correspond with the groupings that can be visually perceived in the two dimensional representation) as well as in query execution to speed up processing queries: with clusters, the system trims the document space by looking only at documents that have a relatively high degree of membership in the cluster that best fits the query.

2.4 Visualization

The user interface has a visualization tool which plots a two dimensional representation of the documents in the search space. LSI is used to obtain a rank-2 decomposition of the n-dimensional search space, which is ultimately a 2xn matrix of points in two dimensions. Our interface plots that matrix, allows the user to move around and inspect documents, and colorizes documents by their degree of membership in any given cluster. The aim of the tool is to allow users to quickly determine the salient characteristics of a search space and to determine, on a very broad level, how the use of an ontology affects the space. The user may also plot a two dimensional representation

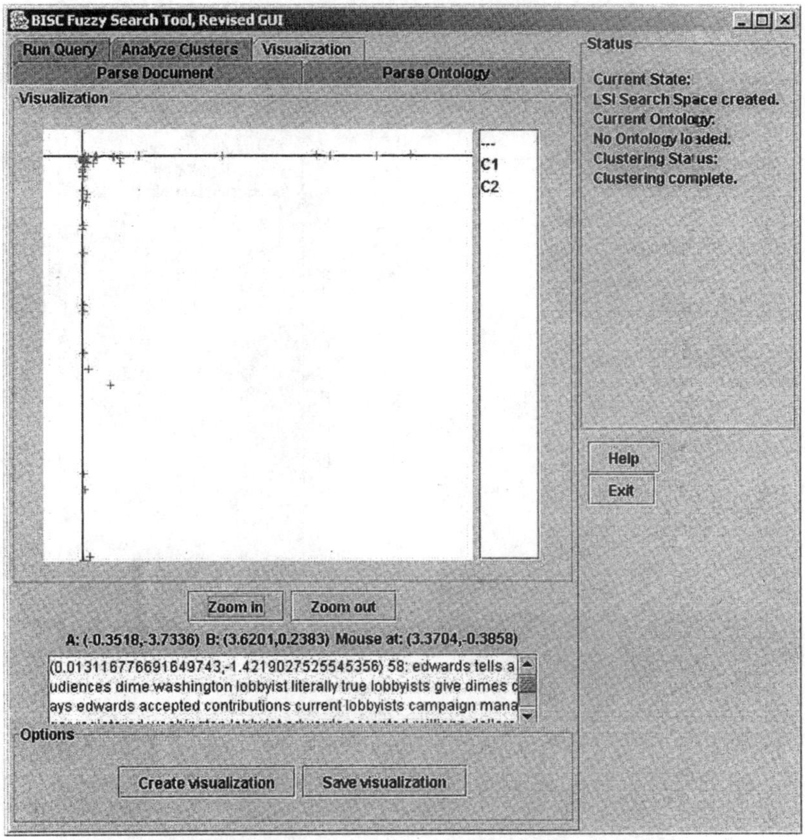

Fig. 2. Two dimensional visualization of a search space containing documents in Spanish and English. English documents cluster on the y axis, while Spanish documents cluster on the x axis.

of terms (which is based on the same underlying search space), in order to determine how certain terms might group together.

2.5 Ontology Generation

A simple tool has been built into the system that takes a search space, finds the most common terms in that search space, and then constructs an ontology out of information available through MIT's Open Mind Common Sense corpus (OMCSNet) found at `http://web.media.mit.edu/~hugo/conceptnet/`. This ontology simply encodes all of the relations to various words and phrases that OMCSNet has for the common terms. Admittedly, this tool is crude, in that it contains no domain specific information and includes irrelevant phrases which are meaningless when broken down into individual terms, it provides a good initial approximation of an ontology for testing.

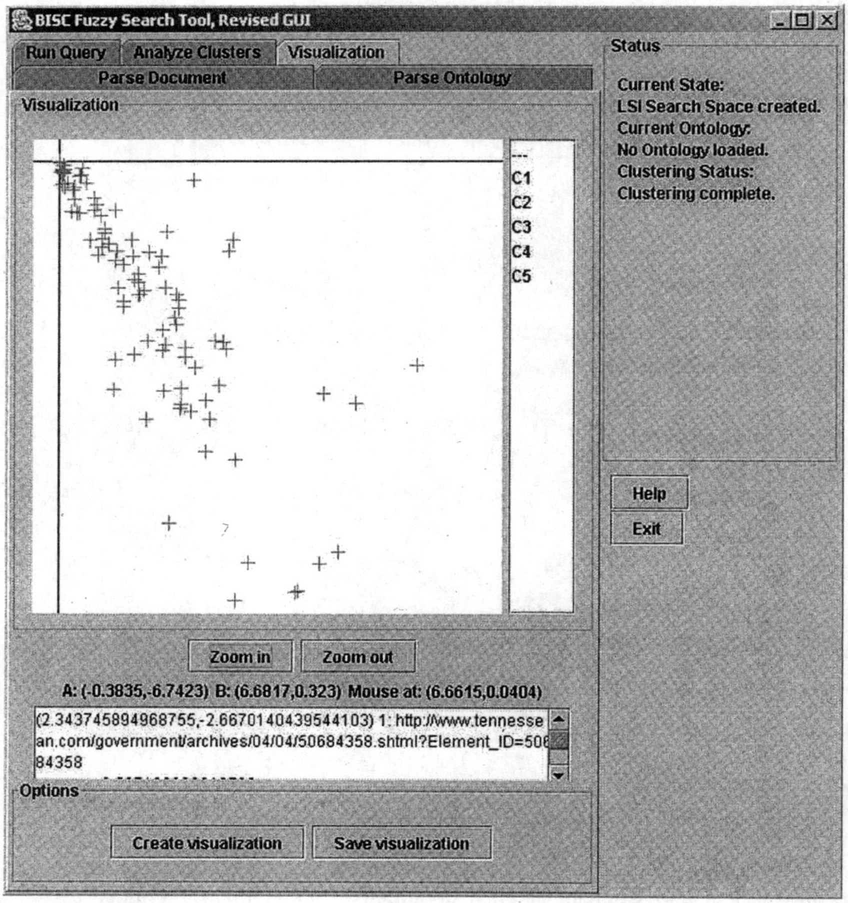

Fig. 3. Two-dimensional visualization of a search space containing documents related to the war in Iraq. No ontological modifications have been made to the search space.

3 Results

To test the operation of our system, we created search spaces with sets of documents drawn from Internet news sites. As an initial test of our two dimensional visualization, we first sampled a number of news sites in two distinct languages: Spanish and English. As we expected, the visualization displayed two completely distinct clusters of documents, as shown in Figure 2. We ran our next set of tests on a body of 124 news articles retrieved from Google™News by searching for the terms "Iraq War". To generate an ontology for testing, we found the most common terms in a set of documents and went to OMCSNet to create an ontology (as mentioned, this functionality is already built into our framework). The ontology we generate thus has

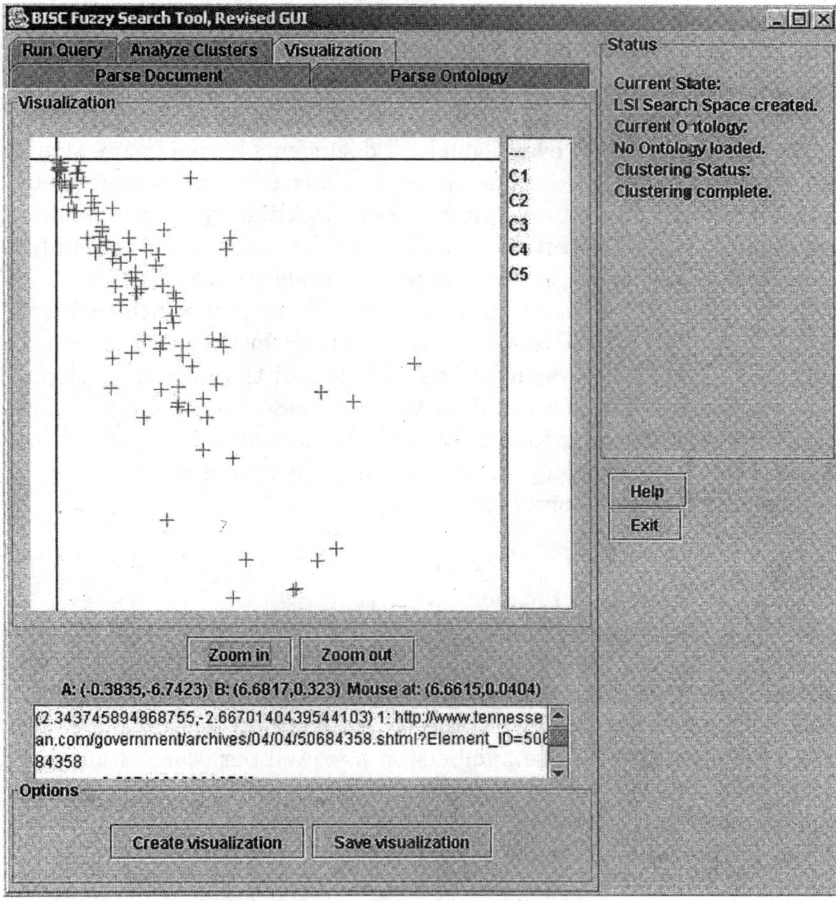

Fig. 4. Two-dimensional visualization of the same search space after "fuzzification" with an English-language ontology based on the most common words in the document set.

no domain-specific information and no notion of abstract concepts; it only encodes the relationship between (hopefully relevant) English terms and other terms that may show up in our search space. Despite this simplicity our expectation was that adjusting term frequency values would make documents which referred to similar topics appear more similar.

3.1 Visualization

We used our visualization tool to get a two dimensional representation of the search space before and after modifying term frequencies with our ontology. Figure 3 shows the visualization before modification, and Figure 4 shows the visualization after modification.

Our OMCSNet ontology only increased term frequencies, and did so regardless of context such that the degree of similarity would only increase between any two documents after modification. Not surprisingly, the visualization after modification showed that all documents were more tightly clustered. We hypothesize that, even though all documents become more similar to each other, topic-related documents see a *greater* effect from the ontological transformation and pull even closer together. Although we were able to inspect visual points to informally verify that similar documents were in fact near each other, we had no systematic way to evaluate the effectiveness of the transformation. If we were to use an advanced ontology which took note of context, we would expect to see a greater impact in the visualization. In general, because of the coarse nature of visuals and the high level of rank reduction we need to obtain a two-dimensional representation, our visualization tool only provides an intuition for how the documents are related and the overall effect of an ontology, but cannot give any systematic evidence for the effectiveness of a given ontology.

3.2 Queries

To measure the effectiveness of an ontology at improving information retrieval for a body of documents, we compared search results from a variety of queries on a given corpora with and without the use of an ontology. The accuracy of our results are subjective; having no objective standard to measure our results against, we cannot give concise numbers on how well our search framework performs short of developing a point-based rubric to manually evaluate, rank and compare search results.

As our primary interest while writing this system was verifying that the system itself performed as expected, we did not develop any ranking system for accuracy, but rather evaluated results of several test queries based on informal observation, comparing the use of different ontologies (including the "null" ontology). Figures 5 and 6 show the results of searching for the term "patriotism" before and after ontologically modifying an "Iraq War" search space.

The top results shown in Figure 5 are documents containing the term "patriotism", and they are separated from lower results by a steep drop in similarity index value. Upon further inspection, our lower-ranked documents do not include the term "patriotism" but still seem to hold some relation to the term: this may be an effect of semantic information captured by LSI, or it may simply be that all documents related to the Iraq War are related to patriotism in some way.

In Figure 6, the first document to contain the word "patriotism" is actually ranked in the middle of the list of query results, but we see that higher-ranking documents do discuss the *display* of patriotism, including phrases such as "flag waving", "supporting veterans", and "national pride". It is of course easy to make up a story to explain why the results in one figure are

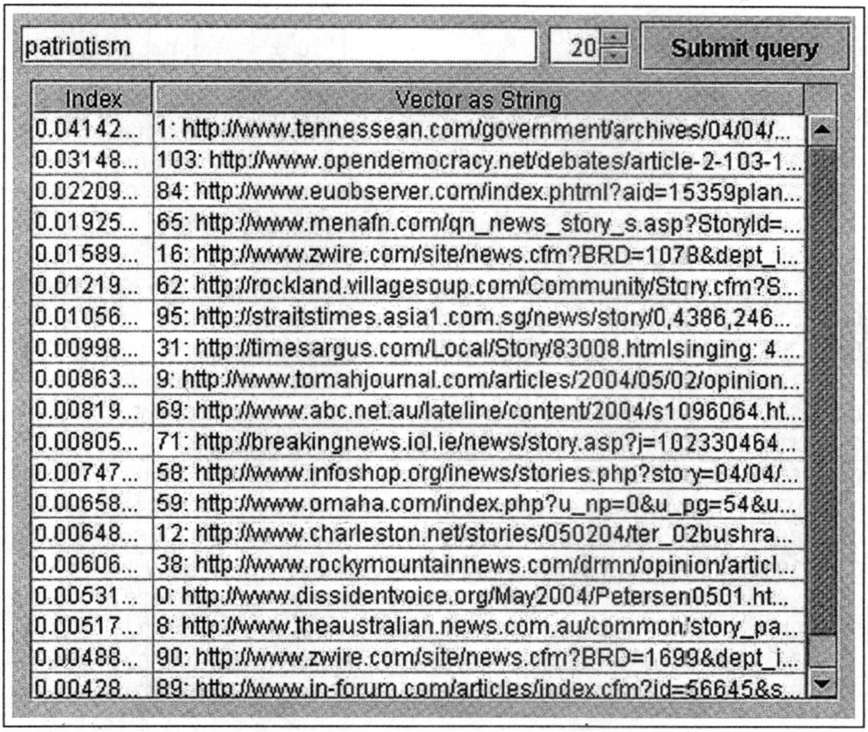

Fig. 5. The result of a search for "patriotism" before modifying the search space.

better than the results of the other: this data is not meant to be evidence that the ontology we used actually improved search results, but rather to demonstrate how an ontology changes the results of our query and how the system allows the user to quickly compare search spaces with and without the help of ontologies.

4 Future Work

We have identified several projects that could be pursued using the framework, either as extensions or as tests performed within the system. Our desire to test some of these ideas motivated the design of this system, but we expect that the framework may prove useful for testing ideas that never occurred to us.

4.1 Hierarchical Conceptual Fuzzy Sets

The framework lends itself to a more advanced notion of "ontology" than we have used in our initial implementation, which focuses simply on direct

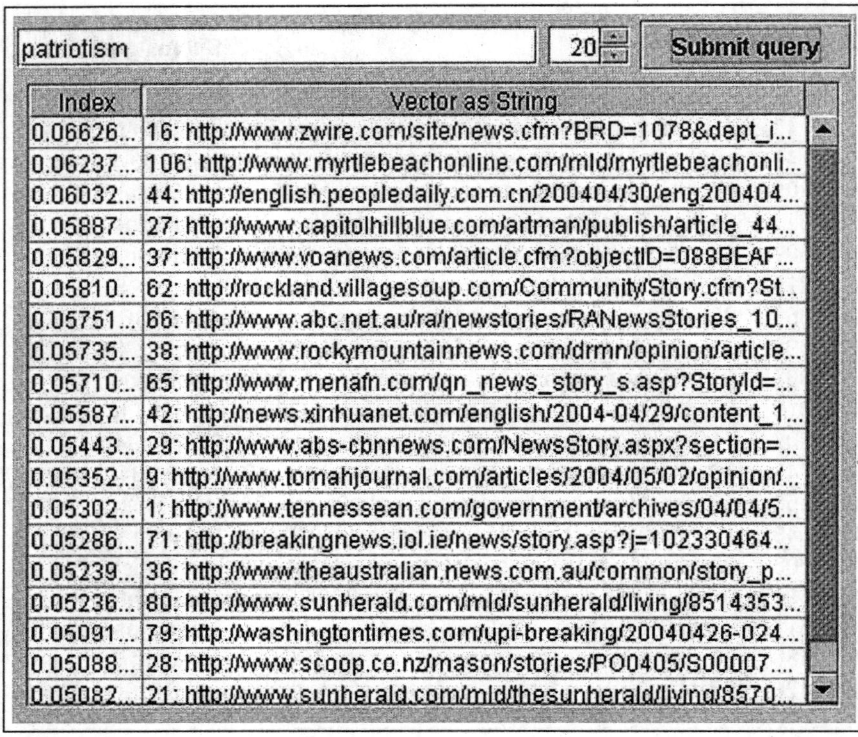

Fig. 6. The result of the same search after modification. Note that all results have received a higher similarity index as a result of uniform clustering and the order of relevance has changed from the query without an ontology.

relationships between terms. To capture semantic information with greater depth, hierarchical conceptual fuzzy sets may prove useful. A hierarchical conceptual fuzzy set network specifies concepts and relations using multiple levels of abstraction. For example, the term `Porsche` would trigger activation of the concept `Sports Cars`, which would in turn activate `Cars`, and then `Moving Vehicles`. In this situation the additional term `Ferrari` would strongly trigger `Sports Car`, while the term `Truck` would strongly trigger the broader `Moving Vehicles`. By more accurately determining the context of words within a document, and thus the "meaning" of the document, the use of a hierarchical conceptual fuzzy set network could further improve the quality of query results.

Extending the framework to test this idea would require an extension to the current ontology parser class, an extension to the current ontology class, and a method to create appropriate "hierarchical ontologies" for testing purposes.

4.2 Tailored Ontologies

Because this tool allows us to compare query results with and without ontological information, it will be useful for testing the effectiveness of various ontologies at capturing semantic structure within a search domain. Possible experiments include:

- Hand craft a set of relations among words related to an academic discipline (i.e. Computer Science), and then use the ontology to search in the domain of technical articles for that discipline.
- Automatically generate an ontology for a search domain based a set of criteria (i.e. term coincidence) and compare results with and without this ontology. Specifically, it may be interesting to evaluate the effectiveness of "fuzzy thesauri" [4] generated from the World Wide Web.
- Test the utility of feedback-driven ontology systems (i.e. the BISC Image Search program). User feedback then be the basis of an interactive system that personalizes context in queries [6] or to create a term-centered (rather than phrase-centered) general knowledge base.
- Automatically create ontologies based on semantic information from a natural language database. [7] While we have already implemented a tool to generate ontologies from MIT's OMCSNet, a context-sensitive ontology from a language database has further applications. That is, with the capacity to parse sentences, it is possible determine which terms and concepts in a document should be activated with a higher degree of accuracy. In the other direction, document-wide term-frequencies along with the activation values of terms and concepts in a conceptual fuzzy set network can aid a natural language parser in resolving ambiguous contexts.

As mentioned in Section 3.2, analysis and evaluation would require a standardized rubric for ranking the quality of results.

4.3 Alternative Frequency Measures

Throughout our program, we have used Term Frequency-Inverse Document Frequency (TF-IDF) measures, but we have not experimented with Non-monotonic Document Frequency (NMDF) measures [5], term ranking algorithms based on evolutionary computing, or other methods for measuring the occurrence of terms within documents. Unfortunately, because indexing methods often rely on detailed information from document parsing, modifications and additions to the modules that deal with indexing will most likely require changes in the parsing modules as well, violating the abstraction barriers and modularity of the framework.

---

[6] For example, a system would detect that its user uses the terms "Bush" and "president" interchangeably and tighten the ontological relationship between these two terms.

[7] Examples include Princeton's WordNet, MIT's OMCSNet, and Berkeley's FrameNet.

4.4 Integration with Google^{TM}Search

We currently parse a manually entered set of web pages or text documents. If we take this idea one step further and try to include some form of automated document search and extraction on the World Wide Web, a natural direction is making use of Google^{TM}'s search engine (via their free API) to download new documents on-the-fly. A typical scenario would have a user searching for documents about "car repair"; the system would fetch a group of documents related to automotive maintenance by using Google^{TM}'s search API, parse them as a document search space, and query within this tight domain of documents (perhaps also using a tailored ontology using automated methods such as Section 4.2's OMCSNet-ontology creation).

4.5 Query Refinement and Expansion

With the World Wide Web (accessed via Section 4.4's methods or some other means) at our fingertips, we should be able to make refinements of queries or expand queries to include other relevant documents and enlarge the size of our search space. For instance, to follow-up and expand on Section 4.4 and integration with Google^{TM}, we can use the following algorithm:

Input query
Use Google^{TM}API to find top n results for the terms in the query
Use OMCSNet to create context-specific ontology based on the most common terms in the top n results
Use OMCSNet ontology to find the top j groups of terms most related to the terms of the query
Use Google^{TM}API to find top n results for each of the groups of terms related to the query
Add all documents retrieved from Google^{TM}to search space
Reorder documents from API search using OMCSNet ontology
Return top k most highly ranked documents

Such a process would in effect expand and refine our search space – by sampling multiple parts of the World Wide Web with the help of an ontology, we are expanding beyond the limited number of terms in the user's query, and by reordering documents with respect to the ontology, we are refining our results and giving higher ranks to documents which are closely related to the query. Abstractly, we are approximating "fuzzification" of our perspective of the World Wide Web with the ontology. If we had re-indexed the entire World Wide Web, documents using ontologically related terms would be similar to each other: although these documents might have no relation to each other that would show up in a Google^{TM}search, this process would ensure that all relevant documents would be fetched and the reordered such that the results would be similar to the results we would expect if we had completely re-indexed.

4.6 Commercial Applications

We designed the system to work well on relatively small corpora, in the range of thousands of documents. If tailored and domain-specific ontologies prove to be a useful technique in improving search results within a restricted domain, these ideas could be applied to a scalable search system[8] based on some of the principles laid out by Brin and Page Brin:1997. Such a system would still probably be best suited to searching restricted (and more structured) domains, rather than the entirety of the World Wide Web, and might be used by libraries and other institutions charged with storing academic or professional knowledge to provide improved search results to their clients.

5 Conclusion

As this ontology based search system is primarily a tool for testing new ideas, this paper's intention is to create awareness of the availability of this tool. For the interested reader, the complete source code for this system, as well as documentation, is available at `http://www-bisc.cs.berkeley.edu/ontologysearch`. Although completely different indexing techniques would be necessary to efficiently apply ontology-based ideas to the task of searching very large corpora, we believe that this system could serve as a fair prototype for a tool to search specific, limited-size corpora (i.e. a set of books or papers in a particular field). If we are able to find and develop a method of capturing semantic significance of documents at this level, this technology then could be expanded into the larger domain of generalized Internet search.

Acknowledgements

The ideas in this project came from the lecture series presented by Dr. Masoud Nikravesh to the undergraduate Berkeley Initiative in Soft Computing research group, and its development was guided by Sergio Guadarrama. The authors would like to thank the BISC group for welcoming undergraduates and fostering their interests.

References

1. Berry, M. W., Browne , M. (1999) Understanding Search Engines: Mathematical Modeling and Text Retrieval (Software, Environments, Tools) SIAM, Philadelphia
2. Bezdek, J. C. (1981) Pattern Recognition with Fuzzy Objective Function Algoritms. Plenum Press, New York.

[8] Such a system would probably not be able to use a vector-space internal representation

3. Brin and Page (1997) The Anatomy of a Large-Scale Hypertextual Web Search Engine
4. De Cock, M., Guadarrama, S., Nikravesh, M. (2004) Fuzzy Thesauri for and from the WWW. Paper prepared for this book.
5. Haveliwala, Gionis, Klein, Indyk (2002) Evaluating Strategies for Similarity Search on the Web
6. Kamvar, Klein, Manning (2002) Spectral Learning
7. Kummamuru, Dhawale, Krishnapuram (2003) Fuzzy Co-clustering of Documents and Keywords
8. Nikravesh, M., Takagi, T., Tajima, M., Shinmura, A., Ohgaya, R., Taniguchi, K., Kazuyosi, K., Fukano, K., Aizawa, A. (2003) Web Intelligence: Conceptual–Based Model. Internal report, Electronics Research Laboratory, College of Engineering, University of California, Berkeley, Memorandum No. UCB/ERL M03/19

Soft Computing for Perception Based Information Processing

Masoud Nikravesh [(1)] and Dae-Young Choi[(1,2)]

[(1)] BISC Program, EECS Department-CS Division
University of California, Berkeley, CA 94720
Nikravesh@cs.berkeley.edu
[(2)] Dept. of MIS, Yuhan College, Koean-Dong, Sosa-Ku,
Puchon City, Kyungki-Do, South Korea
dychoi@green.yuhan.ac.kr

Abstract: Humans have a remarkable capability (perception) to perform a wide variety of physical and mental tasks without any measurements or computations. Familiar examples of such tasks are: playing golf, assessing wine, recognizing distorted speech, and summarizing a story. The question is whether a special type information retrieval processing strategy can be designed that build in perception. Commercial Web search engines have been defined which manage information only in a crisp way. Their query languages do not allow the expression of preferences or vagueness. Even though techniques exist for locating exact matches, finding relevant partial matches might be a problem. It may not be also easy to specify query requests precisely and completely - resulting in a situation known as a fuzzy-querying. It is usually not a problem for small domains, but for large repositories such as World Wide Web, a request specification becomes a bottleneck. Thus, a flexible retrieval algorithm is required, allowing for imprecise or fuzzy query specification or search. In addition, they have problems as follows : (1) large answer set ; (2) low precision; (3) unable to preserve the hypertext structures of matching hyperdocuments; (4) ineffective for general-concept queries. The task is to use user-defined queries to retrieve useful information according to certain measures. In order to handle these problems, we propose the Perception Index (PI) that contains attributes associated with a focal keyword restricted by fuzzy term(s) used in fuzzy queries on the Internet. If we integrate the Document Index (DI) used in commercial Web search engines with the proposed PI, we can handle both crisp terms (keyword-based) and fuzzy terms (perception-based). In this respect, the proposed approach is softer than the keyword-based approach. The PI brings somewhat closer to natural language. It is a further step toward a real human-friendly, natural language-based interface for Internet. It should greatly help the user relatively easily retrieve relevant information. In other words, the PI assists the user to reflect his/her perception in the process of query. Consequently, Internet users can narrow thousands of hits to the few that users really want. In this respect, the PI provides a new tool for targeting queries that users really want, and an invaluable personalized search. In this chapter, we also present the search mechanism based on the integrated index (DI + PI) and fuzzy query based on the integrated index (DI + PI). Moreover, we describe some features of the proposed method and suggest some considerations for implementing the proposed method. The main goal of the perception-based information processes and retrieval system is to design a model for the internet based on user profile with capability of exchanging and updating the rules dynamically and *"do what I mean, not as I say"* and using programming with *"human common sense capability"*.

1 Introduction

Under leadership of DARPA, ARPANET has been designed through close collaboration with UCLA during 1962-1969, 1970-1973, and 1974-1981. Initially designed to keep military sites in communication across the US. In 1969, ARPANET connected researchers from Stanford University, UCLA, UC Santa Barbara and the University of Utah. The Internet community formed in 1972 and the Email is started in 1977. While initially a technology designed primarily for needs of the U.S. military, the Internet grew to serve the academic and research communities. More recently, there has been tremendous expansion of the network both internationally and into the commercial user domain.

There are many publicly available Web search engines, but users are not necessarily satisfied with speed of retrieval (i.e., slow access) and quality of retrieved information (i.e., inability to find relevant information). It is important to remember that problems related to speed and access time may not be resolved by considering Web information access and retrieval as an isolated scientific problem. An August 1998 survey by Alexa Internet (<alexa.com>) indicates that 90% of all Web traffic is spread over 100,000 different hosts, with 50% of all Web traffic headed towards the top 900 most popular sites. Effective means of managing uneven concentration of information packets on the Internet will be needed in addition to the development of fast access and retrieval algorithms (Kabayashi and Takeda 2000).

World Wide Web search engines have become the most heavily-used online services, with millions of searches performed each day. Their popularity is due, in part, to their ease of use. The central tasks for the most of the search engines can be summarize as 1) query or user information request- do what I mean and not what I say!, 2) model for the Internet, Web representation-web page collection, documents, text, images, music, etc, and 3) ranking or matching function-degree of relevance, recall, precision, similarity, etc. ***Table 1*** also compares the issues related to the conventional Database with Internet. Already explosive amount of users on the Internet is estimated over 200 million (***Table 2)***. While the number of pages available on the Internet almost double every year, the main issue will be the size of the internet when we include multimedia information as part of the Web and also when the databases connected to the pages to be considered as part of an integrated Internet and Intranet structure. Databases are now considered as backbone of most of the E-commerce and B2B and business and sharing information through Net between different databases (Internet-Based Distributed Database) both by user or clients are one of the main interest and trend in the future. In addition, the estimated user of wireless devices is estimated 1 billion within 2003 and 95 % of all wireless devices will be Internet enabled within 2005. ***Table 3*** shows the evolution of the Internet, World Wide Web, and Search Engines.

Table 1. Database Vs. Internet

Database	Internet
Distributed	Distributed
Controlled	Autonomous
Query (QL)	Browse (Search)
Precise	Fuzzy/Imprecise
Structure	Unstructured

Table 2. Internet and rate of changes

Jan 1998: 30 Millions web hosts
Jan 1999: 44 Millions web hosts
Jan 2000: 70 Millions web hosts
Feb 2000: +72 Millions web hosts

Dec 1997: 320 Millions
Feb 1999: 800 Millions
March 2000: +1,720 Millions

The number of pages available on the Internet almost doubles every year

Courtois and Berry (Martin P. Courtois and Michael W. Berry, ONLINE, May 1999-Copyright © Online Inc.) published a very interesting paper "Results Ranking in Web Search Engines". In their work for each search, the following topics were selected: credit card fraud, quantity theory of money, liberation tigers, evolutionary psychology, French and Indian war, classical Greek philosophy, Beowulf criticism, abstract expressionism, tilt up concrete, latent semantic indexing, fm synthesis, pyloric stenosis, and the first 20 and 100 items were downloaded using the search engine. Three criteria 1) All Terms, 2) Proximity, and 3) Location were used as a major for testing the relevancy ranking. ***Table 4*** shows the concept of relevancy and its relationship with precision and recall (***Table 5*** and ***Figure 1***). ***Table 6*** shows the summary of the results. The effectiveness of the classification is defined based on the precision and recall (***Tables 4-5*** and ***Figure 1***).

Table 4. Similarity/Precision and Recall

	Relevant	Non-Relevant	
Retrieved	$A \cap B$	$\bar{A} \cap B$	**B**
Not Retrieved	$A \cap \bar{B}$	$\bar{A} \cap \bar{B}$	$\bar{B}$
	A	$\bar{A}$	**N**

N: Number of documents

Table 5. Similarity/Measures of Association

There are five commonly used measures of association in IR :

Simple matching Coefficiet: $|X \cap Y|$

Dice's Coefficiet: $2\dfrac{|X \cap Y|}{|X|+|Y|}$

Jaccard's Coefficiet: $\dfrac{|X \cap Y|}{|X \cup Y|}$

Cosine Coeffciet: $\dfrac{|X \cap Y|}{|X|^{1/2} \times |Y|^{1/2}}$

Overlap Coefficiet: $\dfrac{|X \cap Y|}{\min(|X|,|Y|)}$

Disimilarty Coefficeint: $\dfrac{|X \Delta Y|}{|X|+|Y|} = 1 - \text{Dice's Coefficient}$

$|X \Delta Y| = |X \cup Y| - |X \cap Y|$

Table 6. Results Ranking in Web Search Engines

Criteria	All Terms		Proximity		Location	
First 20 and 100 items	20/100 hits	Mean hits	20/100 hits	Mean hits	20/100 hits	Mean hits
ALTAVISTA	31/13%	22%	11/7%	9%	41/10%	25.5%
EXCITE	18/5%	11.5%	28/5%	16.5%	77/53%	65%
HOTBOT	19/12%	15.5%	40/24%	32%	62/29%	45.5%
INFOSEEK	23/16%	19.5%	14/10%	12%	79/50%	64.5%
LYCOS	8/5%	6.5%	49/26%	37.5%	69/32%	50.5%

Effectiveness is a measure of the system ability to satisfy the user in terms of the relevance of documents retrieved. In probability theory, precision is defined as conditional probability, as the probability that if a random document is classified under selected terms or category, this decision is correct. Precision is defined as portion of the retrieved documents that are relevant with respect to all retrieved documents; number of the relevant documents retrieved divided by all documents retrieved. Recall is defined as the conditional probability and as the probability if a random document should be classified under selected terms or category, this decision is taken. Recall is defined as portion of the relevant retrieved documents that are relevant with respect to all relevant documents exists; number of the relevant documents retrieved divided by all relevant documents. The performance of each request is usually given by precision-recall curve (***Figure 1***). The overall performance of a system is based on a series of query request. Therefore, the performance of a system is represented by a precision-recall curve, which is an average of the entire precision-recall curve for that set of query request.

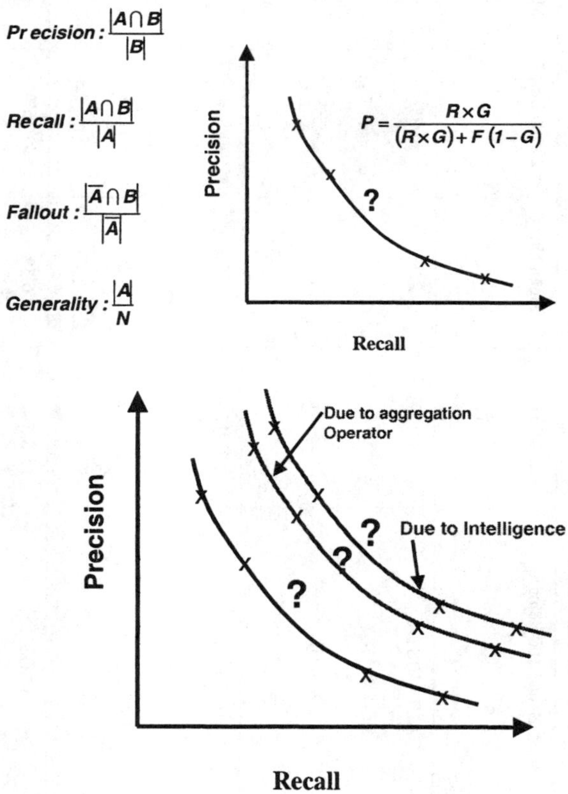

Figure 1. 1.a.) relationship between Precision and Recall, **1.b.)** inverse relationship between Precision and Recall.

To improve the performance of a system one can use different mathematical model for aggregation operator for (A∩B) such as fuzzy logic. This will sift the curve to a higher value as is shown in ***Figure 1.b***. However, this may be a matter of scale change and may not change the actual performance of the system. We call this improvement, virtual improvement. However, one can shit the curve to the next level, by using a more intelligent model that for example have deductive capability or may resolve the ambiguity (***Figure 1.b***).

Many search engines support Boolean operators, field searching, and other advanced techniques such as fuzzy logic in variety of definition and in a very primitive ways (***Table 7***). While searches may retrieve thousands of hits, finding relevant partial matches and query relevant information with deductive capabilities might be a problem. ***Figure 2.*** shows a schematic diagram of model presented by Lotfi A. Zadeh (2002) for the flow of information and decision. What is also important to mention for search engines is query-relevant information rather than generic information. Therefore, the query needs to be refined to capture the user's perception. However, to design such a system is not trivial, however, Q/A systems information can be used as a first step to build a knowledge based to capture some of the common user's perceptions. Given the concept of the perception, new machineries and tools need to be developed. Therefore, we envision that non-classical techniques such as fuzzy logic based-clustering methodology based on perception, fuzzy similarity, fuzzy aggregation, and FLSI for automatic information retrieval and search with partial matches are required.

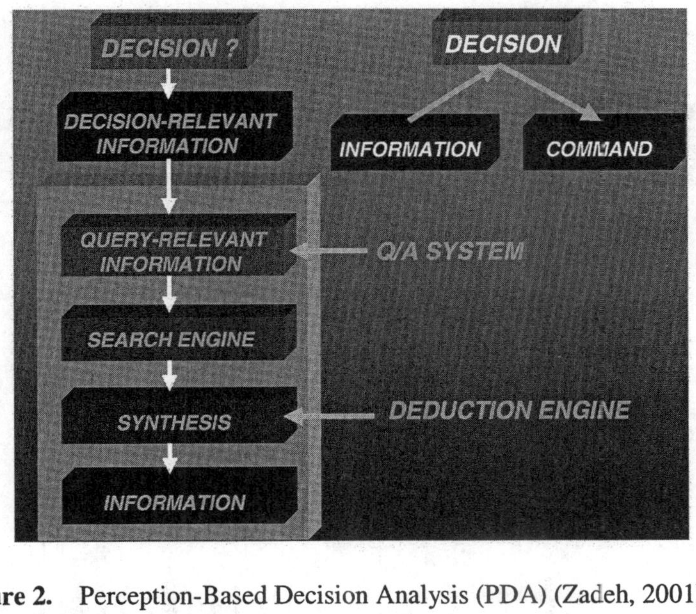

Figure 2. Perception-Based Decision Analysis (PDA) (Zadeh, 2001)

Table 7. Examples of Fuzzy Web Search Engines

Search Engine	Simple Form	Search Logic				Fuzzy Logic in any form	Term Weighting	Sorted Output	Ranked output	Find Like
		Boolean	Proximity	Nesting	Truncation					
Excite!	X	X	X	X		X	X	X	X	X
Alta Vista	X	X	X	X	X			X	X	
HotBot		X	X	X		X	X		X	
Infoseek	X	X	X	X	X	X	X	X	X	
Lycos	X				X	X*			X	
Open Text		X	X	X					X	(X)
Web Crawler	X	X	X	X		X	X	X	X	
Yahoo	X	X	X	X			X			
Google	X	X	*	*	*	X	*	*	*	*
Northern Light Power	X	X	*	*	*	X	*	*	*	*
Fast Search Advanced	X	X	*	*	*	X	*	*	*	*

2 Intelligent Search Engines

Design of any new intelligent search engine should be at least based on two main motivations:

- The web environment is, for the most part, unstructured and imprecise. To deal with information in the web environment what is needed is a logic that supports modes of reasoning which are approximate rather than exact. While searches may retrieve thousands of hits, finding decision-relevant and query-relevant information in an imprecise environment is a challenging problem, which has to be addressed.
- Another, and less obvious, is deduction in an unstructured and imprecise environment given the huge stream of complex information.

Tim Bernres-Lee (1999) in his transcript refers to the fuzzy concept and the human intuition with respect to the Web (Transcript of Tim Berners-Lee's talk to the LCS 35th Anniversary celebrations, Cambridge Massachusetts, 1999/April/14):

Lotfi A. Zadeh (2001a) consider fuzzy logic is a necessity to add deductive capability to a search engine: "Unlike classical logic, fuzzy logic is concerned, in the main, with modes of reasoning which are approximate rather than exact. In Internet, almost everything, especially in the realm of search, is approximate in nature. Putting these two facts together, an intriguing thought merges; in time, fuzzy logic may replace classical logic as what may be called the brainware of the Internet.

...

In my view, among the many ways in which fuzzy logic may be employed, there are two that stand out in importance. The first is search. Another, and less obvious, is deduction in an unstructured and imprecise environment. Existing search engines have zero deductive capability. ... To add a deductive capability to a search engine, the use of fuzzy logic is not an option - it is a necessity."

With respect to the deduction and its complexity, Lotfi's viewpoint (2001a and 2002) is summarized as follows: "Existing search engines have many remarkable capabilities. But what is not among them, is the deduction capability -- the capability to answer a query by drawing on information which resides in various parts of the knowledge base or is augmented by the user. Limited progress is achievable through application of methods based on bivalent logic and standard probability theory. But to move beyond the reach of standard methods it is necessary to change direction. In the approach, which is outlined, a concept which plays a pivotal role is that of a prototype -- a concept which has a position of centrality in human reasoning, recognition, search and decision processes. … The concept of a prototype is intrinsically fuzzy. For this reason, the prototype-centered approach to deduction is based on fuzzy logic and perception-based theory of probabilistic reasoning, rather than on bivalent logic and standard probability theory. What should be underscored, is that the problem of adding deduction capability to search engines is many-faceted and complex. It would be unrealistic to expect rapid progress toward its solution."

During 80, most of the advances of the automatic document categorization and IR were based on knowledge engineering. The models were built manually using expert systems capable of taking decision. Such expert system has been typically built based on a set of manually defined rules. However, the bottleneck for such manual expert systems was the knowledge acquisition very similar to expert system. Mainly, rules needed to be defined manually by expert and were static. Therefore, once the database has been changed or updated the model must intervene again or work has to be repeated anew if the system to be ported to a completely different domain. By explosion of the Internet, these bottlenecks are more obvious today. During 90, new direction has been merged based on machine learning approach. The advantage of this new approach is evident compared to the previous approach during 80. In machine learning approach, most of the engineering efforts goes towards the construction of the system and mostly is independent of the domain. Therefore, it is much easier to port the system into a new domain. Once the system or model is ported into a new domain, all that is needed is the inductive, and updating of the system from a different set of new dataset, with no required intervention of the domain expert or the knowledge engineer. In term of the effectiveness, IR techniques based on machine learning techniques achieved impressive level of the performance and for example made it possible automatic document classification, categorization, and filtering and making these processes viable alternative to manual and expert system models.

Doug B. Lenat both the founder of the CYC project and president of Cycorp (http://www.cyc.com) puts the concept of deduction into perspective and he expresses that both commonsense knowledge and reasoning are key for better information extraction (2001).

Lotfi A. Zadeh (2002) express qualitative approach towards adding deduction capability to the search engine based on the concept and framework of protoforms:

"At a specified level of abstraction, propositions are p-equivalent if they have identical protoforms." "The importance of the concepts of protoform and p-equivalence derives in large measure from the fact that they serve as a basis for knowledge compression."

"A knowledge base is assumed to consist of a factual database, FDB, and a deduction database, DDB. Most of the knowledge in both FDB and DDB is perception-based. Such knowledge cannot be dealt with through the use of bivalent logic and standard probability theory. The deduction database is assumed to consist of a logical database and a computational database, with the rules of deduction having the structure of protoforms. An example of a computational rule is "if Q_1 A's are B's and Q_1 (A and B)'s are C's," then "$Q_1 Q_2$ A's are(B and C)'s, where Q_1 and Q_2 are fuzzy quantifiers and A, B and C are labels of fuzzy sets. The number of rules in the computational database is assumed to be very large in order to allow a chaining of rules that may be query-relevant."

Computational theory of perception (CTP) (Zadeh, 1999 and 2001b; Nikravesh et al., 2001; Nikravesh, 2001a and 2001b) is one of the many ways that may help to address some of the issues presented by both Berners Lee and Lotfi A. Zadeh earlier, a theory which comprises a conceptual framework and a methodology for computing and reasoning with perceptions. The base for CTP is the methodology of computing with words (CW) (Zadeh 1999). In CW, the objects of computation are words and propositions drawn from a natural language.

3 Perception-Based Information Processing for Internet

One of the problems that Internet users are facing today is to find the desired information correctly and effectively in an environment that the available information, the repositories of information, indexing, and tools are all dynamic. Even though some tools were developed for a dynamic environment, they are suffering from "too much" or " too little" information retrieval. Some tools return too few resources and some tool returns too many resources ***(Figure 3).***

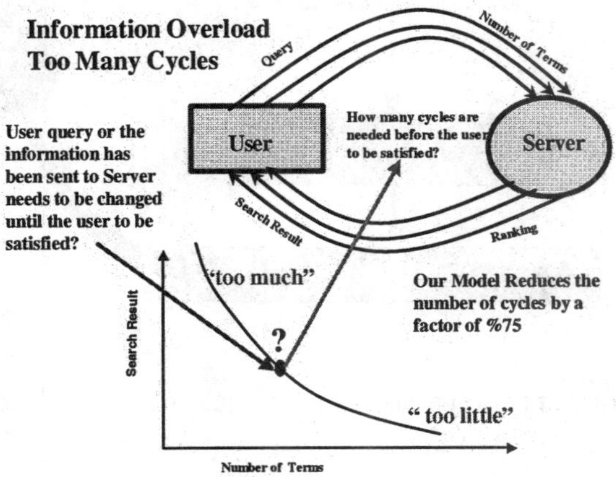

Figure 3. Information overload

The main problem with conventional information retrieval and search such as vector space representation of term-document vectors are that 1) there is no real theoretical basis for the assumption of a term and document space and 2) terms and documents are not really orthogonal dimensions. These techniques are used more for visualization and most similarity measures work about the same regardless of model. In addition, terms are not independent of all other terms. With regards to probabilistic models, important indicators of relevance may not be term -- though terms only are usually used. Regarding Boolean model, complex query syntax is often misunderstood and problems of null output and Information overload exist. One solution to these problems is to use extended Boolean model or fuzzy logic. In this case, one can add a fuzzy quantifier to each term or concept. In addition, one can interpret the AND as fuzzy-MIN and OR as fuzzy-MAX functions. Alternatively, one can add agents in the user interface and assign certain tasks to them or use machine learning to learn user behavior or preferences to improve performance. This technique is useful when past behavior is a useful predictor of the future and wide variety of behaviors amongst users exist.

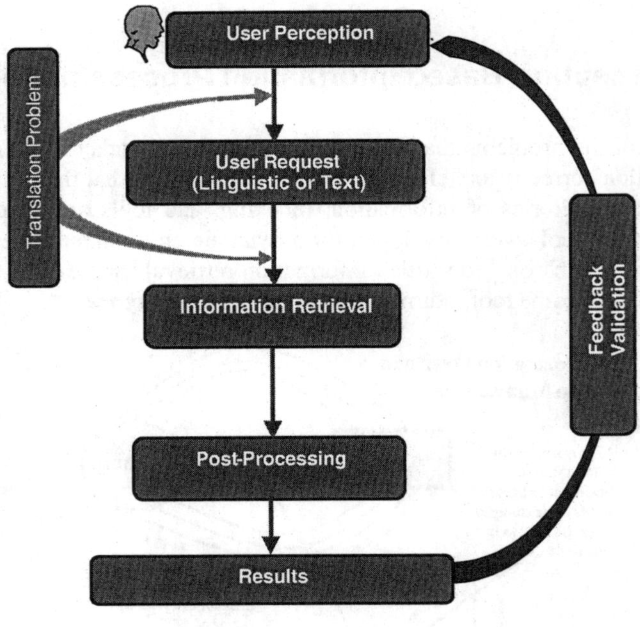

Figure 4.a. Structure of conventional search engine and retrieval technique

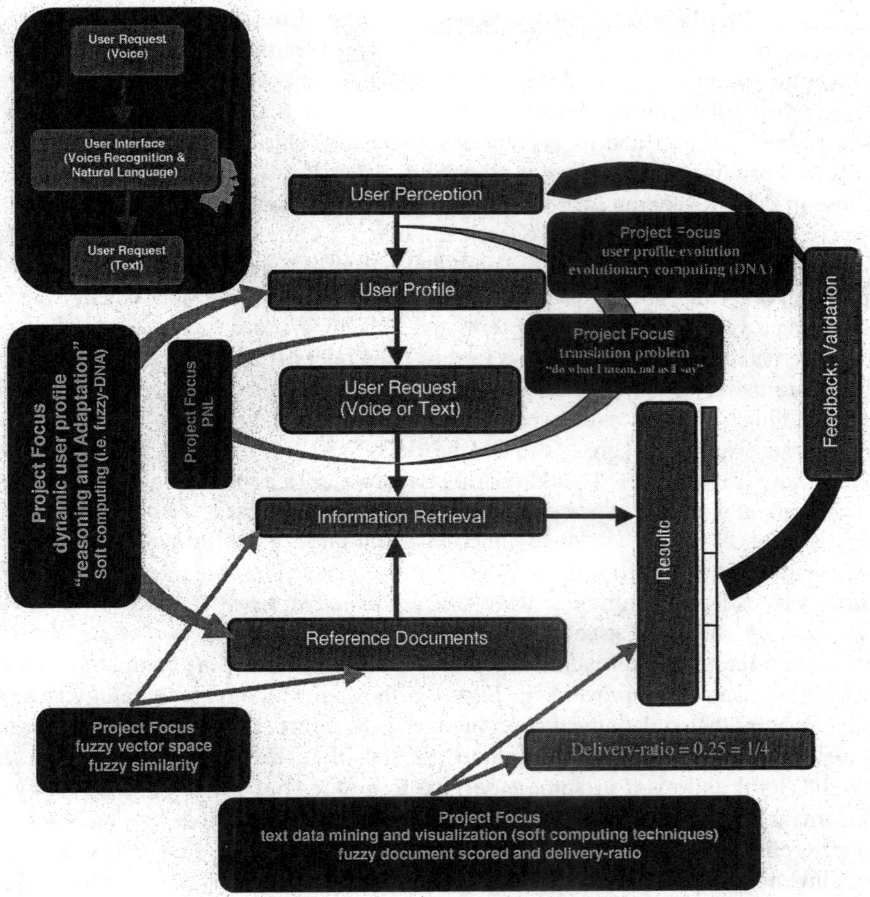

Figure 4. b Structure of search engine and retrieval technique and the problem related to perception and areas that soft computing can be used as a mean for improvement.

In addition, the user's perception, which is one of the most important key features, is oftentimes ignored. For example, consider the word " football". The perception of an American differs from the perception of an European who understands football to mean "Soccer." Therefore, if the search engine knows something about the user and its perception, it might be able to better refine the users results. For this example, there is no need to eliminate American football pages for those in the UK looking for real football information, since this information inclusively exists in user's profile. Search Engines also often return a large list of irrelevant search results due to the ambiguity of search query terms. To solve this problem one can use the following approaches 1) from Users Side/ Client Side by selecting a very specific (unique) term and 2) from Systems Sides/Server by offering alternate query terms for users to refine the query terms. Sources of the ambiguity are mainly due to 1) definition/meaning and as an example-what is the largest building? (for this case, what is the meaning of "largest") and 2) specificity and as an example- where is the GM headquarters? (for this case, what level of specificity is required?). To address this issue, a clarification dialog is required.

The main goal of the perception-based information processes and retrieval system is to design a model for the internet based on user profile with capability of exchanging and updating the rules dynamically and *"do what I mean, not as I say"* and using programming with *"human common sense capability"*. ***Figures 4.a and 4.b*** show the structure of conventional search engine and retrieval technique and the problem related to perception and areas that soft computing can be used as a mean for improvement. ***Figure 5*** shows the automated ontology generation and automated document indexing using the terms similarity based on Fuzzy-Latent Semantic Indexing Technique (FLSI). Often time it is hard to find the "right" term and even in some cases the term does not exist. The ontology is automatically constructed from text document collection and can be used for query refinement. ***Figure 6*** shows documents similarity map that can be used for intelligent search engine based on FLSI, personalization and user profiling. The user profile is automatically constructed from text document collection and can be used for query refinement and provide suggestions and for ranking the information based on pre-existence user profile.

4 Fuzzy Conceptual Model and Search Engine

One can use clarification dialog, user profile, context, and ontology, into a integrated frame work to address some of the issues related to search engines were described earlier. In our perspective, we define this framework as ***Fuzzy Conceptual Matching based on Human Mental Model (Figure 7)***. The Conceptual Fuzzy Set (CFS) model will be used for intelligent information and knowledge retrieval through conceptual matching of both text and images (here defined as "Concept").

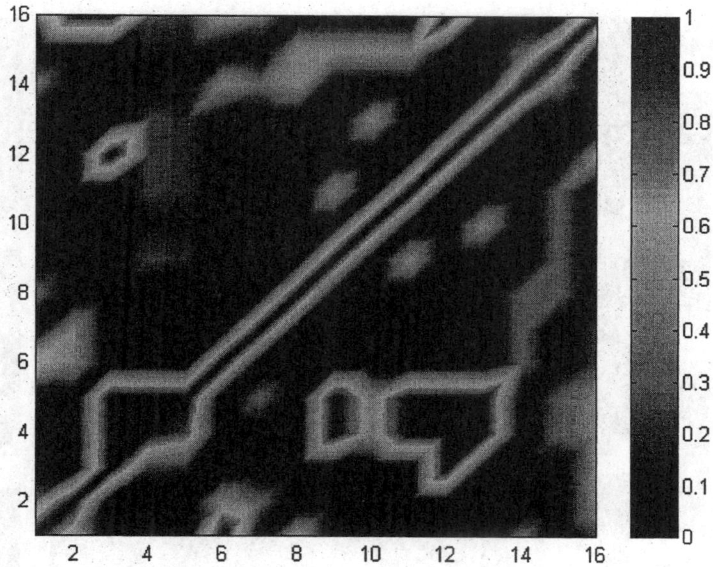

Figure 5. Terms Similarity; Automated Ontology Generation and Automated Indexing

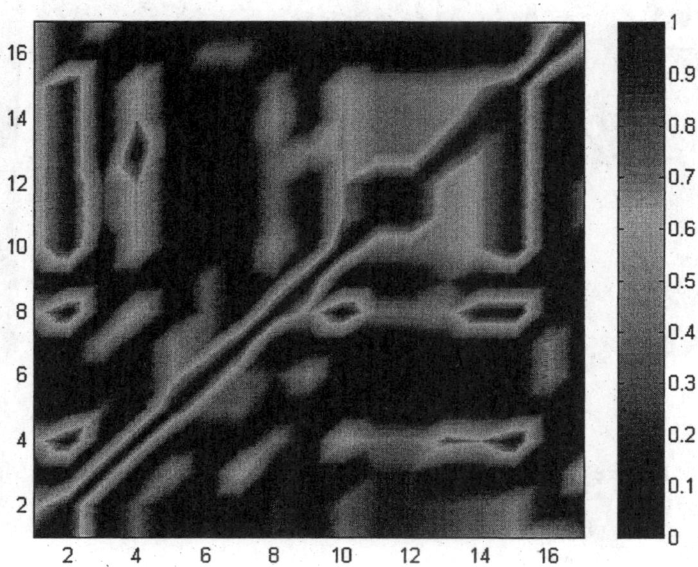

Figure 6. Documents Similarity; Search Personalization-User Profiling

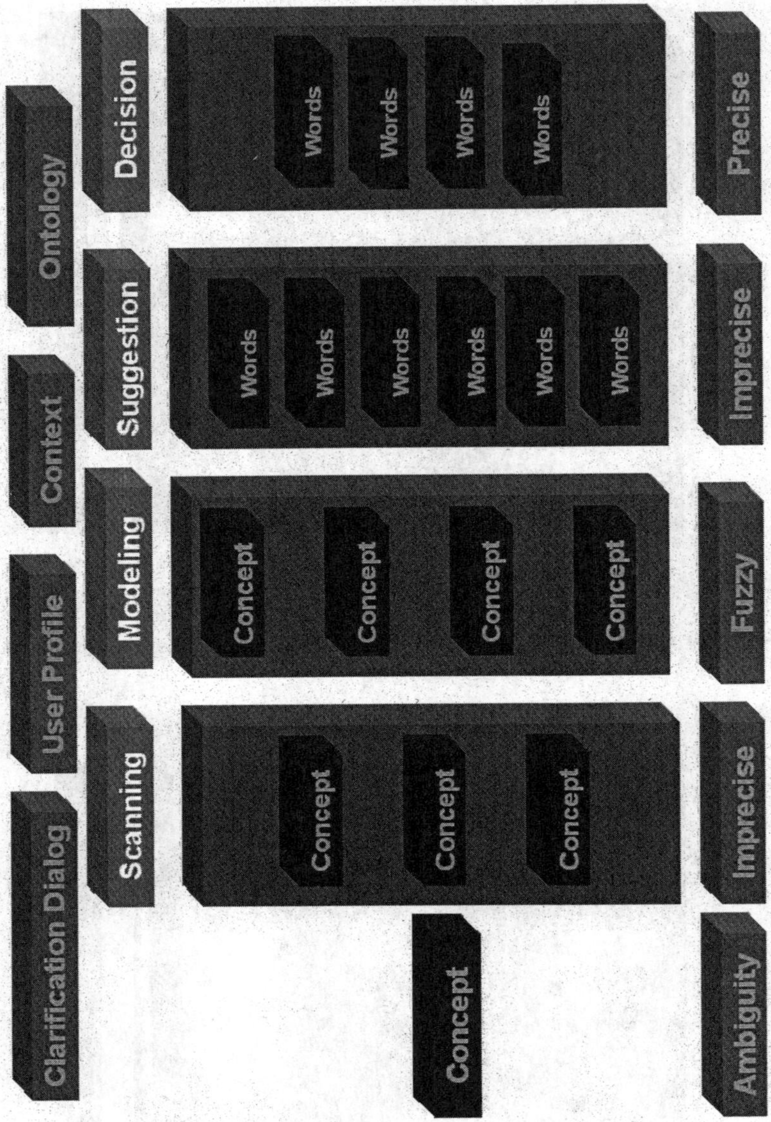

Figure 7. Fuzzy Conceptual Matching and Human Mental Model

The selected query doesn't need to match the decision criteria exactly, which gives the system a more human-like behavior. The CFS can also be used for constructing fuzzy ontology or terms related to the context of search or query to resolve the ambiguity. It is intended to combine the expert knowledge with soft computing tool. Expert knowledge needs to be partially converted into artificial intelligence that can better handle the huge information stream. In addition, sophisticated management work-flow need to be designed to make optimal use of this information. The new model can execute conceptual matching dealing with context-dependent word ambiguity and produce results in a format that permits the user to interact dynamically to customize and personalized its search strategy.

5 Fuzzy Query on the Internet

In this section, we do not attempt to solve 'unable to preserve the hypertext structures of matching hyperdocuments' problem. However, we try to tackle in part the other problems (i.e., 'large answer set', 'low precision', 'ineffective for general-concept queries'). In order to handle these problems, we propose the Perception Index (PI) that contains attributes associated with a focal keyword restricted by fuzzy term(s) used in fuzzy queries on the Internet.

5.1 Integration of Document Index with Perception Index

The central concept of information retrieval is the notion of relevance (Salton 1989). A user with a given query for information tries to find any specific results that he/she really wants. There are several models for specifying the representations used for the documents and the queries, as well as the matching of these representations (Kraft and Petry 1997). The most used model is that of the Boolean query based on set theory. Documents are represented as sets of terms and queries are Boolean expressions on terms. The retrieval mechanism does an exact match by classifying documents that satisfy the Boolean query as being relevant, all other documents as being irrelevant. This model is used by virtually all commercial textual-document retrieval systems. However, it is difficult to overcome the limitations of this model, including the inability to handle properly imprecision and subjectivity. The second model is the vector space model (Salton 1989) where documents and queries are represented as vectors in the space of all possible index terms. The document vectors consist of weights based on term frequencies in the collection, while the query vectors are binary vectors on the terms. The matching is based on a similarity measure between the documents and the query (often involving the cosine of the angle between the query vector and a given document vector). To date, this model leads the others in terms of performance. The third model is the probabilistic model (Salton 1989) where documents are represented as binary vectors. The queries are vectors of terms with weights based on the estimated probability of relevance of documents with those terms. Like the vector space model, the key advantage is the ability to rank documents on the likelihood of relevance. The fourth model is the generalized Boolean model, where fuzzy set theory allows the extension of the classical Boolean model to incorporate weights and partial matches, and adding the idea of document ranking.

The importance of representations of uncertainty in databases is increasing as more complex applications such as CAD/CAM and geographical information systems (GIS) are being undertaken in object-oriented and multi-media databases. Query languages are designed to express the user's retrieval requests in either a crisp manner or not. Much of the work in the database area has been in extending query languages to permit the representation and retrieval of imprecise data (Kacprzyk and Ziolkowski 1986, Nakajima et el. 1993, Petry and Bosc 1996, Rasmussen and Yager 1999, and Testemale 1986). There are some current commercial attempts at providing fuzzy query capabilities as front ends to conventional database systems (Nakajima et el. 1993).

Until now, however, commercial systems including informational retrieval systems (IRS), data base management systems (DBMS), and Web search engines have been defined which manage information only in a crisp way. Moreover, (crisp) traditional query languages do not allow the expression of preferences or vagueness which could be desirable for the following reasons (Kraft and Petry 1997):

- to control the size of the results;
- to express soft retrieval conditions;
- to produce a discriminated answer.

Although the commercial Web search engines such as Yahoo !, Google, Lycos, etc. help Internet users get to good information, they do not properly handle fuzzy query and tend to ignore the importance of fuzzy terms in a query. The expressive power of conventional search engine query interfaces is relatively weak when restricted to keyword-based search (i.e., Document Index (DI)-based search) (Kao et al. 2000).

In Section 5.2, we introduce the integrated index (DI + PI) and suggest a new search mechanism based on the integrated index. In Section 5.3, we describe fuzzy query based on the integrated index. This section is divided into three parts : types of fuzzy query, query processing based on the integrated index, and user interface based on the integrated index. In Section 5.4, we summarize some features of the proposed method. In Section 5.5, we show the effectiveness of our approach. In Section 5.6, we suggest some considerations for implementing the proposed method. We discuss the proposed method in Section 5.7.

5.2 Integration of Document Index with Perception Index

The most important of the tools for information retrieval is the index – a collection of terms with pointers to places where information about documents can be found. The development of effective indexing tools to aid in filtering is one of major classes of problems associated with Web search and retrieval. Removal of spurious information is a particularly challenging problem (Kobayashi and Takeda 2000).

Search engines are the most popular tools that people use to locate information on the Web. A search engine works by traversing the Web via the hyperlinks that connect the Web pages, performing text analysis on the pages it has encountered, and indexing the pages based on the keywords they contain. A user seeking information from the Web would formulate his/her information goal in terms of a few keywords composing a query. A search engine, on receiving a query, would match the query against its Document Index (DI). All of the pages that match the user query will be selected into an *answer set* and be ranked according to how relevant the pages are with respect to the query. Relevancy here is usually based on the number of matching keywords that a page contains (Kao et al. 2000). The DI is generally consisted of keywords that appear in the title of a page or in the text body. Based on the DI, the commercial Web search engines such as Yahoo !, Google, Lycos, etc. help users get to good information. For example, BigBook (or SuperPages) can help users to find 'Italian restaurants within a 1-mile radius from a specific address' (U.S. yellow pages services) (Lidsky and kwon 1997). This proximity search is processed based on crisp query with keywords (i.e., 'Italian restaurants', '1-mile', 'a specific address'). However, they do not properly process fuzzy queries. For example, find '***popular*** national parks in the USA'. In addition, they have problems as follows (Kao et al. 2000):

- large answer set;
- low precision;
- unable to preserve the hypertext structures of matching hyperdocuments;
- ineffective for general-concept queries.

In this section, we do not attempt to solve 'unable to preserve the hypertext structures of matching hyperdocuments' problem. However, we try to tackle in part the other problems (i.e., 'large answer set', 'low precision', 'ineffective for general-concept queries'). In order to handle these problems, we propose a Perception Index (PI). The remarkable human capability to perform a wide variety of physical and mental tasks without any measurements and any computations is derived from the brain's crucial ability to manipulate perceptions – perceptions of distance, size, weight, color, speed, time, direction, force, number, truth, likelihood, and other characteristics of physical and mental objects. Familiar examples of the remarkable human capability are parking a car, driving in heavy traffic, playing golf, riding a bicycle, understanding speech, and summarizing a story (Zadeh 1999). In the computational theory of perceptions (CPT) (Zadeh 1999), words play the role of labels of perceptions and, more generally, perceptions are expressed as propositions in a natural language. Computing with words (CW) techniques are employed to translate propositions expressed in a natural language into what is called the generalized constraint language (GCL). In this language, the meaning of a proposition is expressed as a generalized constraint, X *isr* R, where X is the constrained variable, R is the constraining relation and *isr* is a variable copula in which r is a discrete variable whose value defines the way in which R constrains X (Zadeh 1997 and 1999). Among the basic types of constraints are : possibilistic, veristic, probabilistic, random set, Pawlak set, fuzzy graph and usuality (Zadeh 1999). These perceptions are mainly manipulated based on fuzzy concepts. For processing a fuzzy query, the PI is consisted of attributes associated with a keyword restricted by fuzzy term(s) in a fuzzy query. In this respect, the restricted keyword is named as a focal keyword, whereas attribute(s) associated with the focal keyword may be regarded as focal attribute(s). The PI can be mainly derived from the contents in the text body of a Web page or from the other sources of information with respect to a Web page. For example, the PI may be consisted of distance, size, weight, color, etc. on a keyword in the text body of a Web page. Using the PI, search engines can process fuzzy concepts (terms). In the sequel, if we integrate the DI used in commercial Web search engines with the proposed PI, search engines can process fuzzy queries. For example, consider a fuzzy query that finds '***popular*** national parks in the USA'. In this case, the fuzzy term '*popular*' is processed by using the PI, whereas keywords '*national parks*' and '*USA*' are processed by using the DI. We note that 'in' and 'the' in the above fuzzy query are examples of stop words ignored by search engines (see <www.google.com>).

Table 8. An example of Integrated Index (DI + PI)

Document Index (DI)	IPs	Perception Index (PI)				FPs (Results)
Keywords	URLs	Distance	Size	No. of visitors	...	Targeted URLs

(IPs : Intermediate Pointers; FPs : Final Pointers; URLs : Uniform Resource Locators)

It should be noted that fuzzy term(s) may be regarded as a constraint on a fuzzy query. For example, consider a fuzzy query that finds '***popular*** national parks in the USA'. In this case, the fuzzy term '***popular***' play the role of a constraint on the fuzzy query. In other words, using fuzzy term(s), Internet users can narrow thousands of hits to the few that users really want. In this respect, the PI provides helpful hints for targeting queries that users really want, and an invaluable personalized search.

The expressive power of conventional search engine query interfaces is relatively weak when restricted to keyword-based search (Kao et al. 2000). At present, commercial Web search engines based on the DI (i.e., keyword-based search engines) present limitations in modeling perceptual aspects of humans. In addition, they generally return a bunch of Web pages (or URLs) irrelevant to user's query. Although much Web search engines have been developed, they do not properly handle the fuzzy terms representing human's perception. In addition, they appear to have trouble with returning the targeted results. In order to tackle this problem, we integrate the DI used in commercial Web search engines with the proposed PI. In the proposed method, given a fuzzy query, search engine processes the fuzzy query based on the integrated index (DI + PI) as in **Figure 8**.

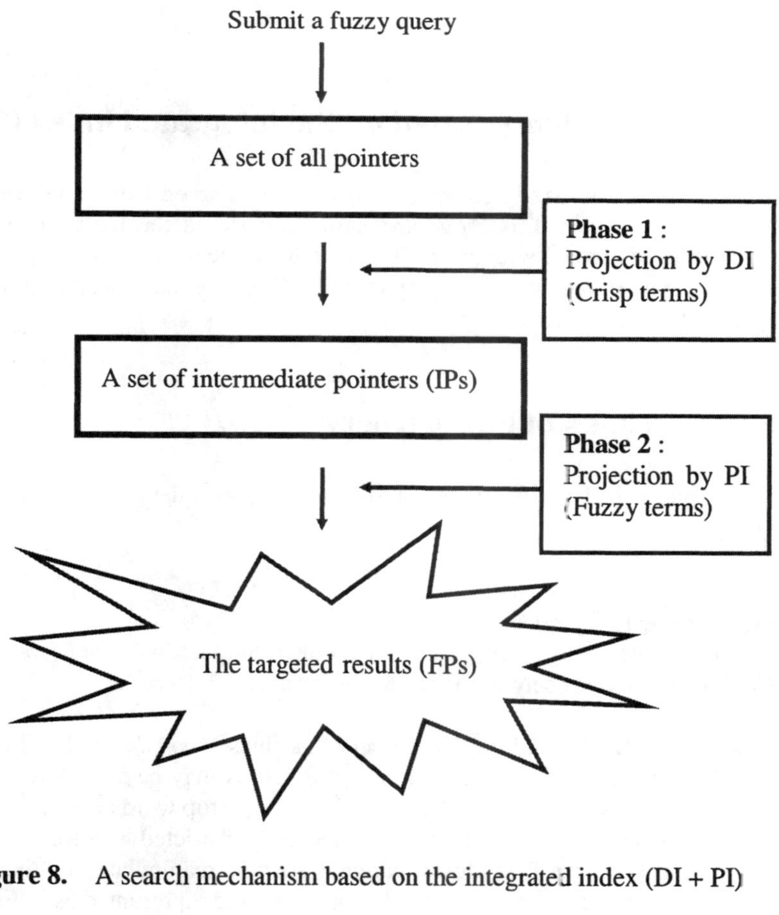

Figure 8. A search mechanism based on the integrated index (DI + PI)

In **Figure 8**, if we submit a query with only crisp terms (keyword-based query), this search engine uses only the phase 1. By applying the DI, the phase 1 performs an elimination-based approach to eliminate the URLs which are impossible to be the answers of the query. In this case, this search engine will return the same results that the existing search engines do. On the other hand, if we submit a query with both crisp terms and fuzzy terms, this search engine uses both phase 1 and phase 2. In this case, by applying the PI, the URLs reflecting fuzzy terms are extracted. More specifically, the phase 2 evaluates the fuzzy terms in detail on the set of intermediate pointers (i.e., the candidate URLs), and then generates the final pointers (FPs) (i.e., targeted results) that user really wants. This search mechanism can be conceptually explained by SQL-like language as follows : *SELECT * FROM {a set of intermediate pointers that satisfies focal keyword(s) in the DI} [WHERE the value(s) of focal attribute(s) in the PI are satisfied by the user]*. We note that commercial Web search engines tend to ignore the importance of *[WHERE]* part. In this approach, the PI may be regarded as a constraint on the DI.

5.3 Fuzzy Query based on the Integrated Index (DI + PI)

We assume that a fuzzy term in a fuzzy query is marked with an asterisk. For example, it is expressed as '***popular** national parks in the USA'. If a query has fuzzy term(s) marked with asterisk(s), search engine displays a PI associated with a focal keyword restricted by fuzzy term(s). Then user can specify values with respect to the fuzzy terms.

5.3.1 Types of Fuzzy Query

Fuzzy query is largely divided into simple fuzzy query and compound fuzzy query.

(1) Simple fuzzy query

The simple fuzzy query does not include conjunction ('*and*') or disjunction ('*or*') connective(s) between fuzzy terms, or negation ('*not*').

Example 1. Consider a fuzzy query that finds '**popular* national parks in the USA'. In this case, the DI, the PI, and stop words may be as follows : DI = {<u>national parks</u>, USA, …}, PI = {<u>No. of visitors</u>, …}, stop words = {in, the}. We note that a focal keyword 'national parks' in the DI is restricted by a fuzzy term 'popular'. In this case, the fuzzy term '*popular*' may be manipulated by the number of visitors (i.e., a focal attribute in the PI) per year, and represented as in **Figure 9**.

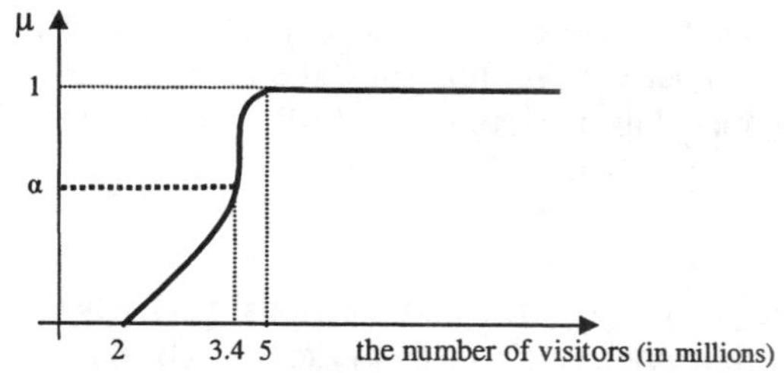

Figure 9. A membership function of '*popular*'

Example 2. Consider a fuzzy query that finds 'national parks ****moderate*** distance from San Francisco'. In this case, the DI, the PI and stop words may be as follows : DI = {national parks, San Francisco, ...}, PI = {distance, ...}, stop words = {from}. We note that a focal keyword 'San Francisco' in the DI is restricted by a fuzzy term 'moderate'. In this case, the fuzzy term '*moderate*' may be manipulated by the degree of distance (i.e., a focal attribute in the PI) from San Francisco, and represented as in **Figure 10**.

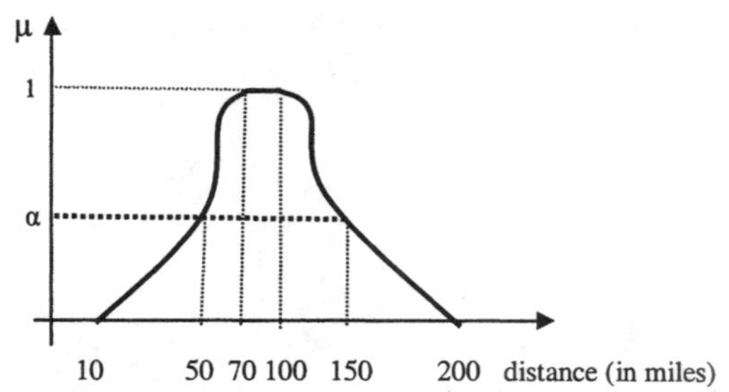

Figure 10. A membership function of '*moderate*'

(2) Compound fuzzy query

The compound fuzzy query includes conjunction ('*and*') or disjunction ('*or*') connective(s) between fuzzy terms, or negation ('*not*').

• *Conjunction ('and')*

Example 3. Consider a fuzzy query that finds 'national parks that ****popular*** *and* ****moderate*** distance from San Francisco'. In this case, the DI, the PI, logical operator, and stop words may be as follows : DI = {national parks, San Francisco, ...}, PI = {No. of visitors, distance, ...}, logical operator = {and}, stop words =

{that, from}. We note that a focal keyword 'San Francisco' in the DI is restricted by fuzzy terms 'popular' and 'moderate'. In this case, the fuzzy terms '*popular*' and '*moderate*' may be manipulated as in **Figures 9** and **10**, respectively.

- ***Disjunction ('or')***

Example 4. Consider a fuzzy query that finds 'national parks that ***popular** or ***moderate** distance from San Francisco'. In this case, the DI, the PI, logical operator, and stop words may be as follows : DI = {national parks, San Francisco, ...}, PI = {No. of visitors, distance, ...}, logical operator = {or}, stop words = {that, from}. We note that a focal keyword 'San Francisco' in the DI is restricted by fuzzy terms 'popular' and 'moderate'. In this case, the fuzzy terms '*popular*' and '*moderate*' may be manipulated as in **Figures 9** and **10**, respectively.

- ***Negation ('not')***

Example 5. Consider a fuzzy query that finds '*not* ***popular** national parks in the USA'. In this case, the DI, the PI, logical operator and stop words may be as follows : DI = {national parks, USA, ...}, PI = {No. of visitors, ...}, logical operator = {not}, stop words = {in, the}. This fuzzy query is similar to Example 1 but the fuzzy term 'popular' is negated. According to **Figures 9**, the negated fuzzy term '*not popular*' may be represented as in **Figure 11**.

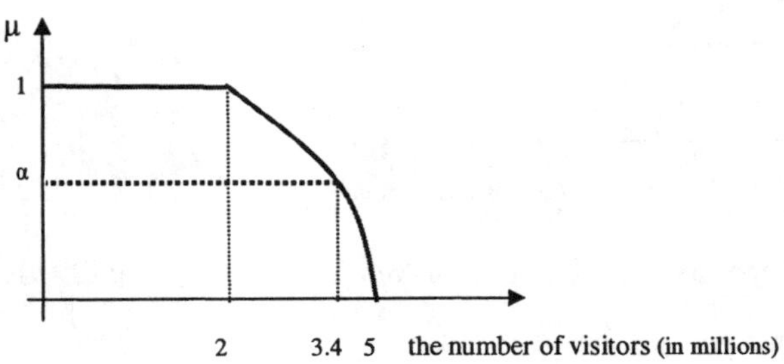

Figure 11. A membership function of '*not popular*'

5.3.2 Query Processing based on the Integrated Index (DI + PI)

Now, we present how this search engine processes fuzzy queries. Let the set of national parks in the USA be A = {A_1, A_2, ..., A_{99}, A_{100}} and each A_i, (i = 1,2, ..., 100) has its own PT(page title) or URL.

Example 6. Consider a crisp query that finds 'national parks in the USA' (Q_1). In this case, the PI is not used. So, this search engine uses only the phase 1 in **Figure 8**. Thus, the integrated index is made as in **Table 9**.

Table 9. A snapshot of Integrated Index (DI + PI) after processing Q_1

Document Index (DI)	IPs	Perception Index (PI)			FPs (Results)
National parks, USA	A_1	Distance	No. of visitors	…	A_1
	A_2	Distance	No. of visitors	…	A_2
	…	…	…	…	…
	A_{99}	Distance	No. of visitors	…	A_{99}
	A_{100} + Irrelevant URLs	Distance	No. of visitors	…	A_{100} + Irrelevant URLs

In the crisp query case, this search engine returns the same results that the existing search engines do. We note that IPs and FPs are equal.

Example 7. Consider a fuzzy query that finds '****popular*** national parks in the USA' (Q_2). In this case, the DI and the PI are used. So, this search engine uses both the phase 1 and the phase 2 in **Figure 8**. We assume that '****popular*** national parks in the USA' are A_p, $A_p \in \{A_1, A_2, \ldots, A_{99}, A_{100}\}$, by using α-cut in **Figure 9**. Thus, the integrated index is made as in **Table 10**.

Table 10. A snapshot of Integrated Index (DI + PI) after processing Q_2

Document Index (DI)	IPs	Perception Index (PI)			FPs (Results)
<u>**National parks**</u>, USA	$\{A_1, \cdots, A_{100}\}$ + Irrelevant URLs	Distance	<u>**No. of visitors**</u>	…	URLs w.r.t $\{A_p\}$

(Focal keyword : National parks; Focal attribute : No. of visitors)

Example 8. Consider a fuzzy query that finds 'national parks ****moderate*** distance from San Francisco' (Q_3). In this case, the DI and the PI are used. So, this search engine uses both the phase 1 and the phase 2 in **Figure 8**. We assume that 'national parks ****moderate*** distance from San Francisco' are A_m, $A_m \in \{A_1, A_2, \ldots, A_{99}, A_{100}\}$, by using α-cut in **Figure 10**. Thus, the integrated index is made as in **Table 11**.

Table 11. A snapshot of Integrated Index (DI + PI) after processing Q_3

Document Index (DI)	IPs	Perception Index (PI)			FPs (Results)
National parks, **San Francisco**	$\{A_1,\cdots, A_{100}\}$ + Irrelevant URLs	**Distance**	No. of visitors	…	URLs w.r.t $\{A_m\}$

(Focal keyword : San Francisco; Focal attribute : Distance)

Example 9. Consider a fuzzy query that finds 'national parks that ****popular*** *and* ****moderate*** distance from San Francisco' (Q_4). In this case, the DI and the PI are used. So, this search engine uses both the phase 1 and the phase 2 in **Figure 8**. Then the query results with respect to Q_4 become $\{A_p\}\cap\{A_m\}$. For instance, let A_p be a set $\{A_1, A_2, A_3\}$ and A_m be a set $\{A_1, A_4, A_5\}$, then $\{A_p\}\cap\{A_m\} = \{A_1\}$. Thus, the integrated index is made as in **Table 12.**

Table 12. A snapshot of Integrated Index (DI + PI) after processing Q_4

Document Index (DI)	IPs	Perception Index (PI)			FPs (Results)
National parks, **San Francisco**	$\{A_1,\cdots, A_{100}\}$ + Irrelevant URLs	**Distance**	**No. of visitors**	…	URLs w.r.t $\{A_p\}\cap\{A_m\}$

(Focal keyword : San Francisco; Focal attributes : Distance and no. of visitors)

Example 10. Consider a fuzzy query that finds 'national parks that ***popular** <u>or</u> ***moderate** distance from San Francisco' (Q_5). In this case, the DI and the PI are used. So, this search engine uses both the phase 1 and the phase 2 in **Figure 8**. Then the query results with respect to Q_5 become $\{A_p\} \cup \{A_m\}$. For instance, let A_p be a set $\{A_1, A_2, A_3\}$ and A_m be a set $\{A_1, A_4, A_5\}$, then $\{A_p\} \cup \{A_m\} = \{A_1, A_2, A_3, A_4, A_5\}$. Thus, the integrated index is made as in **Table 13**.

Table 13. A snapshot of Integrated Index (DI + PI) after processing Q_5

Document Index (DI)	IPs	Perception Index (PI)			FPs (Results)
National parks, **<u>San Francisco</u>**	$\{A_1,\cdots, A_{100}\}$ + Irrelevant URLs	**<u>Dis-tance</u>**	**<u>No. of visi-tors</u>**	…	URLs w.r.t $\{A_p\}\cup\{A_m\}$

(Focal keyword : San Francisco; Focal attributes : Distance and no. of visitors)

Example 11. Consider a fuzzy query that finds '<u>*not*</u> ***popular** national parks in the USA' (Q_6). In this case, the DI and the PI are used. So, this search engine uses both the phase 1 and the phase 2 in **Figure 8**. Then the query results with respect to Q_6 become $\{\sim A_p\}$. For instance, let A_p be a set $\{A_1, A_2, A_3\}$, then $\{\sim A_p\} = \{A_4, A_5, \ldots, A_{99}, A_{100}\}$ if the universal set $A = \{A_1, A_2, \ldots, A_{99}, A_{100}\}$. Thus, the integrated index is made as in **Table 14**.

Table 14. A snapshot of Integrated Index (DI + PI) after processing Q_6

Document Index (DI)	IPs	Perception Index (PI)			FPs (Results)
<u>National parks</u>, USA	$\{A_1,\cdots, A_{100}\}$ + Irrelevant URLs	Dis-tance	**<u>No. of visi-tors</u>**	…	URLs w.r.t $\{\sim A_p\}$

(Focal keyword : National parks; Focal attribute : No. of visitors)

5.3.3 User Interface based on the Integrated Index (DI + PI)

Williams (1984) developed a user interface for information retrieval systems to aid users in formulating a query. The system, *RABBIT III*, supports interactive refinement of queries by allowing users to critique retrieved results with labels such as 'require' and 'prohibit'. Williams claims that this system is particularly helpful to naïve users with only a vague idea of what they want and therefore need to be guided in the formulation/reformulation of their queries or who have limited knowledge of a given database or who must deal with a multitude of databases. This process allows users to refine their queries. In a similar sense, we can refine user's query by means of the phase 2 for processing fuzzy term(s) in **Figure 8**. Thus, search engine will return the targeted results that users really want. An important problem relating to personalization concerns understanding how a machine can help an individual user via suggesting recommendations (Bekin 2000). In our approach, the PI can help the user to specify clearly what he/she really wants. More specifically, the user in the system is asked to specify fuzzy term(s) in a query. In this respect, the PI may be regarded as a recommendation for handling fuzzy term(s) in a query. As a result, search engine returns 'the targeted results'. Now, we describe user interface for phase 1 and phase 2 in **Figure 8**.

(1) User interface for phase 1

Initially, user interface for phase 1 lets user specify his/her queries with only crisp terms (keywords), or both crisp terms and fuzzy terms. If user submits a query with only crisp terms, only user interface for phase 1 is used, and search results are returned based on only the DI. On the other hand, if user submits a query with both crisp terms and fuzzy terms, user interface for phase 2 is also displayed to process the fuzzy terms.

(2) User interface for phase 2

For the fuzzy query on the Internet, the 'easy of use' is important because Internet users are broad spectrum in terms of cultural differences, level of intelligence, etc. In this respect, user interface for phase 2 should provide Internet users with an easy user interface for specifying these fuzzy terms such as '*popular*', '*moderate*', '*big*', etc. In addition, we need to reflect cultural differences. For instance, different people generally use different scales (i.e., feet, miles, meter, etc). Internet users have their own membership functions with respect to fuzzy terms in a fuzzy query, by means of human's perception capability. Consequently, they can give values with respect to fuzzy terms in the user interface for phase 2.

User interface for phase 2 displays a PI associated with a focal keyword. For example, given a fuzzy query that finds '***popular** national parks in the USA', a PI associated with a focal keyword 'national parks' is displayed as shown in **Table 10**. It should be noted that different people may use different conceptual comprehension (fuzzy terms, membership functions, α-cut), with respect to the same situation. It is the user's task in this user interface to examine the suggested attributes in the PI, and to specify the values of the focal attributes reflecting user's query requirements. Using the PI, search results can be restricted within narrow limit. We call it '*target search by fuzzy terms*'. In other words, search engine will return the targeted results that users really want.

Fuzzy terms are specified in user interface for phase 2. For example, they can be expressed as point value, interval value, multiple values, etc.

- ***Point value***

Example 12. In Example 1, given a α-cut, the fuzzy term '*popular*' may be specified by using a focal attribute 'no. of visitors'. More specifically, it is expressed as a point 3.4 (i.e., 'no. of visitors' $\geq$ 3.4 millions).

- ***Interval value***

Example 13. In Example 2, given a α-cut, the fuzzy term '*moderate*' may be specified by using a focal attribute 'distance'. More specifically, it is expressed as an interval (i.e., distance = [50, 150] in miles).

- ***Multiple values***

A veristic variable (Zadeh 1997 and 1999) which can be assigned two or more values in its universe simultaneously will be specified as multiple values.

Example 14. Let U be the universe of natural languages and let X denote the fluency of an individual in English, French and Italian. Then, X *isv* (1.0 English + 0.8 French + 0.6 Italian) means that the degrees of fluency of X in English, French and Italian are 1.0, 0.8 and 0.6, respectively (Zadeh 1997 and 1999).

5.4 Some Features of the Proposed Method

Remark 1. The higher the α in α-cut ($0 \leq \alpha \leq 1$), the smaller the number of the targeted results. This property provides continual incremental result from 'the highest constraint (i.e., $\alpha = 1$)' to 'the lowest constraint (i.e., $\alpha = 0$)'. Consequently, we can achieve 'interactive user control of the query processing' by adjusting the value of α.

Remark 2. If $\alpha = 0$, search results coincide with the results by applying only the DI (i.e., the existing keyword-based search). In this case, the results of phase 1 in **Figure 8** become search results.

Remark 3. Even though the same integrated index (DI+PI) is given, different search results are returned by adjusting the value of α or by using different focal attributes in the PI. In the case of 'using different focal attributes in the PI', for example, consider a fuzzy query that finds '*attractive* car', where '*attractive*' means 'comfortable and fast'. In this case, for the fuzzy term '*attractive*', people may use different focal attributes (i.e., size, speed, etc.) in the PI. In addition, different people may use different conceptual comprehension (fuzzy terms, membership functions, α-cut), with respect to the same situation. Thus, search engine will return the personalized search results that users really want. In the meantime, clustering (i.e., grouping similar documents together to expedite information retrieval) is adaptively determined depending on the value of α or the selected focal attributes in the PI.

Remark 4. Using the PI, Internet users can narrow thousands of hits to the few that users really want.

Remark 5. Using the PI, therefore, we can tackle in part the major problems in commercial Web search engines (i.e., 'large answer set', 'low precision', 'ineffective for general-concept queries').

5.5 Performance Analysis

For comparing with commercial keyword-based search engines, the ratio *[the number of FPs / the number of IPs]* can be used as a measure of performance evaluation on the proposed method. We note that the number of IPs is the result of phase 1 and the number of FPs is the result of phase 2 in **Figure 8**. The smaller the ratio, the better the filtering effect of the proposed method. More specifically, **Table 15** illustrates the problem 'quality of retrieved information' in the commercial Web search engines by showing the results obtained from querying two popular search engines with 6 sample queries ($Q_1 \sim Q_6$ in Subsec. 5.3.2).

Table 15. Example queries and results

Queries	Search engines	No. of hits
Q_1 (Crisp Query)	Yahoo !	returns about 112,000
	Google	returns about 240,000
	The proposed method	returns the same results that the existing search engines do
Q_2 (Fuzzy Query)	Yahoo !	returns about 34,200
	Google	returns about 73,100
	The proposed method	returns URLs w.r.t $\{A_p\}$ (see **Table 10**)
Q_3 (Fuzzy Query)	Yahoo !	returns about 1,380
	Google	returns about 3,960
	The proposed method	returns URLs w.r.t $\{A_m\}$ (see **Table 11**)
Q_4 (Fuzzy Query)	Yahoo !	returns about 1,330
	Google	returns about 2,050
	The proposed method	returns URLs w.r.t $\{A_p\}\cap\{A_m\}$ (see **Table 12**)
Q_5 (Fuzzy Query)	Yahoo !	returns about 1,000
	Google	returns about 2,050
	The proposed method	returns URLs w.r.t $\{A_p\}\cup\{A_m\}$ (see **Table 13**)
Q_6 (Fuzzy Query)	Yahoo !	returns about 29,100
	Google	returns about 62,200
	The proposed method	returns URLs w.r.t $\{\sim A_p\}$ (see **Table 14**)

5.6 Additional Considerations

The work of Lidsky and Kwon (1997) is an opinionated but informative resource on search engines. It describes 36 different search engines and rates them on specific details of their search capabilities. For instance, in one study, searches are divided into five categories : (1) simple searches; (2) custom searches; (3) directory searches; (4) current news searches; and (5) Web content. The five categories of search are evaluated in terms of power and easy of use. Variations in ratings sometimes differ substantially for a given search engine. In the meantime, they chose the respective best search engine according to five categories : (1) search indexes and directories; (2) people finders; (3) business finders; (4) usenet search; and (5) metasearch. The data indicate that as the number of people using the Internet and Web has grown, user types have diversified and search engine providers have begun to target more specific types of users and queries with specialized and tailored search tools. In this respect, for the fuzzy query processing, topic-specific (or domain-specific) requirement is necessary because of the following reasons : (1) commonsense knowledge - the present state of AI is not up to formulating a full commonsense database, but full commonsense knowledge is not necessary (McCarthy 2000). In this respect, for the fuzzy query processing, if we design a search engine based on 'domain-specific' concept, the degree of freedom on fuzzy terms will be highly reduced. In other words, 'domain-specific' concept provides the higher possibility for a well-defined (restricted) condition. For example, given a travel-domain database, consider a fuzzy query that finds '***popular** national parks in the USA'. In this case, the fuzzy term 'popular' is used to restrict 'national parks', not 'music', 'car', etc.; (2) indexing overhead - human indexing (for example, Yahoo !, LookSmart, etc.) is currently the most accurate because experts on popular subjects organize and compile the directories and indexes in a way which facilitates the search process. However, the enormous number of existing Web pages and their rapid increase and frequent updating make the indexing a difficult one or an overhead. If we design a search engine based on 'domain-specific' concept, the indexing overhead on the PI will be highly reduced; (3) storage requirement - comparing with traditional Web search engines, recommending the PI requires the system to maintain more data. If we design a search engine based on 'domain-specific' concept, the storage requirement on the PI will be highly reduced; (4) uneven concentration - if we design a search engine based on 'domain-specific' concept, 'uneven concentration of information packets on the Internet' problem, as described in Section 1, will be highly reduced.

5.7 Remarks

Although the commercial Web search engines such as Yahoo !, Google, Lycos, etc. help Internet users get to good information, they do not properly handle fuzzy query. For example, consider a fuzzy query that finds '***popular*** national parks in the USA'. In this case, '***popular***', 'national parks' and 'USA' are generally processed as keywords in the commercial Web search engines. It should be noted that fuzzy term '***popular***' is a constraint on a focal keyword 'national parks' rather than an independent keyword. In other words, the fuzzy term '***popular***' plays the role of a constraint on the fuzzy query. However, commercial Web search engines tend to ignore the importance of fuzzy terms in a query processing. As a result, search engines return a bunch of page titles (or URLs) irrelevant to user's query. For example, in the case of a fuzzy query that finds '***popular*** national parks in the USA', Yahoo ! returns about 34,200 page titles (or URLs) and Google returns about 73,100 page titles (or URLs). Intuitively, we find that there are so many page titles (or URLs) irrelevant to user's query.

In this section, we present the search mechanism based on the integrated index (DI + PI) and fuzzy query based on the integrated index (DI + PI). Moreover, we describe some features of the proposed method and suggest some considerations for implementing the proposed method.

6 Ranking Algorithm based on Perception Index

In Section 5, we have introduced the Perception Index (PI) that contains attributes associated with a focal keyword restricted by fuzzy term(s) in a fuzzy query.

Ranking algorithms play an important role in Web search engines. Although the existing ranking methods for Web search engines provide users with their own ranking algorithms based on popularity, bookmark, etc., they often tend to return unranked random samples in response to user's query. In order to tackle this problem, we introduce a new ranking algorithm based on the Perception Index (PI). Using the values of focal attributes in the PI, user's search intentions can be explicitly reflected. Consequently, the proposed ranking algorithm ensures consistently high-quality returns in terms of user's search intentions.

6.1 Overview

Increased capabilities of computer hardware and software have created a vast body of machine-readable resources. Typically there is no lack of available information; more often, users, seeking needles in haystacks, are overwhelmed by the quantity of irrelevant information. Often this is caused by a poor query (too vague or too generic; for example, try searching for "computer science") (Ali and McRoy 2000) . Without the context of the query and the relations of the information, a search engine is doomed to return random samples of the Internet. With no ability to control or organize the sprawl on the Internet, how will we ever be able to find the intelligence in all the data, information and knowledge that we presume to be there ? Perhaps the Internet is more like TV. Is it mostly a collection of garbage, gleaned at the lowest common denominator, serving merely to provide eyeballs to advertisers; or is it a free exchange of information that's just too cheap to meter ? (Hoebel and Welty 1999). Despite numerous refinements, most Web search engines still return too many results and random samples of the Internet. In other words, they often give users a bunch of garbage. In this respect, we need a new tool to handle both the removal of spurious results and the random samples of the Internet. In section 5, we have mainly discussed the problem on 'the removal of spurious results'. The PI provides a deductive capability to query language on the Internet. In other words, the useful URLs (targeted URLs) are separated from the useless by the PI. In this section, we will focus on the ranking within the targeted URLs. The Compaq study found that most searchers (68%) look only at the first page of results. This means that ranking algorithm plays an important role in Web search engines. Although the existing ranking methods for Web search engines provide users with their own ranking algorithms based on popularity, bookmark, etc., they often tend to return unranked random samples in response to user's query. In order to tackle this problem, we introduce a new ranking algorithm based on the Perception Index (PI). Using the values of focal attributes in the PI, user's search intentions can be explicitly reflected. For example, consider a fuzzy query that finds '*attractive* car', where '*attractive*' means 'comfortable and fast'. In this case, for the fuzzy term '*attractive*', people may use different focal attributes (i.e., size, speed, etc.) in the PI. In this respect, it provides a user with the personalized ranking based on user's search intentions.

In Section 6.2, we briefly summarize the existing ranking methods. In Section 6.3, we introduce a new ranking algorithm based on the PI, and compare the proposed ranking algorithm with the existing Web ranking methods. We discuss the proposed ranking method in Section 6.4.

6.2 Summary of the existing ranking methods

In conventional information retrieval (IR), a variety of techniques have been developed for ranking retrieved documents for a given query. A textual database can be represented by a word-by-document matrix whose entries represent the frequency of occurrence of a word in a document. Thus, documents can be thought of as vectors in a multidimensional space, the dimensions of which are the words used to represent the texts. In a standard 'keyword-matching' vector system (Salton and McGill 1983), the similarity between two documents is computed as the inner product or cosine of the corresponding two columns of the word-by-document matrix. Queries can also be represented as vectors of words and thus compared against all document columns with the best matches being returned. An important assumption in this vector space model is that the words (i.e., dimensions of the space) are orthogonal or independent. While it has been a reasonable first approximation, the assumption that words are pairwise independent is not realistic.

Recently, several statistical and AI techniques have been used to better capture term association and domain semantics. One such method is latent semantic indexing (LSI) (Berry et al. 1995, Deerwester et al. 1990). LSI is an extension of the standard vector retrieval method designed to help overcome some of the retrieval problems described previously. In LSI the associations among terms and documents are calculated and exploited in retrieval. The assumption is that there is some underlying or 'latent' structure in the pattern of word usage across documents and that statistical techniques can be used to estimate this latent structure. A description of terms, documents, and user queries based on the underlying latent semantic structure is used for representing and retrieving information. The particular LSI analysis described by Deerwester et al. (1990) uses singular value decomposition (SVD), a technique closely related to eigenvector decomposition and factor analysis. SVD takes a large word-by-document matrix and decomposes it into a set of k, typically 100 to 300, orthogonal factors from which the original matrix can be approximated by linear combination. Instead of representing documents and queries directly as vectors of independent words, LSI represents them as continuous values on each of the k orthogonal indexing dimensions derived from the SVD analysis. One advantage of this approach is that queries can retrieve documents even if they have no words in common. The LSI technique captures deeper associative structure than simple term-to-term correlations and clusters and is completely automatic. We can interpret the analysis performed by SVD geometrically. The result of the SVD is a k-dimensional vector space containing a vector for each term and each document. The location of term vectors reflects the correlations in their usage across documents. Similarly, the location of document vectors reflects correlations in term usage. In this space the cosine or dot product between vectors corresponds to their estimated similarity. Retrieval proceeds by using the terms in a query to identify a vector in the space, and all documents are then ranked by their similarity to the query vector. The LSI method has been applied to several standard IR collections with favorable results.

In the Web search engines, however, detailed information regarding ranking algorithms used by major search engines is not publicly available. A simple means to measure the quality of a Web page, proposed by Carriere and Kazman (1997), is to count the number of pages with pointers to the page. Google is a representative Web search engine that uses link information. Its rankings are based, in part, on the number of other pages with pointers to the page. In November 1999, Northern Light introduced a new ranking system, which is also based, in part, on link data (see <http://www.searchenginewatch.com/sereport/ 99/11briefs.html>). In other words, Google and Northern Light rank search results, in part, by popularity. In the meantime, HotLinks ranks search results based on the bookmarks of its registered users. Yahoo's Inktomi-served results aren't ranked by popularity (see <http://websearch.about.com/internet/webserch/library /weekly/aa052199.htm>).

In theory, more popular links indicate more relevant content, but if a user differs from the crowd, simply popularity-based ranking approaches dive deeply into other possibilities on the Web. Consequently, they often give users a bunch of garbage. In addition, they often tend to return unranked random samples in response to user's query. In order to tackle these problems, we introduce a new ranking algorithm based on the Perception Index (PI).

6.3 A new ranking algorithm based on the Perception Index (PI)

As described in Section 6.2, the existing ranking methods can be largely categorized into the following two classes : keyword-based approach and hyperlink-based approach. On the other hand, in Section 5, we have shown that fuzzy terms play the role of a constraint on the fuzzy query and can be expressed by using the focal attributes in the PI. In this Section, we propose a new ranking algorithm based on the PI. It may be regarded as a fuzzy term-based approach.

Although Web search engines generally return large amounts of web pages (or URLs) for a given query, only a small fraction of the returns will actually be relevant to any particular person. Thus, there is the problem of determining what information is of interest to any particular person, while minimizing the amount of search through irrelevant information. In Section 5, we have mainly discussed the problem on 'the removal of spurious results' irrelevant to user's search intentions. The PI provides a deductive capability to query language on the Internet. In other words, the useful URLs (targeted URLs) are separated from the useless by the PI. In this Section, we will focus on the ranking within the targeted URLs. The Compaq study found that most searchers (68%) look only at the first page of results. This means that ranking algorithm plays an important role in Web search engines. Although the existing ranking methods for Web search engines provide users with their own ranking algorithms based on popularity, bookmark, etc., they often tend to return unranked random samples in response to user's query. In order to tackle this problem, we introduce a new ranking algorithm based on the PI. It provides a user with the personalized ranking based on user's search intentions.

Zadeh suggested we can represent linguistic quantifiers as fuzzy subsets of the unit interval (Zadeh 1997). In this representation the membership grade of any proportion $r \in [0, 1]$, $Q(r)$, is a measure of the compatibility of the proportion r with the linguistic quantifier we are representing by the fuzzy subset Q. For example, if Q is the quantifier 'most' then Q(0.9) represents the degree to which 0.9 satisfies the concept 'most'. Yager identified three classes of linguistic quantifiers that cover most of these used in natural language (Yager 1991 and 1996).

(i) A quantifier Q is said to be monotonically nondecreasing if $r_1 > r_2$ then $Q(r_1) \geq Q(r_2)$.
(ii) A quantifier Q is said to be monotonically nonincreasing if $r_1 > r_2$ then $Q(r_1) \leq Q(r_2)$.

(iii) A quantifier Q is said to be unimodal if there exists two values a ≤ b both contained in the unit interval such that for r < a, Q is monotonically nondecreasing, for r > b, Q is monotonically nonincreasing, and for $r \in$ [a, b], Q® = 1.

Figure 12 shows prototypical examples of these quantifiers.

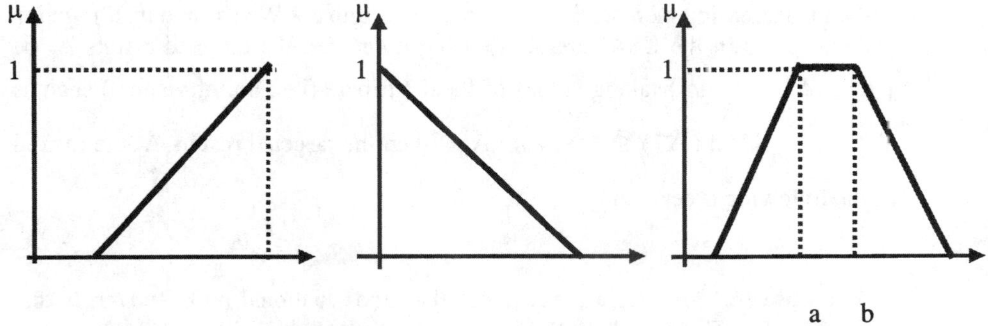

(i) Monotonically nondecreasing (ii) Monotonically nonincreasing (iii) Unimodal

Figure 12. Three types of quantifiers

In a similar way, we can identify three classes of fuzzy terms that cover most of these used in natural language. For example, in Section 5.3.1, we have represented the fuzzy terms 'popular' (monotonically nondecreasing), 'moderate' (unimodal), 'not popular' (monotonically nonincreasing), (see **Figures 9-11**). In this respect, we design a new ranking algorithm based on the PI as follows :

Algorithm 1 : *Ranking for one focal attribute*

(i) Monotonically nondecreasing case

The larger the value of focal attribute in the PI, the higher the rank retrieved documents for a given fuzzy query.

(ii) Monotonically nonincreasing case

The larger the value of focal attribute in the PI, the lower the rank retrieved documents for a given fuzzy query.

(iii) Unimodal case

If an interval of focal attribute determined by α-cut is $[a_i, b_i]$, and let β denote the midpoint between a_i and b_i, then the degree of closeness (nearness) to β can be used as a ranking criterion. In other words, the closer the β, the higher the rank retrieved documents for a given fuzzy query.

Example 15. Consider a fuzzy query that finds '***popular*** national parks in the USA'. In this case, the fuzzy term 'popular' may be represented by a monotonically nondecreasing membership function, (see **Figure 9**)We assume that '***popular*** national parks in the USA' are A_p by using α-cut. Let the targeted results A_p be $\{ A_p^1, A_p^2, ..., A_p^r \}$ taking values of focal attribute (i.e., no. of visitors) such as $Val(A_p^1) \le Val(A_p^2) \le ... \le Val(A_p^r)$, then the targeted results A_p are ranked as the following order : $A_p^r, ..., A_p^2, A_p^1$.

Example 16. Consider a fuzzy query that finds 'national parks ***moderate*** distance from San Francisco'. In this case, the fuzzy term 'moderate' may be represented by a unimodal membership function, (see **Figure 10**). We assume that '***moderate*** national parks in the USA' are A_m by using α-cut. Let an interval of focal attribute (i.e., distance) determined by α-cut be $[a_i, b_i]$, and let β denote the midpoint between a_i and b_i, and let the targeted results A_m be $\{ A_m^1, A_m^2, ..., A_m^s \}$ taking values of the focal attribute such as $Val(A_m^1)$, $Val(A_m^2)$, ..., $Val(A_m^s)$. If the degree of closeness (nearness) to β is the order $Val(A_m^1)$, $Val(A_m^2)$, ..., $Val(A_m^s)$, then the targeted results A_m are ranked as the following order : $A_m^1, A_m^2, ..., A_m^s$.

Example 17. Consider a fuzzy query that finds 'not ***popular*** national parks in the USA'. In this case, the negated fuzzy term 'not popular' may be represented by a monotonically nonincreasing membership function, (see **Figure 11**). We assume that 'not ***popular*** national parks in the USA' are $\sim A_p$ by using α-cut. Let the targeted results $\sim A_p$ be $\{ A_p^1, A_p^2, ..., A_p^t \}$ taking values of focal attribute (i.e., no. of visitors) such as $Val(A_p^1) \le Val(A_p^2) \le ... \le Val(A_p^t)$, then the targeted results $\sim A_p$ are ranked as the following order : $A_p^1, A_p^2, ..., A_p^t$.

Algorithm 2 : ***Ranking for multiple focal attributes***

If we have multiple focal attributes (for instance, 'no. of visitors' and 'distance'), weighting the importance of focal attributes should be considered. For the weighted case, assume that $\theta_1, \theta_2, \ldots, \theta_n$ are ordinal weights. Then we refer to $\Theta = (\theta_1, \theta_2, \ldots, \theta_n)$ as a weighting, where θ_i is the weight of attribute i. Intuitively, the targeted results can be ranked according to the ordinal weights. For a respective focal attribute, the rank retrieved documents for a given fuzzy query can be determined based on the **Algorithm 1**.

Example 18. Consider a fuzzy query that finds 'national parks that ***popular*** *and* ***moderate*** distance from San Francisco'. In this case, the fuzzy terms 'popular' and 'moderate' may be represented by a monotonically nondecreasing membership function and a unimodal membership function, respectively. Using the results of Examples 15 and 16, if the weight of focal attribute 'no. of visitors' is more important than the weight of focal attribute 'distance', then the targeted results are ranked as the following order : $A_p^r, \ldots, A_p^2, A_p^1, A_m^1, A_m^2, \ldots, A_m^s$.

As described in Section 5.3.3, user interface for phase 2 displays a PI associated with a focal keyword. For example, consider a fuzzy query that finds '***popular*** national parks in the USA', a PI associated with a focal keyword 'national parks' is displayed. It is the user's task in this user interface to examine the suggested attributes in the PI, and to specify the values of the focal attributes reflecting user's search intentions. Consequently, search results can be restricted within narrow limit. We call it '*target search by fuzzy terms*'. In other words, search engine will return the targeted results that users really want. Now, if we apply Algorithm 1 and 2, the targeted results can be displayed from the highest rank to the lowest rank.

Although the existing ranking methods for Web search engines also provide users with their own ranking algorithms based on popularity, bookmark, etc., their approaches look like the behind-the-scenes processing. In the proposed approach, user's search intentions can be explicitly reflected by using the values of focal attributes in the PI. In this respect, we can explicitly describe how to rank the search results by means of the proposed approach. Consequently, the proposed approach provides a user with the personalized ranking based on user's search intentions.

7 Challenges and Road Ahead

During the August 2001, BISC program hosted a workshop toward better understanding of the issues related to the Internet (Fuzzy Logic and the Internet-FLINT2001, Toward the Enhancing the Power of the Internet). The main purpose of the Workshop was to draw the attention of the fuzzy logic community as well as the Internet community to the fundamental importance of specific Internet-related problems. This issue is critically significant about problems that center on search and deduction in large, unstructured knowledge bases. The Workshop provided a unique opportunity for the academic and corporate communities to address new challenges, share solutions, and discuss research directions for the future. Followings are the areas that were recognized as challenging problems and the new direction toward the next generation of the search engines and Internet. We summarize the challenges and the road ahead into four categories as follows:

I. Search Engine and Queries:

- Deductive Capabilities
- Customization and Specialization
- Metadata and Profiling
- Semantic Web
- Imprecise-Querying
- Automatic Parallelism via Database Technology
- Approximate Reasoning
- Ontology
- *Ambiguity Resolution through Clarification Dialog; Definition/Meaning & Specificity* User Friendly
- Multimedia
- Databases
- Interaction

II. Internet and the Academia:

- Ambiguity and Conceptual and Ontology
- Aggregation and Imprecision Query
- Meaning and structure Understanding
- Dynamic Knowledge
- Perception, Emotion, and Intelligent Behavior
- Content-Based
- Escape from Vector Space Deductive Capabilities
- Imprecise-Querying
- *Ambiguity Resolution through Clarification Dialog*
- *Precisiated Natural Languages (PNL)*

III. Internet and the Industry:

- XML=>Semantic Web
- Workflow
- Mobile E-Commerce
- CRM
- Resource Allocation
- Intent
- Ambiguity Resolution
- Interaction
- Reliability
- Monitoring
- Personalization and Navigation
- Decision Support
- Document Soul
- Approximate Reasoning
- Imprecise Query
- Contextual Categorization

IV. Fuzzy Logic and Internet; Fundamental Research:

- Computing with Words (CW)
- Computational Theory of Perception (CTP)
- Precisiated Natural Languages (PNL)

The potential Area and applications of Fuzzy Logic for the Internet include:

I. Potential Areas:

- Search Engines
- Retrieving Information
- Database Querying
- Ontology
- Content Management
- Recognition Technology
- Data Mining
- Summarization
- Information Aggregation and Fusion
- E-Commerce
- Intelligent Agents
- Customization and Personalization

I. Potential Applications:

- Search Engines and Web Crawlers
- Agent Technology (i.e., Web-Based Collaborative and Distributed Agents)
- Adaptive and Evolutionary techniques for dynamic environment (i.e. Evolutionary search engine and text retrieval, Dynamic learning and adaptation of the Web Databases, etc)
- Fuzzy Queries in Multimedia Database Systems
- Query Based on User Profile
- Information Retrievals
- Summary of Documents
- Information Fusion Such as Medical Records, Research Papers, News, etc
- Files and Folder Organizer
- Data Management for Mobile Applications and eBusiness Mobile Solutions over the Web
- Matching People, Interests, Products, etc
- Association Rule Mining for Terms-Documents and Text Mining
- E-mail Notification
- Web-Based Calendar Manager
- Web-Based Telephony
- Web-Based Call Centre
- Workgroup Messages
- E-Mail and Web-Mail
- Web-Based Personal Info
- Internet related issues such as Information overload and load balancing, Wireless Internet-coding and D-coding (Encryption), Security such as Web security and Wireless/Embedded Web Security, Web-based Fraud detection and prediction, Recognition, issues related to E-commerce and E-bussiness, etc.

8 Conclusion

Intelligent search engines with growing complexity and technological challenges are currently being developed. This requires new technology in terms of understanding, development, engineering design and visualization. While the technological expertise of each component becomes increasingly complex, there is a need for better integration of each component into a global model adequately capturing the imprecision and deduction capabilities. In addition, intelligent models can mine the Internet to conceptually match and rank homepages based on predefined linguistic formulations and rules defined by experts or based on a set of known homepages. The FCM model can be used as a framework for intelligent information and knowledge retrieval through conceptual matching of both text and images (here defined as "Concept"). The FCM can also be used for constructing fuzzy ontology or terms related to th context of the query and search to resolve the ambiguity. This model can be used to calculate conceptually the degree of match to the object or query.

The expressive power of conventional search engine query interfaces is relatively weak when restricted to keyword-based search (i.e., Document Index (DI)-based search). At present, the keyword-based search engines present limitations in modeling perceptual aspects of humans. In addition, they appear to have trouble with returning the targeted results. In other words, they generally return a bunch of Web pages (or URLs) irrelevant to user's query. In this respect, we need a new tool to handle both the fuzzy query and the removal of spurious results. In order to tackle these problems, we introduce the Perception Index (PI) that contains attributes associated with a focal keyword restricted by fuzzy term(s) in a fuzzy query. If we integrate the Document Index (DI) used in commercial Web search engines with the proposed PI, we can handle both crisp terms (keyword-based) and fuzzy terms (perception-based). In this respect, the proposed approach is softer than the keyword-based approach (i.e., commercial Web search engines). It is a further step toward a real human-friendly, natural language-based interface for Internet. It should greatly help the user relatively easily retrieve relevant information. In other words, the proposed method assists the user to reflect his/her perception in the process of query. As a consequence, Internet users can narrow thousands of hits to the few that users really want. In this respect, the PI provides a new tool for targeting queries that users really want, and an invaluable personalized search. The use of PI provides helpful hints for solving the problems of 'large answer set', 'low precision', 'ineffective for general-concept queries' suffered by most search engines.

Although the existing ranking methods for Web search engines provide users with their own ranking algorithms based on popularity, bookmark, etc., they often tend to return unranked random samples in response to user's query. In theory, more popular links indicate more relevant content, but you will have to determine that for yourself. If you differ from the crowd, simply popularity-based ranking approaches dive deeply into other possibilities on the Web. Consequently, they often give users a bunch of garbage. In order to tackle this problem, we introduce a new ranking algorithm based on the Perception Index (PI). Using the values of focal attributes in the PI, user's search intentions can be explicitly reflected. For example, consider a fuzzy query that finds '*attractive* car', where '*attractive*' means 'comfortable and fast'. In this case, for the fuzzy term '*attractive*', people may use different focal attributes (i.e., size, speed, etc.) in the PI. In this respect, it provides a user with the personalized ranking based on user's search intentions. Consequently, the proposed ranking algorithm ensures consistently high-quality returns in terms of user's search intentions.

Acknowledgement

Funding for this research was provided by the British Telecommunication (BT) and the BISC Program of UC Berkeley. This work was also supported by postdoctoral fellowships program from Korea Science & Engineering Foundation (KOSEF). The Authors thanks Sun-Gyung Jung, plan & control manager/education center of Oracle Korea.

References

S. S. Ali and S. McRoy (2000) Information retrieval, Intelligence (ACM) 11(4) : 17-19.

J. Baldwin, Future directions for fuzzy theory with applications to intelligent agents, in M. Nikravesh and B. Azvine, FLINT 2001, New Directions in Enhancing the Power of the Internet, UC Berkeley Electronics Research Laboratory, Memorandum No. UCB/ERL M01/28, August 200.

J. F. Baldwin and S. K. Morton, conceptual Graphs and Fuzzy Qualifiers in Natural Languages Interfaces, 1985, University of Bristol.

M. J. M. Batista et al., User Profiles and Fuzzy Logic in Web Retrieval, in M. Nikravesh and B. Azvine, FLINT 2001, New Directions in Enhancing the Power of the Internet, UC Berkeley Electronics Research Laboratory, Memorandum No. UCB/ERL M01/28, August 2001.

N.J. Belkin (2000) Helping people find what they don't know, Communications of the ACM 43(8) : 58-61.

H. Beremji, Fuzzy Reinforcement Learning and the Internet with Applications in Power Management or wireless Networks, in M. Nikravesh and B. Azvine, FLINT 2001, New Directions in Enhancing the Power of the Internet, UC Berkeley Electronics Research Laboratory, Memorandum No. UCB/ERL M01/28, August 2001.

M. W. Berry, S. T. Dumais and G. W. O'Brien (1995) Using linear algebra for intelligence information retrieval, SIAM Rev. 37(4) : 573-595.

T.H. Cao, Fuzzy Conceptual Graphs for the Semantic Web, in M. Nikravesh and B. Azvine, FLINT 2001, New Directions in Enhancing the Power of the Internet, UC Berkeley Electronics Research Laboratory, Memorandum No. UCB/ERL M01/28, August 2001.

J. Carriere and R. Kazman (1997) WebQuery : searching and visualizing the Web through connectivity, Proceedings of the sixth international conference on the World Wide Web.

D. Y. Choi, Integration of Document Index with Perception Index and Its Application to Fuzzy Query on the Internet, in M. Nikravesh and B. Azvine, FLINT 2001, New Directions in Enhancing the Power of the Internet, UC Berkeley Electronics Research Laboratory, Memorandum No. UCB/ERL M01/28, August 2001.

S. Deerwester, S. T. Dumais, G. W. Furnas, T. K. Landauer and R. Harshman (1990) Indexing by latent semantic analysis, J. Am. Soc. Inf. Sci. 41(6) : 391-407.

N. Guarino, C. Masalo, G. Vetere, "OntoSeek : content-based access to the Web", IEEE Intelligent Systems, Vol.14, pp.70-80 (1999)

K.H.L. Ho, Learning Fuzzy Concepts by Example with Fuzzy Conceptual Graphs. In 1st Australian Conceptual Structures Workshop, 1994. Armidale, Australia.

L. Hoebel and C. Welty (1999) Garbage collection, Intelligence (ACM) 10(2) : 48.

J. J. Hopfield, "Neural networks and physical systems with emergent collective computational abilities", Proceedings of the National Academy of Sciences U.S.A., Vol.79, pp.2554-2558 (1982)

J. J. Hopfield, "Neurons with graded response have collective computational properties like those of two-state neurons, Proceedings of the National Academy of Sciences U.S.A., Vol.81, pp.3088-3092 (1984)

A. Joshi and R. Krishnapuram, Robust Fuzzy Clustering Methods to Support Web Mining, in Proc Workshop in Data Mining and Knowledge Discovery, SIGMOD, pp. 15-1 to 15-8, 1998.

B. Kao, J. Lee, C. Y. Ng and D. Cheung (2000) Anchor point indexing in Web document retrieval, IEEE trans. on SMC (part C) 30(3) : 364-373.

J. Kacprzyk and A. Ziolkowski, Retrieval from databases using queries with fuzzy linguistic quantifiers, Fuzzy logic in knowledge engineering (Edited by Prade H and Negoita C. V), Verlag TUV Rheinland, 1986.

M. Kobayashi, K. Takeda, "Information retrieval on the web", ACM Computing Survey, Vol.32, pp.144-173 (2000)

B. Kosko, "Adaptive Bi-directional Associative Memories," Applied Optics, Vol. 26, No. 23, 4947-4960 (1987).

B. Kosko, "Neural Network and Fuzzy Systems," Prentice Hall (1992).

D. H. Kraft and F. E. Petry (1997) Fuzzy Information systems : managing uncertainty in databases and information retrieval systems, Fuzzy sets and systems 90(2) : 183-191.

R. Krishnapuram et al., A Fuzzy Relative of the K-medoids Algorithm with application to document and Snippet Clustering , in Proceedings of IEEE Intel. Conf. Fuzzy Systems-FUZZIEEE 99, Korea, 1999.

T. B. Lee , Transcript of Tim Berners-Lee's talk to the LCS 35th Anniversary celebrations, Cambridge Massachusetts, 1999/April/14

D. B. Lenat, From 2001 to 2001: Common Sense and the Mind of HAL; A chapter from Hal's Legacy: 2001 as Dream and Reality (http://www.cyc.com/publications.html)

Lidsky D and Kwon R (1997) Searching the net, PC magazine Dec. 2 : 227-258.

T. P. Martin, Searching and smushing on the Semantic Web – Challenges for Soft Computing, in M. Nikravesh and B. Azvine, FLINT 2001, New Directions in Enhancing the Power of the Internet, UC Berkeley Electronics Research Laboratory, Memorandum No. UCB/ERL M01/28, August 2001.

J. McCarthy (2000) Phenomenal data mining, Communications of the ACM 43(8) : 75-79.

H. Nakajima, T. Sogoh and M. Arao (1993) Development of an efficient fuzzy SQL for a large scale fuzzy relational database, Proc. 5th IFSA world congress : 517-530.

M. Nikravesh and B. Azvine, FLINT 2001, New Directions in Enhancing the Power of the Internet, UC Berkeley Electronics Research Laboratory, Memorandum No. UCB/ERL M01/28, August 2001.

M. Nikravesh, Fuzzy Logic and Internet: Perception Based Information Processing and Retrieval, Berkeley Initiative in Soft Computing, Report No. 2001-2-SI-BT, September 2001a.

M. Nikravesh, BISC and The New Millennium, Perception-based Information Processing, Berkeley Initiative in Soft Computing, Report No. 2001-1-SI, September 2001b.

S. K. Pal, V. Talwar, and P. Mitra, Web Mining in Soft Computing Framework: Relevance, State of the Art and Future Directions, to be published in IEEE Transcations on Neural Networks, 2002.

F. Petry and P. Bosc, Fuzzy databases : principles and applications, Kluwer, Norwell, MA, 1996.

G. Presser, Fuzzy Personalization, in M. Nikravesh and B. Azvine, FLINT 2001, New Directions in Enhancing the Power of the Internet, UC Berkeley Electronics Research Laboratory, Memorandum No. UCB/ERL M01/28, August 2001.

D. Rasmussen and R. R. Yager (1999) Finding fuzzy and gradual functional dependencies with summarySQL, Fuzzy sets and systems 106(2) : 131-142.

G. Salton, Automatic text processing : the transformation, analysis and retrieval of information by computer, Addison-Wesley, Reading, MA, 1989.

G. Salton and M. J. McGill (1983) Introduction to modern information retrieval, McGraw-Hill.

E. Sanchez, Fuzzy logic e-motion, in M. Nikravesh and B. Azvine, FLINT 2001, New Directions in Enhancing the Power of the Internet, UC Berkeley Electronics Research Laboratory, Memorandum No. UCB/ERL M01/28, August 2001.

A. M. G. Serrano, Dialogue-based Approach to Intelligent Assistance on the Web, in M. Nikravesh and B. Azvine, FLINT 2001, New Directions in Enhancing the Power of the Internet, UC Berkeley Electronics Research Laboratory, Memorandum No. UCB/ERL M01/28, August 2001.

S. Shahrestani, Fuzzy Logic and Network Intrusion Detection, in M. Nikravesh and B. Azvine, FLINT 2001, New Directions in Enhancing the Power of the Internet, UC Berkeley Electronics Research Laboratory, Memorandum No. UCB/ERL M01/28, August 2001.

T. Takagi and M. Tajima, Proposal of a Search Engine based on Conceptual Matching of Text Notes, in M. Nikravesh and B. Azvine, FLINT 2001, New Directions in Enhancing the Power of the Internet, UC Berkeley Electronics Research Laboratory, Memorandum No. UCB/ERL M01/28, August 2001.

T. Takagi, A. Imura, H. Ushida, and T. Yamaguchi, "Conceptual Fuzzy Sets as a Meaning Representation and their Inductive Construction," International Journal of Intelligent Systems, Vol. 10, 929-945 (1995).

T. Takagi, A. Imura, H. Ushida, and T. Yamaguchi, "Multilayered Reasoning by Means of Conceptual Fuzzy Sets," International Journal of Intelligent Systems, Vol. 11, 97-111 (1996).

T. Takagi, S. Kasuya, M. Mukaidono, T. Yamaguchi, and T. Kokubo, "Realization of Sound-scape Agent by the Fusion of Conceptual Fuzzy Sets and Ontology," 8th International Conference on Fuzzy Systems FUZZ-IEEE'99, II, 801-806 (1999).

T. Takagi, S. Kasuya, M. Mukaidono, and T. Yamaguchi, "Conceptual Matching and its Applications to Selection of TV Programs and BGMs," IEEE International Conference on Systems, Man, and Cybernetics SMC'99, III, 269-273 (1999).

C. A. Testemale, Database system dealing with incomplete or uncertain information and vague queries, Fuzzy logic in knowledge engineering (Edited by Prade H and Negoita CV), Verlag TUV Rheinland, 1986.

M. Williams (1984) What makes rabbit run ?, J. man-mach. Stud. 2a (1) : 333-352.

Wittgenstein, "Philosophical Investigations," Basil Blackwell, Oxford (1953).

R. Yager, Aggregation Methods for Intelligent Search and Information Fusion, in M. Nikravesh and B. Azvine, FLINT 2001, New Directions in Enhancing the Power of the Internet, UC Berkeley Electronics Research Laboratory, Memorandum No. UCB/ERL M01/28, August 2001.

R. Yager (1991) On linguistic summaries of data, In knowledge discovery in databases, Piatetsky-Shapiro G and Frawley B (Eds.), MIT Press : 347-363.

R. Yager (1996) Database discovery using fuzzy sets, Int. J. of. Intelligence systems 11: 691-712.

J. Yen, Incorporating Fuzzy Ontology of Terms Relations in a Search Engine, in M. Nikravesh and B. Azvine, FLINT 2001, New Directions in Enhancing the Power of the Internet, UC Berkeley Electronics Research Laboratory, Memorandum No. UCB/ERL M01/28, August 2001.

L. A.Zadeh, The problem of deduction in an environment of imprecision, uncertainty, and partial truth, in M. Nikravesh and B. Azvine, FLINT 2001, New Directions in Enhancing the Power of the Internet, UC Berkeley Electronics Research Laboratory, Memorandum No. UCB/ERL M01/28, August 2001 [2001a].

L.A. Zadeh, A Prototype-Centered Approach to Adding Deduction Capability to Search Engines -- The Concept of Protoform, BISC Seminar, Feb 7, 2002, UC Berkeley, 2002.

L. A. Zadeh, " A new direction in AI – Toward a computational theory of perceptions, AI Magazine 22(1): Spring 2001, 73-84

L. A. Zadeh, From Computing with Numbers to Computing with Words-From Manipulation of Measurements to Manipulation of Perceptions, IEEE Trans. On Circuit and Systems-I Fundamental Theory and Applications, 45(1), Jan 1999, 105-119.

L. A. Zadeh (1997) Toward a theory of fuzzy information granulation and its centrality in human reasoning
and fuzzy logic, Fuzzy sets and systems 90(2) : 111-127.

L. A. Zadeh (1999) From computing with numbers to computing with words – From manipulation of measurements to manipulation of perceptions, IEEE trans. on circuit and systems 45(1) : 105-119.

L. A. Zadeh (1983) A computational approach to fuzzy quantifiers in natural language, Co put. Math.Appl. 9 : 149-184.

Y. Zhang et al., Granular Fuzzy Web Search Agents, in M. Nikravesh and B. Azvine, FLINT 2001, New Directions in Enhancing the Power of the Internet, UC Berkeley Electronics Research Laboratory, Memorandum No. UCB/ERL M01/28, August 2001.

Y. Zhang et al., Fuzzy Neural Web Agents for Stock Prediction, in M. Nikravesh and B. Azvine, FLINT 2001, New Directions in Enhancing the Power of the Internet, UC Berkeley Electronics Research Laboratory, Memorandum No. UCB/ERL M01/28, August 2001.

Table 3. Understanding and History of Internet, World Wide Web and Search Engine;

Search Engine and Internet	Date	Developer	Affiliation	Comments
ARPANET	1962 –1969 1970-1973 1974-1981	UCLA	Under Leadership of DARPA	Initially designed to keep military sites in communication across the US. In 1969, ARPANET connected researchers from Stanford University, UCLA, UC Santa Barbara and the University of Utah. Internet community formed (1972). Email started (1977).
ALOHANET	1970		University of Hawaii	
USENET	1979	Tom Truscott & Jim Ellis Steve Bellovin	Duke University & University of North Carolina	The first newsgroup.
ARPANET	1982-1987	Bob Kahn & Vint Cerf	DARPA & Stanford University	ARPANET became “Internet”. Vinton Cerf “Father of the Internet”. Email and Newsgroups used by many universities.
CERT	1988-1990	Computer Emergency Response Team		Internet tool for communication. Privacy and Security. Digital world formed. Internet worms & hackers. The World Wide Web is born.
Archie through FTP	1990	Alan Ematage	McGill University	Originally for access to files given exact address. Finally for searching the archive sites on FTP server, deposit and retrieve files.
Gopher	1991	A team led by Mark MaCahill	University of Minnesota	Gopher used to organize all kinds of information stored on universities servers, libraries, non-classified government sites, etc. Archie and Veronica, helped Gopher (Search utilities).
World Wide Web "alt.hypertext	1991	Tim Berners-Lee	CERN in Switzerland	The first World Wide Web computer code. "alt.hypertext." newsgroup with the ability to combine words, pictures, and sounds on Web pages
Hyper Text Transfer Protocol (HTTP).	1991	Tim Berners-Lee	CERN in Switzerland	The 1990s marked the beginning of World Wide Web which in turn relies on HTML and Hyper HTTP. Conceived in 1989 at the CERN Physics Laboratory in Geneva. The first demonstration

				December 1990. On May 17, 1991, the World Wide Web was officially started, by granting HTTP access to a number of central CERN computers. Browser software became available-Microsoft Windows and Apple Macintosh
	1992			The first audio and video broadcasts:"MBONE." More than 1,000,000 hosts.
Veronica	1993	System Computing Services Group	University of Nevada	The search Device was similar to Archie but search Gopher servers for Text Files
Mosaic	1993	Marc Andeerssen	NCSA (the National Center for Supercomputing Applications); University of Illinois at Urbana Champaign	Mosaic, Graphical browser for the World Wide Web, were developed for the Xwindows/UNIX, Mac and Windows.
World Wide Web Wanderer; the first Spider robot	1993	Matthew Gary	MIT	Developed to count the web servers. Modified to capture URLs. First searchable Web database, the Wandex.
ALIWEB	1993	Martiijn Koster	Now with Excite	Archie-Like Indexing of the Web. The first META tag
JumpStation, World Wide Web Worm .	1993		NASA	Jump Station developed to gathere document titles and headings. Index the information by searching database and matching keywords. WWW worm index title tags and URLs.
Repository-Based Software Engineering (RBSE) Spider	1993		NASA	The first relevancy algorithm in search results, based on keyword frequency in the document. Robot-Driven Search Engine Spidered by content.
	1994			Broadcast over the M-Bone. Japan's Prime Minister goes online at www.kantei.go.jp. Backbone traffic exceeds 10 trillion bytes per month.
Netscape and Microsoft's Internet	1994-1998	Microsoft and Netscape	Microsoft and Netscape	Added a user-friendly point-and-click interface for browsing

Explorer				
Netscape	1994	Dr. James H. Clark and Marc Andreessen		The company was founded in April 1994 by Dr. James H. Clark, founder of Silicon Graphics, Inc. and Marc Andreessen, creator of the NCSA Mosaic research prototype for the Internet. June 5, 1995 - change the character of the World Wide Web from static pages to dynamic, interactive multimedia.
Galaxy	1994	Administered by Microelectronics and computer Technology Corporation	Funded by DARPA and consortium of technologies companies and original prototype by MADE program.	Provided large-scale support for electronic commerce and links documents into hierarchical categories with subcategories. Galaxy merged into Fox/News in 1999.
WebCrawler	1994	Brian Pinkerton	University of Washington	Search text of the sites and used for finding information in the Web. AOL purchased WebCrawler in 1995. Excite purchased WebCrawler in 1996.
Yahoo!	1994	David Filo and Jerry Yang	Stanford University	Organized the data into searchable directory based on simple database search engine. With the addition of the Google, Yahoo! Is the top-referring site for searches on the Web. It led also the future of the internet by changing the focus from search retrieval methods to clearly match the user's intent with the database.
Lycous	1994	Michael Mauldin	Carnegie Mellon University	New features such as ranked relevance retrieval, prefix matching, and word proximity matching. Until June 2000, it had used Inktomi as its back-end database provide. Currently, FAST a Norwegian search provider, replaced the Inktomi.
Excite	1995	Mark Van Haren, Ryan McIntyre, Ben Lutch, Joe Kraus, Graham Spencer, and Martin Reinfried	Architext Sofware	Combined search and retrieval with automatic hypertext linking to document and includes subject grouping and automatic abstract algorithm. IT can electronically parse and abstract from the web.
Infoseek	1995	Steve Kirsch	Infoseek	Infoseek combined many functional elements

		(now with Propel)		seen in other search tools such as Yahoo! And Lycos, but it boasted a solid user-friendly interface and consumer-focused features such as news. Also speed in which indexed Web sites and then added them to its live search database.
AltaVista	1995	Louis Monier, with Mike Burrows	Digital Equipment Corporation	Speed and the first "Natural Language" queries and Boolean operators. It also proved a user-friendly interface and the first search engine to add a link to helpful search tips below search field to assist novice searchers.
MetaCrawler	1995	Erick Selberg and Oren Etizinoi	University of Washington	The first Meta search engine. Search several search engines and reformat the results into a single page.
SavvySearch	1995	Daniel Dreilinger	Colorado State University	Meta Search which was included 20 search engines. Today, it includes 200 search engine.
Inktomi and HotBot	1994-1996	Eric Brewer and Paul Gauthier	University of California-Berkeley Funded by ARPA	Cluster inexpensive workstation computers to achieve the same computing power as expensive super computer. Powerful search technologies that made use of the clustering of workstations to achieve scaleable and flexible information retrieval system. HotBot, powered by Inktomi and was able to rapidly index and spider the Web and developing a very large database within a very short time.
LookSmart	1996	Mr Evan Thornley	LookSmart	Delivers a set of categorized listing presented in a user friendly format and providing search infrastructure for vertical portals and ISPs.
AskJeeves	1997	Davis Warthen and Garrett Gruener	AskJeeves	It is built based on a large knowledge base on pre-searched Web sites. It used sophisticated, natural-language semantic and syntactic processing to understand the meaning of the user's question and match it to a 'question template" in the knowledge

				base.
GoTo	1997	Bill Gross	Indealab!	Auctioning off search engine positions. Advertisers to attach a value to their search engine placement.
Snap	1997	Halsey Minor, CNET Founder	CNET, Computer Network	Redefining the search engine space with a new business model; "portal" as first partnership between a traditional media company and an Internet portal.
Google	1997-1998	Larry Page and Sergey Brin	Stanford University	PageRank™ to deliver highly relevant search results based on proximity match and link popularity algorithms. Google represent the next generation of search engines.
Northern Light	1997	Team of librarians, software engineers, and information industry	Northern Light	To Index and classify human knowledge and has two database 1) contains an index to the full text of millions of Web pages and 2) includes full-text articles from a variety of sources. It searches both Web pages and full-text articles and sorts its search results into folders based on keywords, source, and other criteria.
AOL, MSN and Netscape	1998	AOL, MSN and Netscape	AOL, MSN and Netscape	Search service for the users of services and software
Open Directory	1998	Rick Skrenta and Bob Truel	dmoz	Open directory

Direct Hit	1998	Mike Cassidy	MIT	Direct Hit is dedicated to providing highly relevant Internet search results. Direct Hit's highly scalable search system leverages the searching activity of millions of Internet searchers to provide dramatically superior search results. By analyzing previous Internet search activity, Direct Hit determines the most relevant sites for your search request.
FAST Search	1999	Isaac Elsevier	FAST; Norwegian Company- All the Web	High-capacity search and real-time content matching engines based on the All the Web technology. Using Spider technology to index pages very rapidly. FAST can index both Audio and Video files.

Distributed Architecture for Modeling and Simulation of Autonomous Multi-agent Multi-Physics Systems[1]

Prasanna Sridhar and Mo Jamshidi

Autonomous Control Engineering (ACE), University of New Mexico – Albuquerque

moj@cybermesa.com

Abstract: The need for Modeling and Simulation (M&S) is seen in many diverse applications such multi-agent systems, robotics, control systems, software engineering, complex adaptive systems, homeland security, and many others. In this paper we introduce an architecture for distributed simulation of multi-agent systems called Virtual Laboratory (V-Lab®), based on discrete event system specification (DEVS). V-Lab® is a test bed for many control algorithms and allows the user to demonstrate the working of several soft-computing methodologies like fuzzy logic, learning automata, neural networks, genetic algorithms, etc. applied to multi-agent systems. DEVS defines a framework for discrete event simulation and V-Lab® defines a framework for distributed simulation for multi-agent autonomous systems.

Keywords: Distributed Simulation, Model Continuity, Soft-computing, DEVS, V-Lab®, I-DEVS, CORBA, RMI, HLA

1. Introduction

Modeling of a real-world system is the first step in simulation. Simulation models may be based on physical, mathematical, or logical representations; expert rules;

[1] This work was supported, in part, by NASA Grant NAG 2-1547.

empirical data, etc. in order to describe the behavior of the system being modeled. Simulation is the process of generating the behavior of the model using computing systems (computer, algorithm or human mind). In addition to its use as a tool to better understand and optimize performance and/or reliability of systems, simulation is also extensively used to verify the correctness of designs. Since the data available in the real world is abundant, we define a specification of the conditions called experimental framework, under which the system is observed. System, experimental framework, models and simulators are basic entities of modeling and simulation.

During the design and implementation of a simulator, various techniques and strategies may be adopted to model the behavior of a given system. Simulators are designed using either *continuous* or *discrete-event* [1] techniques to simulate a given system. Continuous simulators are characterized by the extensive use of mathematical formulae, which describe how a simulated component responds when subjected to various conditions. The discrete event modeling provides a general framework for time-oriented simulations of systems. Within the context of discrete-event simulation, an event is defined as an incident, which causes the system to change its state in some way. What separates discrete-event simulation from continuous simulation is the fact that the events in a discrete-event simulator can occur only during a distinct unit of time during the simulation - events are not permitted to occur in between time units. Discrete Event System Specification (DEVS) is a formalism used to create simulation model for discrete event systems.

2. Statement of Problem

As the simulation process becomes more complex and large, it is necessary to divide them into smaller pieces, which are manageable. Applying layered pattern to the design of simulation breaks the simulation into several interconnected layers, which gives *modularity*, *distributability*, and *separability*. Such a layered approach of simulation is provided by V-Lab® [2]. In order to fully utilize and implement the control algorithms for multi-agent systems, an environment for simulation has to be first created. V-Lab® provides a robust environment for testing various control algorithms on distributed systems. Unlike other technology [3], V-Lab® is not confined to specific simulation, but is generic to any multi-agent simulation. In order to fully understand V-Lab® and its modules, we first introduce some definitions and concepts of DEVS.

2.1 Glossary of Terms

V-Lab: [Virtual Laboratory for Autonomous Agents] A distributed simulation environment for modeling and simulation of multi-agent multi-physics systems using discrete event systems [3].

DEVS: [Discrete Event System Specification] A formalism to create simulation models for discrete event systems [4-5].

I-DEVS: [Intelligent DEVS] A fusion of Soft-computing methodologies and discrete event system specification.

Agents: An agent is any entity that can be viewed as perceiving its environment through sensors and acting upon its environment through effector [6].

Autonomous Agents: These are computational systems that inhabit some complex dynamic environment, sense and act autonomously in this environment, and by doing so, realize a set of goals or tasks for which they are designed [6].

2.2 DEVS

The Discrete Event System Specification (DEVS) modeling and simulation environment, DEVSJAVA®, was developed by the Arizona Center for Integrative Modeling and Simulation, headed by Zeigler and Sarjoughian. It was created to provide a robust and generic environment for modeling and simulation applications employing single workstation, distributed, and real-time platforms. The DEVSJAVA 2.7 [4,5] environment provides Java classes that encapsulate all the functionality that is needed to create a module which is fully capable of being connected to other modules in a meaningful relationship, regardless of which machines these modules are located on.

Layered on top of the DEVS environment are the models that a developer would create to compose a simulation. These models are divided into two categories: atomic and coupled. Atomic models compose the functionality of the basic units in a simulation. Using these *atomic* models as building blocks, *coupled* models build up the simulation by linking them together. In addition to containing atomic models, coupled models may also be used as building blocks in other coupled models. Simulations using DEVS are collections of models composed in a hierarchical fashion. For instance, a DEVS coupled model ABC, such as that in Figure 1, can be constructed from an atomic and a coupled model, A and BC, respectively. BC is itself a coupled model that is constructed from two atomic models, B and C. The ABC model clearly has a hierarchical construction from its elements A, BC, B and C.

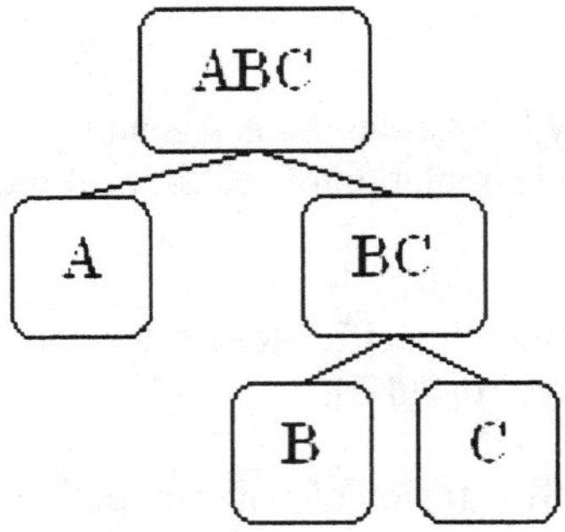

Figure 1. Hierarchical tree for model ABC

The hierarchical specification defines which models are included as sub-models for any given coupled model, but it does not define how these sub models interconnect with the parent model or with each other. This information is defined in the form of ports and couplings. Each model may have an arbitrary number of both in-ports and out-ports that can be coupled to the input and output ports on other models.

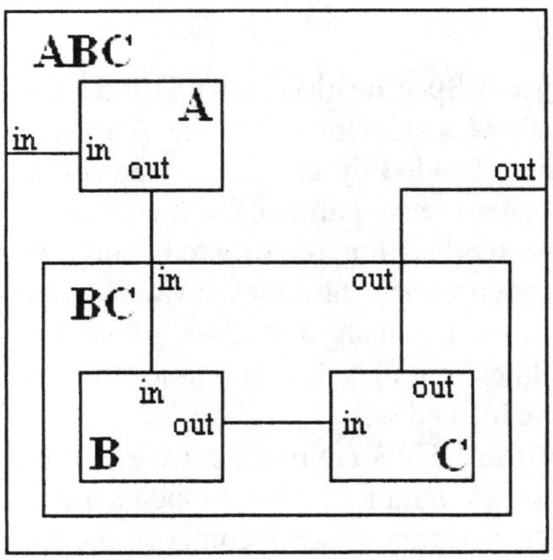

Figure 2. Coupling relation for model ABC

Figure 2 illustrates one possible connection that the ABC model could have. In this example, the input into ABC is coupled with the input in A. In effect, this transfers all messages coming into the ABC model on its in port to the in port on the atomic model A. The output of A is then coupled to the input of the BC model and the output for BC is coupled to the output of ABC. Similarly, since BC is also a coupled model, coupling information for BC would redirect any input coming into the BC model to the in port of the atomic model B. Messages passing out of B would be sent to the in port of C and messages passing out of C would pass out of the BC model and then out of the ABC model itself. Models need not be limited to a single input and output, however, and can have any number of input and output ports.

Formally, an atomic model is represented by a structure M=<X, S, Y, δint, δext, λ, ta> [7] such that X is the set of input values; S is the set of possible states; Y is the set of output values; δint: S→S is the internal state transition; δext: S×Q×X→S | Q={e | 0 < e < sigma}, such that sigma is time to the next internal transition, is the external state transition; λ:S→Y is the output function and ta is the time advance function.

3. Architecture for Distributed Simulation of Multi-agents: V-Lab®

The V-Lab® environment consists of 4 distinct software layers, as Figure 3 illustrates, and each of these layers fills a specific role in the simulation. The foundation of the simulation consists of the operating system and the network code needed to operate the networking hardware, which in turn allows machines to communicate over a network. Using this functionality, a middleware such as the Remote Method Invocation (RMI)[8], Common Object Request Broker Architecture (CORBA)[9], High Level Architecture (HLA)[10] or even sockets acts to solve the problem of how to use the network to connect different portions of a simulation together. While any middleware provides a useful tool for software interconnection, it does not provide the architecture needed to arrange components of a simulation into discrete structures. Using the DEVS environment, V-Lab® defines an appropriate structure in which to organize the elements of DEVS for a distributed agent based simulation. It separates the main components into different categories and defines the logical structure in which they communicate. It also provides the critical objects needed to control the flow of time, the flow of messages, and the base class objects designers will need to create their own V-Lab® modules.

Just as the middleware defines the core functionality of module intercommunication, and DEVS defines the hierarchical and compositional organization of the modules, V-Lab® defines the logical structure in which to implement these modules. Each successive layer defines a more specific organizational structure for the simulation than the last. However, each layer also restricts the domain of problems that can be addressed by the architecture. DEVS may be a valid option for constructing a simulation with millions of cells, but V-Lab® is not. Specifically,

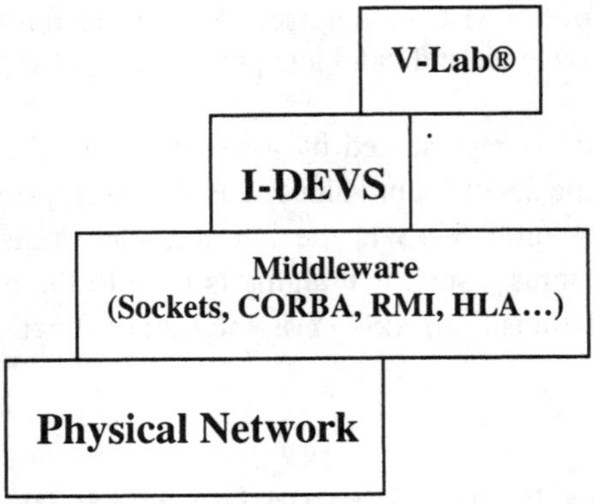

Figure 3. Distributing simulation layers using DEVS

V-Lab® is an architecture that defines a logical structure for simulations with a relatively small number of agents interacting in complex ways. Likewise, in DEVSJAVA 2.7 the user has few restrictions when specifying inter-module communication whereas with V-Lab®, multi-agent simulations require conformance to V-Lab®'s structured communication protocol. Following the rules arising from the architecture defined by V-Lab® allows simulation designers to create a simulation that is modular, extendable and allows for the re-use of pieces of a simulation in future simulations by providing a level of indirection between components of the simulation. Critical backbones of V-Lab® will be tools from soft computing (SC) paradigms like fuzzy logic (FL), neural networks (NN), genetic algorithms (GA) and stochastic learning automaton (SLA). DEVS and SC together constitute what we call IDEVS, intelligent discrete-event systems specification to be detailed in next section.

4. Modules of V-Lab®

Modular architecture of V-Lab® is as presented in Figure 4.

The different modules of V-Lab® are Agent, Controller, Physics, Dynamics, Terrain, SimMan and SimEnv. V-Lab® provides plug-and-play for agents to be added and removed. The architecture looks like *Mediator* pattern in "Design Patterns" [11]. Each of these modules is implemented as a DEVS atomic or coupled model. Essential parts of V-Lab are SimEnv/SimMan, I-DEVS components and Distributed DEVS. But it also provides user an environment to write other modules like Dynamics, Terrain and Physics to suit his simulation

Agents: *If the simulation is robotic, then each robot or its dynamics can be thought of as an agent. The equations or the dynamics governing the motion of the robot is implemented as atomic or coupled model in DEVS. Each of these agents*

can have their own Control Algorithm. This is where the V-Lab® would play an important role as test bed for several control algorithms. Specifically, any of the soft computing methodologies can be implemented and tested using this architecture. Since V-Lab® is modular, user can de-couple the existing control algorithm and plug a new control (like a classic PD, PID...) method for his simulation. Later on we see Neural-Network DEVS (NN-DEVS) which can be one of the control algorithm for V-Lab® Simulation.

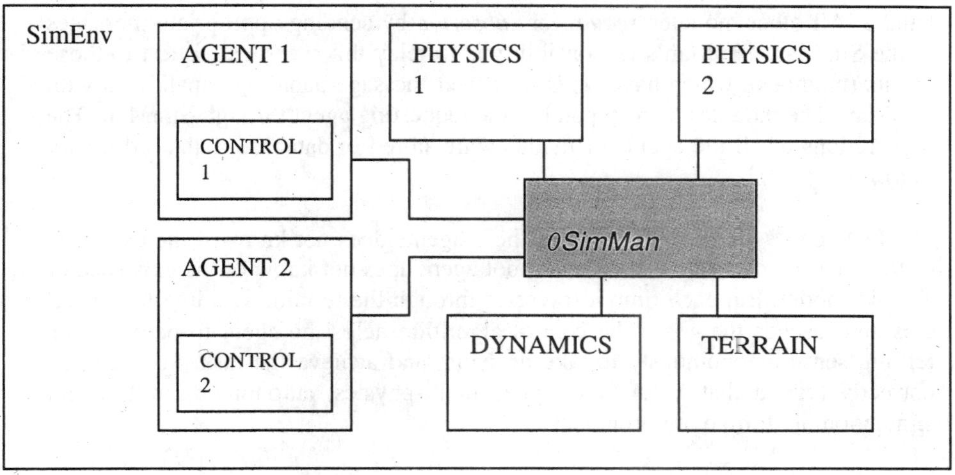

Figure 4. Architecture of V-Lab

Physics: *Several physics model can exist in a single simulation. For example, in robotic simulation, gravity, friction, impact of obstacle collision, force (wind, water...), acceleration/deceleration, many others can be considered. Simulation involving such a complex system of physics equations is essentially multi-physics simulation, which can be demonstrated using V-Lab®.*

Terrain: *This model consists of type of terrain in which the agents traverse and their control algorithm is tested. For example, in robotic simulation, an autonomous agents might come across several obstacles, valleys, elevations, etc., which it has to avoid and traverse to the goal based on the decision making rules from the control algorithm. Terrain generation and terrain traversal algorithms are considered in this model. Again several soft-computing methods like fuzzy logic are applied for terrain traversal [12].*

5. SimEnv/SimMan: Heart of V-Lab:

SimEnv (Simulation Environment) is the high level coupled model. V-Lab Simulation kicks off by starting SimEnv which instantiates all other modules. SimMan (Simulation Manager) acts as message relay for all other modules. With SimMan, V-lab has the property of separability, i.e., the modules do not know the existence of other modules. For example, Agent 1 doesn't know whether Agent 2 exists or Physics module exists. The modules simply *publish* the messages that they can handle. All other modules *request* or *subscribe* by sending appropriate messages to the SimMan. SimMan's responsibility is to relay this message request to those agents (modules), which have *registered* their message handling capabilities with SimMan. The modules then respond to the requesting agent through SimMan. The registrations of all the agents or modules are stored in database (balanced binary tree).

In V-Lab® architecture, each of these agents does not know of the existence of the other models. For example, a robot agent does not know of the existence of Terrain model. But each time it traverses through the terrain, avoiding the obstacles and reaches the goal. The control algorithm helps the agent to perceive the terrain, sense autonomously to take decision, and achieve the task. So it can be correctly argued that V-Lab® provides multi-physics, autonomous multi-agent simulation platform or environment.

6. MODEL CONTINUITY

Model Continuity [13] refers to the ability to use the same model of system throughout its design phase. Such a method helps in designing and testing a system through phases or in a step-wise process. It's a kind of iterative process in which model from the previous phase would act as proof-of-concept for development of model for the next phase. In order to implement V-Lab® we adopt such a continuity feature to develop complete simulation environment from the initial design as shown in Figure 5.

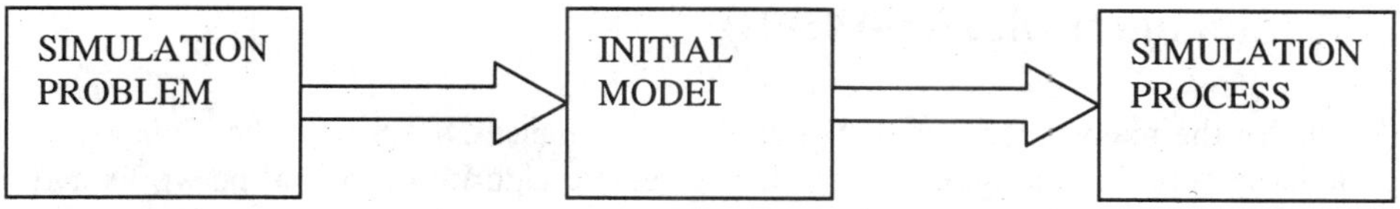

Development of I-DEVS:

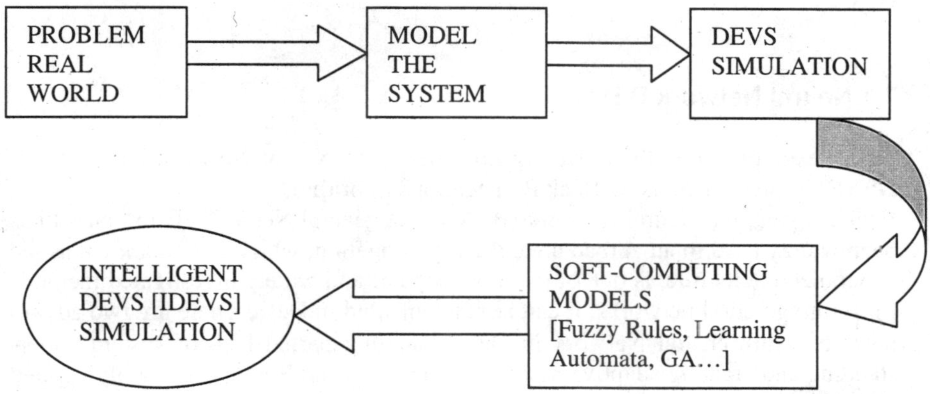

Development of V-Lab®:

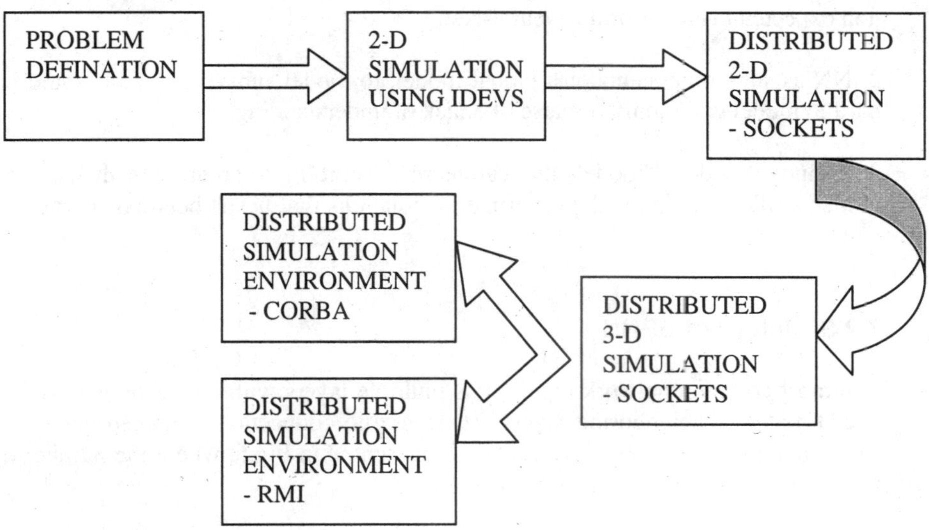

Figure 5. Model Continuity

7. Intelligent DEVS (I-DEVS)

One of the main objectives of V-Lab®. is to enhance DEVS with the tools available by soft computing, e.g. fuzzy logic, genetic algorithms, neural networks and stochastic learning automaton by introducing them in discrete-event simulation environment. In this section four paradigms are introduced within DEVS. We denote this intelligent DEVS as I-DEVS [14]. Here, we will cover Neural Network-DEVS.

7.1 Neural Network DEVS

As an example of I-DEVS we consider here, the Neural Network-DEVS (NN-DEVS) implemented using Back Propagation Algorithm.
Back Propagation Neural Network (BPNN) is a general Neural Network and it has been widely used in an abroad area. Back-propagation, which is also known as the generalized delta rule, is one of the most popular and widely investigated methods for training neural networks. It can be implemented in Devs. There are two advantages of using Neural Network in Devs: handling partial lack of system understanding and creating adaptive models (models that can learn). It is mainly applied in three-areas [15]:

1. Concurrent simulation, where results of a Neural Network (NN) model are compared with results of a less realistic but validated common model to avoid a non expected behavior of the Neural-Net.

2. NN as sub-components of a global model, to model subsystems that would be hard to model commonly because of a lack of understanding.

3. Adaptive models, "models that can learn", according to an error feedback such model would be able to adapt runtime to situations that hasn't been taken into account.

7.2 Structure of BPNN

The most common network topology is multiple layers with connections only between nodes in neighboring layers. There are no connections between nodes located in a common layer. Its structure is presented in Fig.6, where the number of hidden layers can be one or more than one.

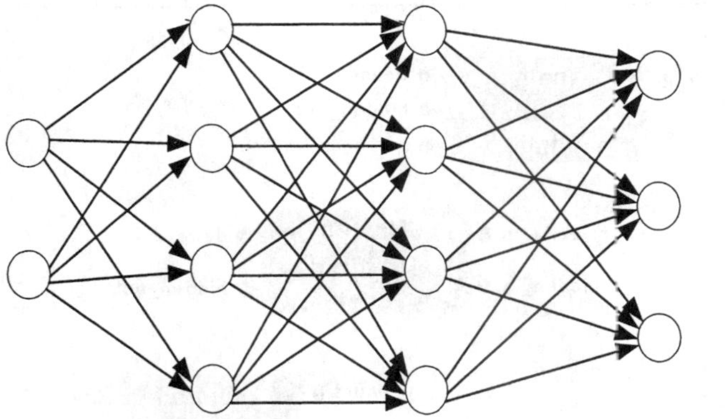

Figure 6. A Typical Neural Network with 2 Hidden Layers

There are two passes in BPNN:
Pass 1: Forward Pass - Present inputs and let the activations flow until they reach the output layer.
Pass 2: Backward Pass - Error estimates are computed for each output unit by comparing the actual output (Pass 1) with the target output. Then, these errors flow from the output layer to the hidden layers. Error estimates are used to adjust the weights in the hidden layer and the input layer.
The foundation of the back-propagation learning algorithm is the nonlinear optimization technique of gradient descent on the sum of squared differences between the actual output in the output nodes and the desired output. The detail about it can be found in several neural network books.

7.3 Implementation of BPNN in Devs

A 4 layer BPNN can be implemented in DEVS, corresponding to the topology of Fig.6. It is composed of input, hidden and output models. Each atomic model includes forward and backward computation (see Figure 7). It has training and testing phases. You can provide training data to inputs and set a stopping criterion to get a desired performance. After the training phase the trained weights can be used to test data. It can be extended to include more hidden layers by adding more hidden models. It can also be decreased to 3 layer BPNN if it is just composed of input and output models.

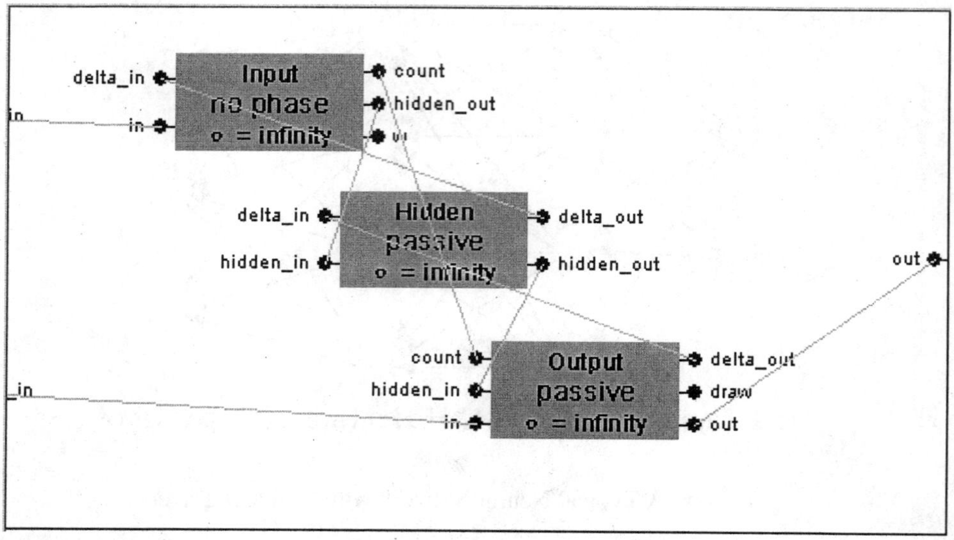

Figure 7. DEVS Atomic model implementing the 4-Layer NN

Other elements of I-DEVS like fuzzy logic-DEVS, GA-DEVS, SLA-DEVS can be found in [14].

8. DISTRIBUTED SIMULATION

The original plan of the V-Lab® project was to integrate CORBA into DEVS for distributed simulation. CORBA is an international standard and is language independent. Additionally, in CORBA, object references are used and methods are invoked remotely so the object stays on the server (client). This is in contrast to sockets where the whole object or enough to reconstruct it at the other end has to be sent. This will impact performance. Also with CORBA, the user does not have to know details about the server, specifically its IP address or socket port number as is necessary with the current socket paradigm. In other words, the remote objects are located through a central service so there should be better location transparency. However, CORBA has proven to be somewhat cumbersome, so it is not as appealing compared to sockets in terms of simplicity. Other possibilities exist, such as Java RMI. In fact, the layered architecture of V-Lab® allows DEVS to be implemented over a wide variety of middleware services, including recent innovations such as peer-to-peer protocols such as JXTA

8.1 Technologies for Middleware:

Sockets: We employed Java Sockets in-order to demonstrate the working of V-Lab® on distributed systems. DEVS provides capability for distributed simulation using sockets (see Figure 8).

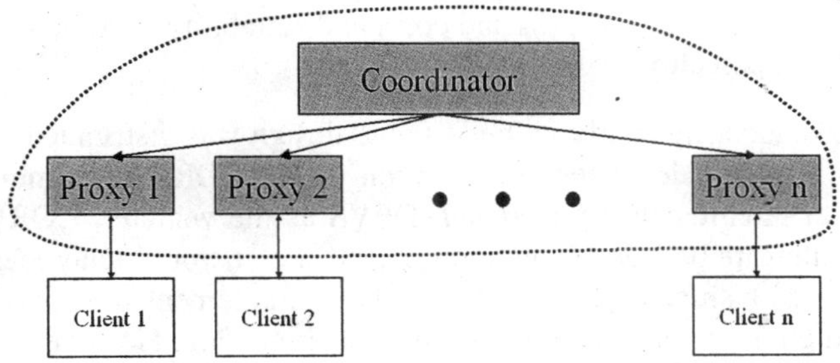

Figure 8. Architecture of DEVS for Distributed Simulation

The user who is interested in distributed simulation need not worry about the inner working of these classes such as *proxy* and *Coordinator.* One needs to do is instantiate coordinator server on the server host and then instantiate a client for each component on the remote host(s). Here is a sample example of the start-up code for the server

for the server

```
public static void main(String[] args)
{
Robot robo = new Robot();  // the top level coupled model
new coordServer(robo, Integer.MAX_VALUE);
// note:  2nd argument is the number of simulation iterations
}
```

And the corresponding start-up client code:

```
   public static void main(String[] args) {
new clientSimulator(new RobotDyn("RobotDyn",1,25,25,Math.PI/4,1,1));
           new clientSimulator(new Controller("Controller"));
new clientSimulator(new Plotxy("PLOT XY"));
           new clientSimulator(new Terrain("Terrain"));
           new clientSimulator(new IRSensor("IR", 3 , 1 , 0 , 0 , 15 ));
           new clientSimulator(new CellGridPlot("Motion Plot",1,100,100));
   }
```

The `clientSimulator` creates the TCP socket and connects to the IP address of the server provided.

RMI/CORBA/HLA: Several other middleware technologies are currently being worked on for distributed simulation such as Remote Method Invocation (RMI), CORBA, HLA and others. The pros and cons, and suitability of these middlewares are discussed in this section.

One of the tasks identified early in the V-Lab® design was distributed simulation. The V-Lab® proposal describes the intention to put CORBA (Common Object Request Broker Architecture) underneath DEVS as the vehicle. CORBA would act as the middleware between simulation objects, i.e. it would relay messages between models and simulators. DEVS has been implemented to execute over CORBA in an environment for real-time system simulation. However, the distributed capability via CORBA is not supported in the latest version of DEVSJAVA® 2.7 [5]. This version is based on a new family of JAVA packages collectively referred to as GenDEVS. Instead of employing CORBA as middleware, GenDEVS provides distributed simulation with TCP/IP sockets as the middleware.

Drawbacks of Sockets: Unfortunately, straight GenDEVS distributed simulation has a limitation that all messages must be strings. This makes it difficult for the theme example because it makes extensive use of double values and arrays in the messages. In a non-distributed environment, it is no problem to simulate with these typed messages because they are simply objects: the simulator just passes object references between models and the models can share them since they share one address space. But when running in the as-is

GenDEVS distributed environment, typed messages will not work because it expects all messages to be strings. In order to eliminate this *encoding* and *decoding* of messages, we have to use some techniques like *Object Serialization* [16]. What if there are technologies, which provide this Object serialization and user need not have to worry about IP address, encoding and decoding mechanisms? ; CORBA or RMI would be good solution.

JAVA RMI: RMI enables the programmer to create distributed Java technology-based to Java technology-based applications, in which the methods of remote Java objects can be invoked from other Java virtual machines. RMI has its own native Object Request Broker (ORB) and eliminates the need to write an IDL (Interface definition Language) unlike in CORBA. RMI removes most of the drawbacks of the socket [16] and provided services similar to CORBA. Only drawback of RMI is its language dependability on JAVA. CORBA plays a good role as middleware where inter-operability is need between different programming languages on server and client. Since the simulation is completely based on GenDEVS, which is implemented in JAVA, RMI can be good choice as middleware as compared to CORBA [17,18]. HLA is not an international standard for middleware technology yet. A more detailed insepection on HLA, CORBA and RMI can be found in several research papers [3,19,20].

9. Experimental Results

We wish to coordinate all elements of IDEVS by performing a multi-agent distributed robotic simulation in a 2-D environment. We call it *Theme Example* here. The objective of this example is to demonstrate and test the various modules of the IDEVS within the proposed V-Lab® architecture in a multi-physics multi-agent distributed simulation. This example allowed one to test some soft computing methods for autonomous agents in DEVSJAVA® 2.7 environment. Autonomous control algorithms are used to control the maneuvering of the rovers and avoid the obstacles to reach the goal position.

The simulation procedure was divided into several modules (see Figure 9) each having specific functionality. Each of these modules can be thought of as a DEVS atomic or coupled module. For Example, a "Robot Dynamics" model was an atomic model, which implemented the kinematics of the robot. Having such a modular approach enables easier implementation of distributed simulation across several machines and also helps to update any modules as and when required.

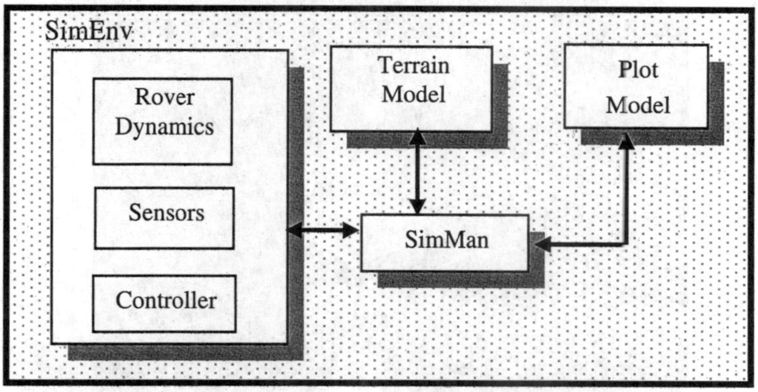

Figure 9. Block Diagram of different modules of *Theme Example*

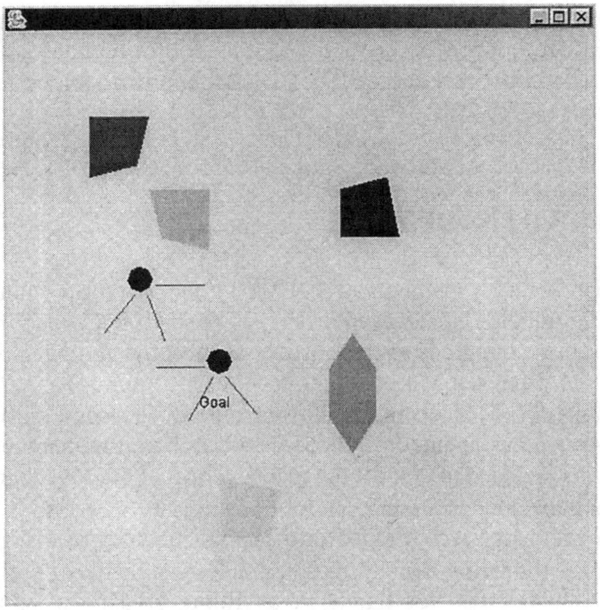

Figure 10. Graphical 2-D Display of Robotic Simulation

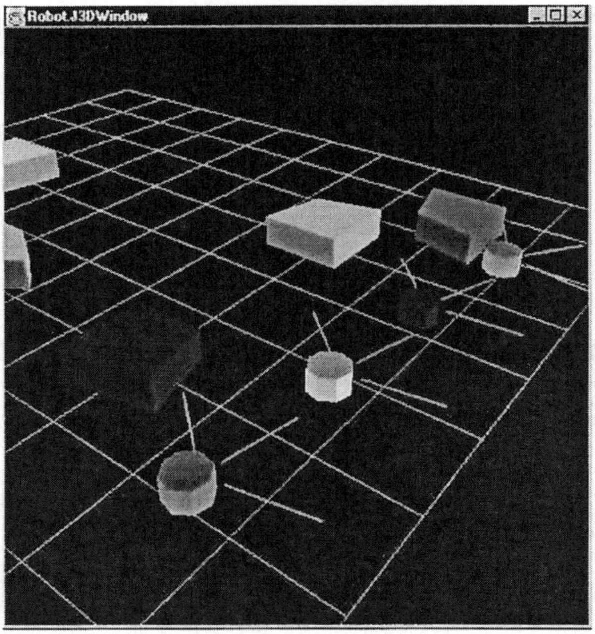

Figure 11. 3-D Display of Robotic Simulation

Figure 10 shows the *Theme Example* results (robotic simulation) of 2 rovers reaching a goal position avoiding the obstacles (convex polygons). Each rover has three Infra-red (IR) sensors through which they detect the distance of obstacle, decelerate and avoid the obstacles based on control algorithms. Fuzzy Rules were written to demonstrate the control of rovers. Several Robots can be added to the simulation creating swarms of robots and each robot can have many sensors. This 2-D simulation runs on distributed machines using sockets.

Figure 11 shows the 3-D display of the *Theme Example* with several rovers (with IR-sensors) maneuvering to the goal avoiding the polygonal obstacles. The 3-D visualization was implemented using JAVA-3D.

10. CONCLUSIONS

In this paper, we described the necessity for simulation environment for complex simulation and a detailed overview of architecture of V-Lab®, which solves such a complex simulation problem. Distributability and Modularity are key features of V-Lab®, which makes simulation more manageable. Such an environment or platform provides user to test several different intelligent or soft computing paradigms with decision and control being one of the applications. With several competing technologies for distributed computing are available, one has to look into standards, international acceptance, ease of use, and suitability to V-Lab®. CORBA or RMI are good choice for development of distributed simulation for V-Lab®. The control algorithms simulated will be tested on hardware platform using ActivMedia Robotics Pioneer II robots.

11. ACKNOWLEDGEMENTS

The author wishes to sincerely thank additional members of the V-Lab® team at the ACE Center, especially John Kitzinger, Shahab Sheikh-Baei, Yan Wang, Tao Song, Shan Xie, Scott Beatty and Umesh Dole. Their dedication has resulted a wonderful work in short time.

REFERENCES

[1] Banks, J, Carson II, J.S., Nelson, B.L., Nicole D.M., "Discrete-event System Simulation", Prentice Hall, 2001.

[2] El-Osery, A., J. Burge, M. Jamshidi, A. Saha, M. Fathi and M. Akbarzadeh-T. "V-Lab – A Distributed Simulation and Modeling Environment for Robotic Agents – SLA-

Based Learning Controllers," *IEEE Transactions on Systems, Man and Cybernetics*, Vol. 32, No. 6, pp. 791-803, 2002

[3] Pasquarelli, A., "Technologies for Distributed Simulation: CORBA and HLA", www.ssgrr.it/en/ssgrr2000/papers/193.pdf , Center of Information Technologies of Alenia Marconi Systems, Rome.

[4] Zeiglar, B. P., Praehofer, H., Kim, T.G., "Theory of Modeling and Simulation", Second edition, Academic Press, Boston, 2000.

[5] Arizona Center for Integrative Modeling and Simulation, "DEVSJAVA 2.7software", http://www.acims.arizona.edu/SOFTWARE/software.shtml

[6] Honavar, V. "Intelligent Agents and Multiagent Systems". Based on material from Intelligent Agents. A tutorial presented at IEEE CEC 99.

[7] Zeiglar, B.P. and Sarjoughian, H., "Introduction to DEVS Modeling and Simulation with JAVA: A Simplified Approach to HLA-Compliant Distributed Simulations", ACIM, *www.acims.arizona.edu*

[8] Wollrath, A. and Waldo, J. "Trail: RMI", http://java.sun.com/docs/books/tutorial/rmi/index.html

[9] The Object Management Group, "CORBA BASICS," electronic document, http://www.omg.org/gettingstarted/corbafaq.htm

[10] US Department of Defense, "High Level Architecture", https://www.dmso.mil/public/transition/hla/

[11] Gamma, E., Helm, R., Johnson, R., and Vlissides, J., "Design Patterns: Elements of Reusable Object-Oriented Software", Addison Wesley Publication, Oct 1994.

[12] Jamshidi M., A. Zilouchian, *Intelligent Control Systems using Soft Computing Methodologies,* CRC Press, Boca Raton, FL, 2001.

[13] Hu, X and Zeiglar, B.P "An Integrated Modeling and Simulation Methodology for Intelligent Systems Design and Testing", *Proc. of PERMIS'02*, Gaithersburg, Aug, 2002.

[14] Jamshidi, M., Sheikh-Bahaei, S., Kitzinger, J., Sridhar, P., Xia, S., Wang, Y., Liu, L., Tunstel, E. Jr, Akbarzadeh, M., El-Osery, A., Fathi, M., Hu, X., and Zeigler, B. P., "A Distributed Intelligent Discrete-Event Environment for Autonomous Agents Simulation", Chapter 11, "Applied Simulation", Kluwer Publications 2003.

[15] Filippi, J-B. Bisgambiglia, P., Delhom, M., "Neuro-Devs, an Hybrid Environment to Describe Complex Systems", ESS '2001 13th European Simulation Symposium and Exhibition

[16] Mahmoud, Q.H., "Transporting Objects over Sockets", http://developer.java.sun.com, Dec 2001.

[17] Curtis, D., "Java, RMI and CORBA", White Paper, www.omg.org/library/wpjava.html, 1997

[18] Reilly, D., "Java RMI and CORBA: A Comparison of Two Competing Technologies", http://www.javacoffeebreak.com/articles/rmi_corba/

[19] Buss, A., Jackson, L., "Distributed Simulation Modeling: A Comparison of HLA, CORBA, and RMI", Proceedings of 1998 Winter Simulation Conference.

[20] Mahmoud, Q. H., "Distributed Programming with Java", Chapter 12, Manning Publications Co., Sept 1999.

Fuzzy Thesauri for and from the WWW

Martine De Cock[1], Sergio Guadarrama[2], and Masoud Nikravesh[3]

[1] Fuzziness and Uncertainty Modelling Research Unit
Dept. of Applied Mathematics and Computer Science
Gent, Belgium
email: martine.decock@ugent.be

[2] Dept. of Artificial Intelligence and Computer Science,
Universidad Politécnica de Madrid
Madrid, Spain
email: sguada@dia.fi.upm.es

[3] Berkeley Initiative in Soft Computing (BISC)
Computer Science Division, Dept. of EECS
University of California
Berkeley CA 94704, USA
email: nikravesh@cs.berkeley.edu

Abstract. We revisit some "old" strategies for the automatic construction of fuzzy relations between terms. Enriching them with new insights from the mathematical machinery behind fuzzy set theory, we are able to put them in the same general framework, thereby showing that they carry the same basic idea.

1 Introduction and Motivation

The remarkable growth of the World Wide Web (WWW) since its origin in the 1990's calls for efficient and effective tools for information retrieval. Attempting to deal with the overwhelming amount of information provided on billions of webpages nowadays does not necessarily imply that we have to develop entirely new technologies from scratch. In the 1970's and 1980's initial research was performed on the retrieval of information from modest text collections, using fuzzy relations to represent associations between terms on one hand, and between terms and documents on the other. Since then the fuzzy mathematical machinery (i.e. fuzzy logical operators, fuzzy similarity measures, operations with fuzzy relations etc.) has come of age.

In contrast to structured databases, most of the information on the WWW is developed to be read and interpreted by human beings rather than by machines. This information, presented in various ways such as natural language, images and video, is often referred to as unstructured or semistructured, terms which apply on the level of individual documents. It seems necessary to build some kind of structure to be able to perform an efficient and effective search. In its easiest form, this structure is an index consisting of terms and pointers to documents containing those terms, or a document–term relation viewed as a matrix in which each element corresponds to the number of occurences of the term in the document. The use of an index or a document–term relation makes the search more efficient than having to go through each document for each keyword based query. However in this approach, documents will not be returned as search results if they do not contain the exact keywords of the query. To satisfy users who expect search engines to come up with "what they mean and not what they say", more sophisticated techniques are needed to enhance the structure which was built automatically on the document collection.

In an approach which even dates back to [7], on top of the document–term relation, one or more term–term relations are provided. The terms are assumed to denote concepts, and the relations between them represent associations such as synonymy, specification, and generalization. Information retrieval certainly does not have a monopoly on these structures; in fact they seem to pop up in many domains that require the semantical representation of language (such as knowledge discovery using association rules, and natural language processing techniques such as machine translation). Throughout the years these kind of structures have been given different names, such as thesauri, taxonomies, or ontologies, sometimes linked to the domain in which they are applied or the relations they represent. We refer to [13] for a discussion and overview of some of this terminology, and the role of ontologies for the semantic web.

Like many relations in real life, relations between terms (or concepts) are a matter of degree. Some terms are related, some terms are not, and in between there is a gradual transition from not "being related" to "being related". Furthermore a term a can be related to a term b, and related to another term c to a lower degree. Therefore it seems intuitively more justifiable to represent associations between terms by $T^2 \rightarrow [0,1]$ mappings (T being the universe of terms), i.e. by fuzzy relations, instead of traditional relations. These fuzzy relations can be used for query expansion: instead of only documents containing exact keywords from the query, also those containing related terms are retrieved. This can be done with or without the knowledge of the user. In the former case lists of related keywords are presented to the user, which he can choose from to refine his query if he is not satisfied with the results obtained so far. Another interesting application for which fuzzy term–term relations are useful, is the clustering of web documents. This helps to present search results in an alternative way, different from the commonly used linearly ordered list.

In early and even in contemporary approaches to information retrieval, these term–term relations are assumed to be given, to be made by an expert. Although this approach is feasable for smaller document collections and has even been applied for parts of the WWW (the open directory project, Yahoo) it can hardly be called a flexible and efficient way for a large and constantly evolving collection of documents such as the WWW. As a more recent trend one tries to build relations for a broader domain from already available domain specific relations (e.g. for the medical domain, for a technical domain) and/or combining them with (multilingual) dictionaries and dictionaries of synonyms ([2], [6], [18]).

Although dictionaries in all languages over the world might be on of the biggest and oldest efforts of mankind directed towards the construction of relations between words, they do certainly not reflect at the same speed the continuous evolution of the assocation between words (concepts) in the human mind. At the time we are writing this paper, probably no current dictionary gives any evidence for a relation between "Schwarzenegger" and "governor", although there has recently grown a strong association between these two words because of the recall election in the state of California, USA (October 7, 2003). This relation is reflected in a high number of documents on the internet dealing with both words: on October 17, 2003, i.e. 10 days after the election, Google claims to have about 932,000 results for the query *governor Schwarzenegger*, compared to 3,410,000 results for *Schwarzenegger* and 15,600,000 results for *governor*.

This paper deals with techniques that can get this kind of useful information which is out there on the WWW fully automatically and unsupervised. Great advantages of an automated process for the construction of fuzzy relations between words include objectivity, high speed and low cost. Indeed such a process is not influenced by background knowledge or point of views of a few experts: the only data given to the system is a collection of hyperlinked webpages that reflect knowledge of all the people involved in making them. In theory, machines can process these documents at a lower cost and a higher speed than human experts, to come up with fuzzy relations overnight that reflect the most recent trends in society. The important question that remains however is if they can be as effective as humans in doing so.

As a possible starting point for such a study, in this paper we gather shattered research ideas for the automatic construction of fuzzy relations between terms. Putting them in a more general framework already sheds new light on the matter (among other things that many researchers individually are doing very similar work, apparently unaware of the existence of a general framework). As such we give an idea about where the theoretical research on the construction of fuzzy thesauri has taken us so far, about a decade after the origin of the WWW. We hope that, by contributing to a solid common starting point, we can speed up further research.

2 Basic Concepts

Throughout this paper let $\mathcal{T}$ denote a triangular norm (t-norm for short), i.e. an increasing, commutative and associative $[0,1]^2 \to [0,1]$ mapping satisfying $\mathcal{T}(1,x) = x$, for all x in $[0,1]$. Furthermore let $\mathcal{S}$ denote a triangular conorm (t-conorm for short), i.e. an increasing, commutative and associative $[0,1]^2 \to [0,1]$ mapping satisfying $\mathcal{S}(0,x) = x$, for all x in $[0,1]$.

Finally let $\mathcal{I}$ denote an implicator, i.e. a $[0,1]^2 \to [0,1]$–mapping $\mathcal{I}$ decreasing in its first, and increasing in its second component, and satisfying $\mathcal{I}(0,0) = 1, \mathcal{I}(1,x) = x$, for all x in $[0,1]$. Examples of these so–called fuzzy logical operators can be found in modern text books on fuzzy set theory and fuzzy logic (see e.g. [17]). Recall that the $\mathcal{T}$–intersection and the $\mathcal{S}$–union of fuzzy sets A and B in X is defined as

$$(A \cap_{\mathcal{T}} B)(x) = \mathcal{T}(A(x), B(x))$$

$$(A \cup_{\mathcal{S}} B)(x) = \mathcal{S}(A(x), B(x))$$

for all x in X. The cardinality of a fuzzy set A in a finite universe X is defined as

$$|A| = \sum_{x \in X} A(x)$$

Since fuzzy relations play such an important role in our framework, we recall some basic notions about them. A fuzzy relation R from a universe X to a universe Y is a fuzzy set in $X \times Y$. The inverse of R is a fuzzy relation from Y to X defined by

$$R^{-1}(y,x) = R(x,y)$$

for all x in X and y in Y. For all y in Y, the R–foreset of y is the fuzzy set Ry defined by

$$Ry(x) = R(x,y)$$

for all x in X.

If R is a fuzzy relation from X to Y, and S is a fuzzy relation from Y to Z, then the composition [22], the subproduct, and the superproduct ([3], [5]) of R and S are fuzzy relations from X to Z, respectively defined as

$$(R \circ_{\mathcal{T}} S)(x,z) = \sup_{y \in Y} \mathcal{T}(R(x,y), R(y,z)) \tag{1}$$

$$(R \triangleleft_{\mathcal{I}} S)(x,z) = \inf_{y \in Y} \mathcal{I}(R(x,y), S(y,z)) \tag{2}$$

$$(R \triangleright_{\mathcal{I}} S)(x,z) = \inf_{y \in Y} \mathcal{I}(S(y,z), R(x,y)) \tag{3}$$

for all x in X and z in Z. The square product of R and S is defined by

$$(R \triangleleft_{\mathcal{I}} S) \cap_{\mathcal{T}} (R \triangleright_{\mathcal{I}} S)) \tag{4}$$

If X and Y are non–empty, finite sets, a fuzzy relation R from X to Y can be represented as a $|X| \times |Y|$–matrix in which the rows correspond to the elements of X, and the columns to the elements of Y. The value on the i-th row and the j-th column of the matrix ($i \in \{1, ..., |X|\}$, $j \in \{1, ..., |Y|\}$) is the degree of relationship between the element x_i of X and the element y_j of Y corresponding to the i-th row and the j-th column respectively, i.e.

$$R_{ij} = R(x_i, y_j)$$

We use the same notation for a fuzzy relation and its representation as a matrix, i.e.

$$R = [R_{ij}]$$

Transposing the matrix corresponds to taking the inverse of the fuzzy relation, i.e.

$$R^T_{ji} = R^{-1}(y_j, x_i) = R(x_i, y_j) = R_{ij}$$

Throughout this paper we use D to denote a non–empty, finite set of documents and T a non–empty, finite set of terms. n is the number of documents and m is the number of terms, i.e. $|D| = n$ and $|T| = m$.

Definition 1 (Fuzzy document–term relation). A fuzzy document–term relation W is a fuzzy relation from D to T.

The symbol W refers to the fact that it contains the data collected from the WWW that will serve as our background knowledge for the construction of fuzzy thesauri. $W(d, t)$ can be obtained in a probabilistic manner by counting frequencies in the so–called TF–IDF approach (see e.g. [4] for a detailed explanation), though one can easily imagine other ways as well, such as the use of the scores that an existing search engine gives a document d when queried for a term t. From this fuzzy document–term relation, we will construct a fuzzy term–term relation.

Definition 2 (Fuzzy thesaurus). A fuzzy thesaurus or fuzzy term–term relation R is a fuzzy relation from T to T.

Meaningful relations between concepts can be of many different natures (see e.g. [20]). In this paper we will study synonym relations, as well as narrower and broader term relations, which are also the cases mentioned in [13].

3 The W^TW–approach

As mentioned above, the data available to us is a fuzzy document–term relation W represented as a $n \times m$–matrix. We are looking for a fuzzy term–term relation, i.e. a $m \times m$–matrix giving us information about the degree of association between terms. From a mathematical point of view, one way to obtain it, is the normalized matrix product W^TW, i.e.

$$R(t_1, t_2) = \frac{1}{n} \sum_{d \in D} W(d, t_1) \cdot W(d, t_2) \tag{5}$$

for all t_1 and t_2 in T. This is closely related to the (sup–$\mathcal{T}$) composition of the fuzzy relations W^{-1} and W in which product is replaced by a t–norm in general, and supremum instead of average is used to aggregate over all documents. Considering sup–$\mathcal{T}$ composition of fuzzy relations as a kind of matrix product goes back to the early days of fuzzy set theory (e.g. [22]).

Kohout et al suggested to use compositions of the document–term relation and its inverse but do not mention the sup–$\mathcal{T}$ composition in [11]. Instead they turn to the subproduct, the superproduct and an alternative version of the square product, replacing the infimum by taking the average over all documents. They use them to generate a specification relation ("more specific than"), a generalization relation ("broader than") and a synonym relation between terms respectively.

(1), (4) and (5) are symmetrical, while (2) and (3) are not in general. Note that

$$(W^{-1} \circ_{\mathcal{T}} W)(t,t) \leq \sup_{d \in D} W(d,t)$$

Hence if t does not have a high $W(d,t)$ value for any of the documents, the association between t and itself by means of (1) will be low! Using (5) the degree of association between t and itself is

$$\frac{1}{n} \sum_{d \in D} W(d,t)^2$$

which will also be small in the aforementioned case. If we are dealing with an implicator satisfying

$$x \leq y \Rightarrow \mathcal{I}(x,y) = 1, \text{for all } x \text{ and } y \text{ in } [0,1]$$

(such as residual implicators) then $(W^{-1} \triangleleft_{\mathcal{I}} W)(t,t)$, $(W^{-1} \triangleright_{\mathcal{I}} W)(t,t)$ and the square product are 1, regardless whether we aggregate by means of infimum or average. This makes (4) a better choice for synonymy then (1) or (5). (2) and(3) can however give rise to counterintuitive results when the first part of the implication is low, e.g. for (2) when $W(d,t_1)$ is low for all documents. Indeed in this case the resulting implication values tend to be high (you can derive everything from a premise which is close to false). In other words a term that is not important for any of the documents will be registered as more specific than all of the other terms! This problem with sub- and superproducts is known. In [5] a patch is provided by taking the intersection of the subproduct with the composition.

4 From Terms to Fuzzy Sets

In the most well spread approach to the automatic construction of fuzzy thesauri, a term is transformed into a fuzzy set. For every term t, the W–foreset of t is the fuzzy set Wt in the universe D of documents, defined as $Wt(d) = W(d,t)$. In other words Wt is the fuzzy set of documents relevant or related to term t. Finding a degree of association between two terms is now shifted to finding a degree of relatedness between the corresponding fuzzy sets. To this end a similarity measure Sim or an inclusion measure Inc is applied. These measures can be defined in different ways among which, for A and B fuzzy sets in D,

$$\text{Sim}(A, B) = \frac{|A \cap_{\mathcal{T}} B|}{|A \cup_{\mathcal{S}} B|}$$

$$\text{Inc}(A, B) = \frac{|A \cap_{\mathcal{T}} B|}{|A|}$$

It is of course assumed that A and B are not empty, i.e. that every term is related to at least one document to a degree greater than 0.

Many authors suggested (often independent of each other's work) the use of $\text{Sim}(Wt_1, Wt_2)$ to construct a symmetrical fuzzy relationship between t_1 and t_2, as well as the use of $\text{Inc}(Wt_1, Wt_2)$ to construct specification and generalization relations, mostly with t–norm $\mathcal{T}$ = min and t–conorm $\mathcal{S}$ = max. E.g. Gotlieb and Kumar [7] use Sim (but in the context of crisp sets). Miyamoto et al ([14], [15]) propose the idea for Inc and Sim, and Ogawa et al [19] use Sim. Recently there seems to be a boom of the same idea resurfacing over and over again ([9], [10], [12], [21]). [6] presents a variant where terms are transformed into crisp sets of their synonyms instead of fuzzy sets of documents. Miyamoto [16] proposes an another extension in which the neighborhood of a term is not necessarily the document in which it occurs, but it can also be a section of a document such as the surrounding words.

This idea of assessing the relatedness of fuzzy sets of documents is in the same spirit as the use of fuzzy relational products. Indeed the value of the (sup–$\mathcal{T}$) composition of W^{-1} and W in (t_1, t_2) measures the compatibility of Wt_1 and Wt_2 using compatibility measure Com, while the subproduct and the superproduct rely on the inclusion measure Inc_2:

$$\text{Com}(A, B) = \sup_{x \in D} \mathcal{T}(A(x), B(x))$$

$$\text{Inc}_2(A, B) = \inf_{x \in D} \mathcal{I}(A(x), B(x))$$

5 Associations

The idea of association rule mining already dates back to Hájek et al (see e.g.[8]). Its application for market basket analysis gained high popularity soon after the re–introduction by Agrawal et al [1] at the beginning of the 1990's. The straightforwardness of the underlying ideas as well as the increasing availability of transaction data from shops certainly helped to this end. In the context of association rule mining, data is represented in a table. The rows correspond to objects (e.g. transactions, patients,...) while the columns correspond to attributes (e.g. items bought in a transaction, symptoms,...). Ones and zeros in the data matrix denote whether or not the object has a specific attribute (whether or not *cheese* was purchased in the 5th transaction, whether or not patient *John* has *fever*,...). In this way, we can think of an object as a set of attributes, but we can also think of an attribute as a set of objects (namely those having that attribute). The purpose of association rule mining is to detect rules of the form

$$A \Rightarrow B$$

in the data, indicating that an object containing attribute A is likely to contain B as well (e.g. *cheese* $\Rightarrow$ *bread*).

The support measure of an association rule checks its statistical significance, while the confidence measure considers how many of the objects that have attribute A, have attribute B as well. If we think of documents as objects and terms as attributes (i.e. the data table is a document–term matrix), then an association rule $A \Rightarrow B$ indicates that a document containing term A is likely to contain term B as well. The support and confidence measure correspond respectively to Com (taking the average instead of the supremum over all documents) and Inc. The use of measures of association for the construction of fuzzy thesauri was already proposed in [20].

6 Conclusion and Future Work

We have shown that (1) composition of fuzzy document–term relations, (2) application of similarity, inclusion, and compatibility measures on fuzzy sets of documents, as well as (3) generation of association rules containing terms as attributes, nicely fit into the same framework. As such it becomes clear that each of these — at first sight different — techniques carries the same basic idea, hence increases the credibility of the power of this idea.

Note however that the general framework (and a fortiori all of the individual techniques discussed above) relies on a fuzzy document–term relation as a starting point. One of the main problems when developing a fuzzy set theoretical application is the definition of the membership functions of the fuzzy sets involved. It is exactly this problem that we are again faced with here. Traditionally one relies on a probabilistic approach to generate the document–term matrix, i.e. by counting frequencies of words within a document and over the set of all documents. In the future we want to move on to new approaches that might capture the semantics better, such as the use of term–sentence, sentence–paragraph, and paragraph–document relations.

Acknowledgements

This paper emerged from the cooperation during a visit of Martine De Cock and Sergio Guadarrama at Berkeley Initiative in Soft Computing, University of California, Berkeley, CA/USA. The authors would like to thank the BISC-group for providing a friendly and inspiring environment. Martine De Cock would like to thank the Fund For Scientific Research – Flanders (FWO) for funding her research.

References

1. Agrawal, R., Imielinski, T., Swami, A. (1993) Mining Association Rules between Sets of Items in Large Databases. Proceedings of ACM SIGMOD International Conference on Management of Data, 207–216
2. Akrivas, G., Wallace, M., Andreou, G., Stamou, G., Kollias, S. (2002) Context-Sensitive Semantic Query Expansion. Proceedings of ICAIS 2002 (IEEE International Conference on Artificial Intelligence Systems), 109–114
3. Bandler, W., Kohout, L. (1980) Fuzzy Relational Products as a Tool for Analysis and Synthesis of the Behaviour of Complex Natural and Artificial Systems. Fuzzy Sets: Theory and Application to Policy Analysis and Information Systems (Wang, S. K., Chang, P. P. eds.) Plenum Press, New York and London, 341–367
4. Berry, M. W., Browne , M. (1999) Understanding Search Engines: Mathematical Modeling and Text Retrieval (Software, Environments, Tools) SIAM, Philadelphia
5. De Baets, B. (1995) Oplossen van vaagrelationele vergelijkingen: een ordetheoretische benadering. PhD thesis (in Dutch), Ghent University
6. Fernandez Lanza, S., Graña Gil, J., Sobrino Cerdeiriña, A. (2002) A Spanish e-Dictionary of Synonyms as a Fuzzy Tool for Information Retrieval. Proceedings of ESTYLF-2002, 31–37
7. Gotlieb, C. C., Kumar, S. (1968) Semantic Clustering of Index Terms. Journal of the Association for Computing Machinery, 15(4):493–513
8. Hájek, P., Havránek T. (1978) Mechanizing Hypothesis Formation Mathematical Foundations for a General Theory. Springer–Verlag, http://www.cs.cas.cz/ hajek/guhabook/
9. Haruechaiyasak, C., Shyu, M.-L., Chen, S.-C., Li, X. (2002) Web Document Classification Based on Fuzzy Association. Proceedings of COMPSAC2002 (26th Annual International Computer Software and Applications Conference), 487–492
10. Intan, R., Mukaidono, M. (2003) A Proposal of Fuzzy Thesaurus Generated by Fuzzy Covering. Proceedings of NAFIPS 2003 (22nd International Conference of the North American Fuzzy Information Processing Society), 167–172
11. Kohout, L. J., Keravnou, E., Bandler, W. (1983) Information Retrieval System Using Fuzzy Relational Products for Thesaurus Construction. Proceedings of IFAC Fuzzy Information, 7–13
12. Martínez-Trinidad, J. F., Ruiz-Shulcloper, J. (2001) Fuzzy Clustering of Semantic Spaces. Pattern Recognition 34:783–793
13. McGuinness, D. L. (2003) Ontologies Come of Age. Spinning the Semantic Web. Bringing the World Wide Web to Its Full Potential (Fensel D., Hendler J., Lieberman H., Wahlster W., eds.) The MIT Press, Cambridge, Massachusetts, 171–194

14. Miyamoto, S., Miyake, T., Nakayama, K. (1983) Generation of a Pseudothesaurus for Information Retrieval Based on Cooccurrences and Fuzzy Set Operations. IEEE Transactions of Systems, Man, and Cybernetics 13(1):62–70
15. Miyamoto, S., Nakayama, K. (1986) Fuzzy Information Retrieval Based on a Fuzzy Pseudothesaurus. IEEE Transactions of Systems, Man, and Cybernetics 16(2):278–282
16. Miyamoto, S. (2003) Proximity measures for terms based on fuzzy neighborhoods in document sets. International Journal of Approximate Reasoning 34:181–199
17. Novák, V., Perfilieva, I., Močkoř, J. (1999) Mathematical Principles of Fuzzy Logic. Kluwer Academic Publishers
18. Nikravesh, M., Takagi, T., Tajima, M., Shinmura, A., Ohgaya, R., Taniguchi, K., Kazuyosi, K., Fukano, K., Aizawa, A. (2003) Web Intelligence: Conceptual–Based Model. Internal report, Electronics Research Laboratory, College of Engineering, University of California, Berkeley, Memorandum No. UCB/ERL M03/19
19. Ogawa Y., Morita, T., Kobayashi, K. (1991) A fuzzy document retrieval system using the keyword connection matrix and a learning method. Fuzzy Sets and Systems 39:163–179
20. Reisinger, L. (1974) On Fuzzy Thesauri. COMPSTAT 1974: Proceedings in Computational Statistics, 199–127
21. Widyantoro D. H., Yen, J. (2001) Incorporating Fuzzy Ontology of Term Relations in a Search Engine. Proceedings of the 2001 BISC International Workshop on Fuzzy Logic and the Internet, 155–160
22. Zadeh, L. A. (1975) The Concept of a Linguistic Variable and its Application to Approximate Reasoning I, II, III. Information Sciences 8:199–249, 301–357, 9:43–80

Consumer Profiling Using Fuzzy Query and Social Network Techniques

F. Olcay Cirit, Masoud Nikravesh, and Sema E. Alptekin

Berkeley Initiative in Soft Computing (BISC), EECS Department
University of California, Berkeley CA 94720, USA

Abstract. Web communities possess the unprecedented ability to map out with ease the networks of communication linking their users. This social connectivity information combined with the instant-feedback nature of web interactivity creates the potential for advanced, automated consumer profiling systems to be used for targeted advertisements and other commercial purposes. This research proposes one method for consumer profiling inspired by social network theory that is based on the BISC Decision Support System. Real-world applications and possible ethical concerns are explored in some detail.

1 Introduction

Conventional advertisements such as billboards lack interactivity: every car passing through a given stretch of highway sees the same billboard, and while it is possible – after adjusting for seasonal variation and other factors – to measure an increase in overall sales following the launch of an ad campaign, it is difficult to assess accurately the impact that such a campaign may have had on an *individual* sale.

In order to increase the effectiveness of conventional advertising, marketers may profile likely consumers of their products, a process that involves building an intuitively or statistically informed macroscopic description of a target consumer group. Traditionally, profiles are based on age, gender, geographic location, income level, and other data collected through means such as surveys, commercial databases, and point-of-purchase information. Guided by a profile, a marketer decides on the type of advertisement that would best appeal to the target audience, and then on the best venues for airing the advertisement so as to reach the target audience effectively.

In contrast to conventional advertisements, web and online advertisements possess a high degree of interactivity: every visitor to a website can see a *personalized* set of ads, and any click-through on a web ad is easily associated with an ensuing sales transaction. Furthermore, even if a user doesn't click on an advertisement immediately but visits the advertised site at a later time, internet technologies such as cookies and web bugs allow his or her advertisement viewings to be tracked to later sales.

We present a method by which databases tracking internet users or members of online communities could be enriched with social connectivity information for the purpose of enhanced, automated profiling and targeted web advertising.

Section 2 briefly introduces the topic of social networks and presents basic research results from the field that ground and motivate the research in this paper.

Section 3 discusses the properties of web-based and other online communities that make them ideal for tracking social connectivity, and describes how such communities could go about tracking social connectivity information in a form that yields to profiling.

Section 4 describes a method for automated consumer profiling using social connectivity information and the BISC DSS, the Berkeley Initiative in Soft Computing's Decision Support System.

Section 5 addresses possible ethical and legal concerns surrounding the implementation of a system based on the principles outlined herein.

2 Social Networks

2.1 Motivating Research

Social network theory is the study of social connections between people in organizations and in society at large. It traces its roots at least as far back as the 19th century.

Decades of research in social networks have lent support to observations of *homophily*, the tendency of people to be friends with others similar to them. For example, the majority of a typical person's friends tend to be of a similar income level and/or education level [3]. Another example of homophily is the tendency of people with social ties to prefer the same brands of goods [6]. These and other instances of homophily should be unsurprising to most, and indeed beg the following question: if people tend be similar to their friends, how could an examination of a person's friends possibly represent an improvement over existing profiling techniques?

The best answer is that matching up friends with friends for the purpose of profiling fills in the gaps in a traditional profile. To provide an example, if one's friends are members of the chess club, then all other things being equal, one is more likely to be a potential candidate for the chess club than an average person. One would have a much easier time identifying potential chess club members through friendship links than through divination on income level, gender, and other sundry data.

More than just simple maps of friendship and correspondence, social networks describe the channels through which ideas, fashion trends, gossip, and diseases flow in a society. *Innovation diffusion*, as this flow is called, refers to the way in which new products or ideas come to be accepted by society [3]. When a new class of product is first introduced, acceptance is initially slow: only the so-called early acceptors will buy it, but once they are sold on the product, they will convince others to buy it as well, accelerating the product's

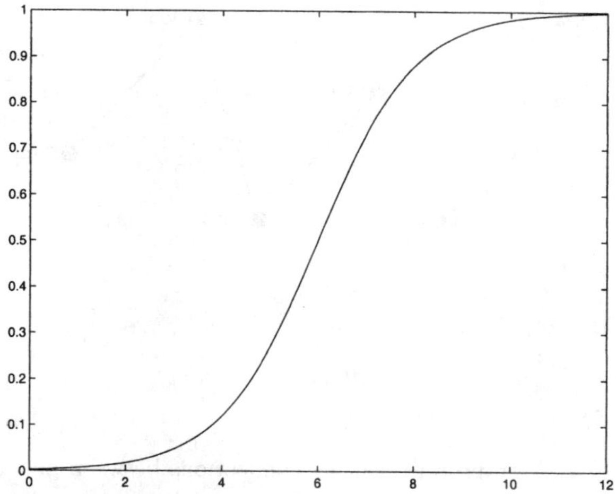

Fig. 1. The logistic function over time: a model for market saturation.

market penetration. The acceptance rate gradually slows as the market for the product becomes saturated (see Figure 1).

Arabie and Wind [2] discuss a good example of network marketing, *MCI's Friends and Family*^{TM}program that in effect provides discounts to MCI customers who encourage their friends and family to sign with MCI. Such methods enlist the customers to participate in the marketing of the product. The targeting method that will be presented here similarly leverages the natural diffusion patterns of a social network to target those nearest the early acceptors first, and in a sense follows the frontier of acceptance through social profiling and targeted advertising.

2.2 Conventions

Social network theory borrows many of its formalisms from graph theory. While a complete discussion is beyond the scope of this paper, a few concepts merit introduction. A detailed discussion of social network methods can be found in Hanneman [4].

A social network can be conceptualized as a *graph* in which the nodes represent people and the connections between the nodes represent connections between people (see Figure 2). To perform calculations on this graph, it can be converted into an equivalent *adjacency matrix* (see Figure 3) such that the rows and columns of the matrix represent the people in the network, and a connection from person A to person B is represented by a value of 1 in row A, column B of the matrix. A lack of a connection between A and B is reflected by a zero in the corresponding matrix cell. A weighted connection that expresses the strength of the relationship between A and B can be represented by a value between 0 and 1.

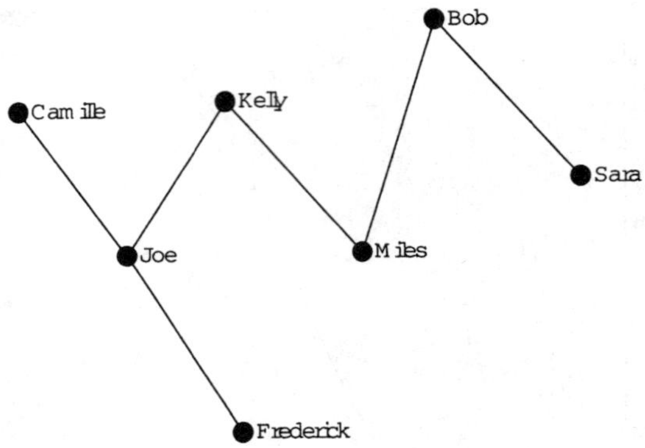

Fig. 2. A graph representing connections between people

	C	K	B	J	M	S	F
C	0	0	0	1	0	0	0
K	0	0	0	1	1	0	0
B	0	0	0	0	1	1	0
J	1	1	0	0	0	0	1
M	0	1	1	0	0	0	0
S	0	0	1	0	0	0	0
F	0	0	0	1	0	0	0

Fig. 3. A matrix representing representing the graph in Figure 2.

3 Online Communities

In past years, when researchers in the field of social network analysis needed to map out a social network, they relied on a technique known as *snowball sampling*, which involves asking a person to fill out a form listing the people with whom he or she corresponds, and then asking each of those people to provide similar lists, and so forth. A moment's contemplation will confirm that snowball sampling can be a tedious and time-consuming undertaking.

In a review of social network studies on computer mediated communication systems, Rice [7] raises the point that "computer-monitored usage and network data are potentially more *accurate* than corresponding self-report data." Rice is referring to the ability of computer communication systems to construct with ease maps of interactions between users, and a methodology for constructing such maps will be developed in this section with some rigor.

3.1 Constructing the Model

The basic principle behind constructing models of social connectivity in online communities is to measure the flow of communication over time between users. To draw upon psychology for justification, the *immediacy principle* [5] holds that "people are drawn toward persons and things that they like, evaluate highly, and prefer." Therefore it is reasonable to assume that users who correspond frequently are closer friends than those who correspond infrequently. Measurements of online correspondence, then, can be thought of as a reliable indicator of affinity between users.

Web-based communities such as Yahoo!™and Friendster™, as well as online communities such as America Online™, are all examples of computer communication systems having the potential to collect correspondence information merely by observing system usage. These online communities also serve vast amounts of advertisements to their users, and would benefit greatly from the improved targeting that social profiling could afford.

Communication channels that could be mapped in such a way include public chat and discussion forums, e-mail, instant message services, online game parlors, and potentially any other conceivable form of on-line communication.

Let us consider the case of tracking chatrooms. The flow of information in a chatroom is such that when one users posts a message, all of the other chat users receive that message. The tracking system can then record a movement of, for example, 50 words of information from user X to all other users in the chatroom. Tracking any other communication service simply involves tallying the number of words of text sent between users involved in an exchange – the content of the communication need not be tracked at all. When tabulating totals, the amount of textual flow between a sender and a receiver can be scaled by the total output of the sender during the measurement interval, resulting in a normalized measure of affinity between the sender and the receiver that can be organized into a weighted adjacency matrix like that described in section 2.2.

Instant messaging services and sites such as Friendster have in common the ability of users to enumerate their contacts. AOL instant messenger (AIM) and Yahoo messenger, for instance, let the user create a list of their "buddies" with whom they can communicate easily. Similarly, Friendster lets its users build a list of their friends by searching an online database for people they know. While these buddy and friend lists could be mapped directly onto adjacency matrices, the matrices so derived would necessarily be unweighted. Weighting connections is important because, although AIM users may have many buddies in their lists, they may correspond only infrequently with the majority of those buddies, the bulk of correspondence being with a small subset.

3.2 Integrating Disparate Modalities

Since web communities have already embraced non-text-based forms of communication such as voice and video chat, a natural question that arises is how to integrate measurements of user interaction over disparate modalities into a unified metric. According to psychometric studies performed by Mehrabian, when two people communicate, their opinions of like or dislike toward each other are determined thus: 7% by the words of the speaker, 38% by the vocal emphases of the speaker, and 55% by the speaker's facial expressions [5].

Therefore, text-based communication, consisting entirely of words and bereft of vocal emphasis and facial expression, is only 7% as strong an indicator of affinity as video chat, which consists of words, vocal emphasis, and facial expressions. Voice chat, by the same reasoning, is worth 7+38=45% as much as video chat.

How then, to compare a thirty-minute voice or video chat to a 2,000 word text exchange? There remains the apples and oranges problem of unit clash. It is only necessary to determine how fast, on average, users of voice and video chat services talk, to produce a rough measure equating minutes of voice chat time to words of an equivalent text chat. Then the weighting ratios mentioned above could be used to assign more importance to video than to text chat.

4 Automated Consumer Profiling

Given a large, advertising-driven web community that maintains a database on its users tracking both socially enriched and normal data, the problem of profiling and targeted advertising is this: how to decide which users in the database are more likely than others to be interested in a given product. The approach developed here makes use of a fuzzy query engine that is described in section 4.1. Section 4.2 describes the profiling process itself.

4.1 The BISC DSS

In its current state, the Berkeley Initiative in Soft Computing's Decision Support System (DSS) allows fuzzy queries on a database. A fuzzy query consists of a series of fuzzy-valued properties paired with weightings that determine the relative importance of each property in making a match. For example, in a date-matching application of the DSS, relevant properties include age, weight, height, etc. A date-matching fuzzy query might specify, among other things, the ideal age of the date in fuzzy terms, along with the relative importance of age when making a match.

Given a fuzzy query, the DSS returns a listing of entries in its database ranked by how well they match the fuzzy query. The user can browse these entries, and later refine the fuzzy query by manually adjusting the ranking of the entries. The DSS then applies a genetic algorithm to tune the query so as to match the new ranking.

4.2 The Targeting Process

Although the DSS as described above seems designed with hand-crafted queries and query refinements in mind, the process described in this section adapts it for automated profiling and profile refinement.

Assuming that one starts with no idea of how to target the advertisement, the best one can do at first is to show the ad to randomly selected users of the web community. Then, one can construct an initial pool of users who have clicked on the advertisement and gone on to buy the product. If the product being sold is the Acme Widget, we can call this initial group of users the Acme Widget User's Club, or AWUC.

The task of profiling and targeting can then be seen as a search for potential new members of the AWUC. We would like to use the DSS to perform a fuzzy search of the database to find these new members, but how to construct a meaningful query?

As it turns out, the manner in which the initial query is constructed is not particularly important, but a simple way to construct a query given the initial club is to compare the club members to each other, and then to compare the club as a whole to the rest of the community as a whole. Put in concrete terms, for each property that we can search using the DSS, we can construct an average for the club, A_c, and an average for everyone in the community, A_e. It is also useful to calculate a standard deviation for the club, D_c. The ideal value for the fuzzy property is then set to A_c, and relative importance is set to

$$k * |A_c - A_e| / D_c$$

where k is a constant scaling value. Setting the relative importance in this way emphasizes those features that differentiate members of the club from the community as a whole.

The next step is to incorporate the social network information into the profiling process. To do this, we can model the flow of information through the network using matrices. We choose a row vector n that is of the same width as the socially weighted adjacency matrix M such that

$$n[a] = 1$$

for every AWUC member a, and for which all other entries are zero. Then, we can compute a score vector s such that

$$s = n * M^p$$

Where p is a pre-determined power. A formal interpretation of s is that it is the image of the AWUC club n on the fuzzy relation M representing the social connectivity of the community. For ease of conceptualization, however, s can be thought of as modeling the "buzz" that each person might hear about Acme Widgets, or as the social nearness of each person to the members of the AWUC. If $p = 1$, this nearness is only defined for direct friends. As p

increases, more and more distant friend relationships affect the score. For the simple unweighted graph in figure 2, if Joe is the sole member of the AWUC and $p = 1$, Camille, Kelly, and Frederick will all have an s-score of 1.

The scores in vector s are entered into the database as a property of each user. Because of the nature of the s-score calculation, the members of the AWUC will all have the highest s-scores in the community, and so the initial query to the BISC DSS will be constructed with a high ideal s-score and large relative importance on the s-score. The initial query is now ready to be presented to the BISC DSS.

The BISC DSS will return a ranked listing of the users that shows how closely each user matches the profile of the AWUC. This ranking is then used to guide the display of advertisements, so that users closer to the top of the list who are not already members of the club are scheduled to receive targeted advertising for Acme Widgets. In addition, a certain smaller percentage of users should be selected randomly to receive advertisements to ensure good coverage and to monitor the usefulness of the profile.

As new users join the Acme Widget User's Club, the profile can be refined. Rather than going back to the equations used for constructing the initial profile, we can use the query refinement capability of the BISC DSS. The new ranking given to the DSS will re-rank the users based on how early they joined the AWUC. This puts new members lower on the ranking than older members, and leaves the rankings of non-AWUC-members unspecified. The DSS will then use a genetic algorithm to find an adjusted query that matches the new ranking, after which the advertising schedule is recalculated. This process is repeated often as the web community database is updated and new members join the community and/or the AWUC.

4.3 Analysis

Once an initial group of early acceptors joins the AWUC, the profiling process targets first the friends of the early acceptors, and then the friends of their friends, and so forth. In such a manner, the profiling system rides the frontier of product acceptance, always attempting to sell to those nearest the sold.

The genetic algorithm refinement process will eventually supercede any inaccuracies or inadequacies of the initially constructed profile, and may converge to a relatively stable query even as calculated s-scores and other variables in the database fluctuate. On the other hand, if there is a considerable difference between the early acceptors and those who are the last to buy a product, the genetic algorithm is flexible enough to adjust for this difference incrementally over the course of the product's market life.

The queries produced by the DSS during the profiling process are useful to marketers in that they constitute ready-made, human-readable profiles of the target market.

5 Ethical and Legal Concerns

As the techniques described in this paper represent a slight departure from established practice among web communities, this section discusses potential ethical and legal concerns that may arise should such a profiling system ever be implemented.

5.1 Privacy

It is quite likely that the use of social profiling techniques by large Web portals would be met with resistance by consumer advocacy groups and internet users at large. For example, many people might be averse to the idea of their web community tracking the length and frequency of inter-user interactions, even if the the content of such interactions is not screened, and even if the resulting measurements are used only for the purpose of targeted advertising.

A quick perusal of the privacy policies of companies mentioned in this paper reveals that Yahoo's privacy policy in its current form [1] *could* allow for some form of social profiling based on usage of the Yahoo! Groups feature. This is not to suggest, however, that Yahoo! or its affiliates engages in such practices.

Supposing that targeted advertisements based on social information come into widespread use, web users may become wise to the workings of the targeting system, opening the door for unintended privacy violations. For example: if John Doe inexplicably begins to see advertisements for blood sugar monitors while accessing his preferred web community, he may conclude with some measure of confidence that at least one of his close online friends is diabetic. In order to prevent the accidental spread of sensitive health information, care must be taken to avoid socially targeting advertisements for health products and other items of a possibly private nature.

5.2 Legal Issues

So long as a web community that wishes to perform social profiling informs users of its intentions through a privacy policy, it can avoid the bulk of potential litigation. There will inevitably arise considerable grassroots opposition to the practice of social profiling, just as there has been opposition to technologies such as cookies and web bugs in the past. It is worth noting, though, that both cookies and web bugs are now widespread and considered standard practice for gathering information on web users.

On the other hand, maintaining information on the duration of inter-user communication, as is suggested by this paper, could open up a web community to subpoena by law enforcement officials. This is, however, something that internet companies as well as telephone companies already deal with frequently, and would not likely represent an undue burden given that the process of releasing pertinent information could easily be streamlined.

6 Conclusions and Future Work

The ideas contained herein remain untested for the simple reason that testing them would require a massive internet presence and traffic stream, or a partnership with an entity of like proportions. It is instructive, however, to consider possible extensions of this method.

More sophisticated data mining techniques could greatly improve the usefulness of the targeting. If the BISC DSS were extended to use fuzzy-rule-based queries instead of simple fuzzy queries, the profiles produced through genetic refinement would be of much greater interest to marketers. If a certain product appealed to two or more distinct groups, a profile based on fuzzy rules could accurately capture the bimodal nature of the target population, whereas a simple fuzzy query could only approximate it.

We thank Dr. Martine De Cock and the BISC group members for their help with this research.

References

1. *Yahoo! Privacy Policy.* http://privacy.yahoo.com.
2. P. Arabie and Y. Wind. Marketing and social networks. In J. Galaskiewicz and S. Wasserman, editors, *Advances in Social Network Analysis*, pages 254–274. SAGE Publications, London, 1994.
3. A. Degenne. *Introducing the Social Network.* SAGE Publications, London, 1999.
4. R. Hanneman. *Introduction to Social Network Methods.* self published at: http://faculty.ucr.edu/ hanneman/SOC157/TEXT/TextIndex.html, 2001.
5. A. Mehrabian. *Silent Messages.* Wadsworth Publishing Company, Belmont, California, 1971.
6. P. Reingen, B. Foster, J. Brown, and S. Seidman. Brand congruence in interpersonal relations: a social network analysis. *Journal of Consumer Research*, 11:771–783, 1984.
7. R. Rice. Network analysis and computer-mediated communication systems. In J. Galaskiewicz and S. Wasserman, editors, *Advances in Social Network Analysis*, pages 167–206. SAGE Publications, London, 1994.

A Trial to Represent Dynamic Concepts

Kazushi Kawase, Tomohiro Takagi, and Masoud Nikravesh[(1)]
Department of Computer Science, Meiji University
1-1-1 Higashi-Mita, Tama-ku, Kawasaki-shi, Kanagawa-ken 214-8571, Japan
kkawase@cs.meiji.ac.jp, takagi@cs.meiji.ac.jp
BISC Program, EECS Department, University of California, Berkeley
URL: http://www-bisc.cs.berkeley.edu/ Email: Nikravesh@eecs.berkeley.edu

Abstract: We consider the expression and recognition of dynamic concepts by assigning the movement patterns learned in a recurrent neural net as symbols. We then develop a method to express more abstract dynamic concepts by combining them with symbols and connecting several recurrent neural networks. Application of the method to actual recognition cases, such as ball bouncing and dance movement (i.e. dancing), demonstrated its effectiveness. These experiments showed the ability of the method to deal with dynamic concepts that are difficult to describe because of vagueness.

1. Introduction

Video retrieval uses various techniques including video segmentation, indexing, and similarity calculation. Retrieval based on keyword input by the user is classified as keyword- based retrieval, and the retrieval based on images or sketches is classified as content- based retrieval.
Keyword-based video-segment retrieval is based on keyword input by the user. If the annotation for the segments in a database is automatically created by the system, it may not match very well. To overcome this problem, there has been much research on automatic annotation. R. Tusch et al.[3]made content-based video queries possible using a combination of low -level (physical) and high-level (semantic) video indexing. W. Zhow et al. [2] focused on the inductive learning of decision trees and proposed a method in which the characteristics of a scene are recognized and used to annotate the scene. For example, A. Kuchinsky et al.[5] built a system combining manual and automatic annotation, making semantic

online video classification possible.

In this paper, we describe a method in which recognized concepts describing the movements of objects in a video scene using recurrent neural networks. In the next phase of our work, we will use these concepts to annotate video scenes.

2. Expression of Dynamic Concepts

2.1. Dynamic concepts

We use the term "dynamic concepts" to represent various things. For example, they can be simple descriptions of movement, such as "fast" or "slow". They can be abstract, such as "happiness" or "sorrow". We previously proposed conceptual fuzzy sets (CFSs) [1][2], which conform to Wittgenstein's concept [Wittgenstein, "Philosophical Investigations," Basil Blackwell, Oxford(1953)] to represent the meanings of concepts. In a CFS, the meaning of a concept is expressed using other concepts. The method described here is based on this idea: the meaning of a dynamic concept is expressed using other dynamic concepts.

Time is also involved in a dynamic concept, as shown in Figure 2. For example, the concepts "slow" and "low acceleration" may help express the concept "sorrow". However, simply looking at the change in these two concepts over time does not clearly define the concept "sorrow". Dynamic concepts involve space and time fuzziness, so expressing them is complicated.

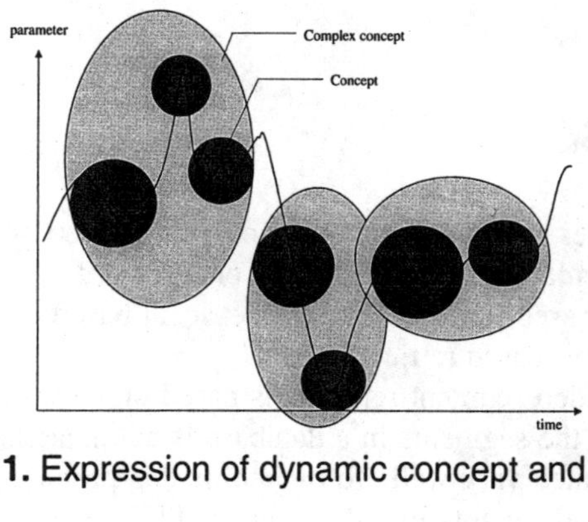

Figure 1. Expression of dynamic concept and time

2.2 Approach

This is the reasons we use recurrent neural networks (henceforth RNNs) to recognize dynamic concepts. That is, we express one or more dynamics concepts using an RNN and express more complex concepts by combining RNNs. Each dynamic concept is expressed as a memorized pattern in an RNN. The outputs of RNNs become the inputs of another RNN, enabling us to express ever more complicated dynamic concepts.

2.3 Features

The proposed expression of dynamic concepts has three major features.

*1) **Relate data patterns to symbols** :*A dynamic concept is physically represented as data that changes over time, as shown by the curve in Figure 1. This data pattern is expressed as a symbol, which can be handled in high-level brain functions. That is, a dynamic concept is related to the symbol for a data pattern. The symbol corresponds to the point where energy in the RNNs is minimal. More abstract concepts are expressed by connecting RNNs which relates more complicated data patterns to symbols.

*2) **Express concepts using other concepts:*** A dynamic concept can be expressed as an individual component, and concepts can be expressed by combining other concepts.

*3) **Cope with vagueness:*** The region of the meaning of a dynamic concept is fuzzy. For example, the boundary between "sorrow" and "happy" is not clear. Consequently, a method, such as the object-oriented paradigm, in which subclasses are defined using binary digits, cannot express dynamic concepts well. Our method can better cope with vagueness.

3. Outline of the recognition system

We developed a system for representing and recognizing movements according to the expression of dynamic concepts described above. As shown Figure 4, it consists of an image processing unit and a recognition unit. In the image processing unit, the characteristic values are extracted by measuring the changes between video frames. Using those characteristic values as inputs and the concepts it has learned, the recognition unit recognizes the dynamic concepts. The recognition unit consists of several RNNs connected to each other.

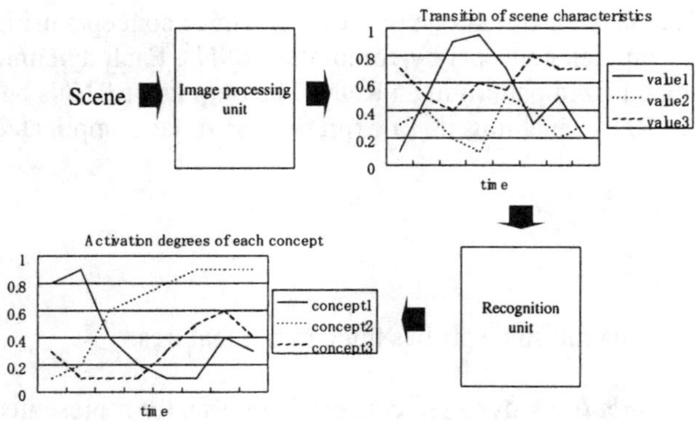

Figure 2. Movement recognition using expression of dynamic concepts

4. Expression and recognition of a simple concept

To evaluates our proposed method we first videotaped a ball bouncing as a simple example of dynamic concepts. We then used our method to describe the movement.

4.1. Recognized Concepts

Assume that there are two ways a ball can bounce: "heavy" and "light". We define each as a concept determined by the time series of ball heights before and after the first bounce. Both concepts are expressed by an RNN whose input is the ball height extracted from the image, as shown in Figure 3.

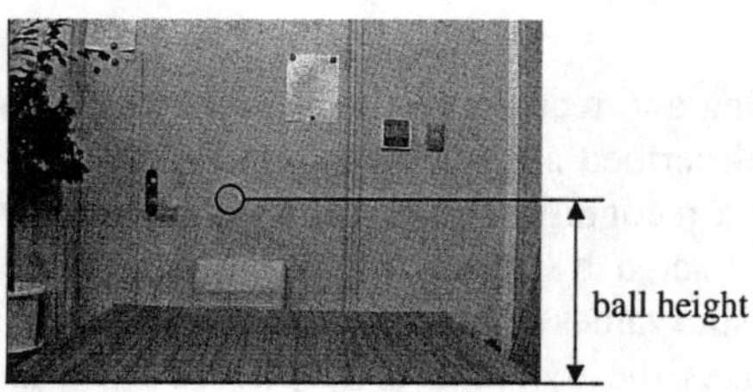

Figure 3. Height of a ball in image

4.2 Recognition Results

The transitions in the activation values are shown in Figure 4. Correct recognition is not ensured until the moment the ball first bounces in the 5-6th time unit. As shown in Figure 4 (a), the activation value of "light" is higher than that of "heavy" until about the 5th time unit, then the activation value of "heavy" becomes larger.

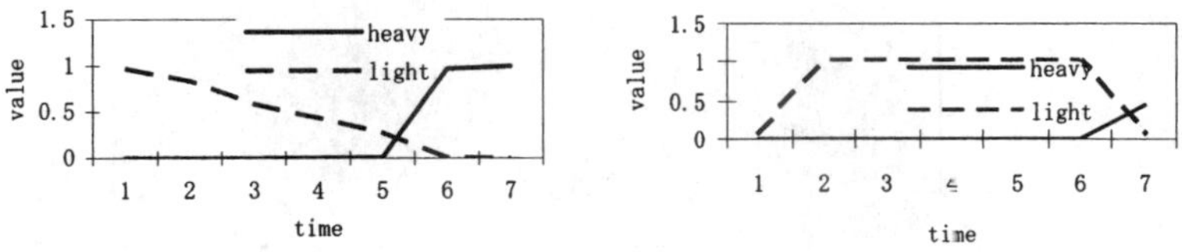

a) Heavy bounce b) Light bounce

Figure 4. Transition in activation values for "heavy" and "light" bounces

5. Expression and recognition of complex concept

We then evaluated our proposed method by using a more complex movement: three features based dancing model (the use of sorrow, enthusiastic, and happy expressions) [7].

5.1 Recognized concepts

In dancing, emotions are expressed through the total flow of body movement. It is thus differs from gestures and sign languages, which use simple patterns. The meanings of the expressions are very vague, so they are expressed by using other concepts.

Each motive (to word motive represent body/mind expression) expresses a feeling that can be distinguished from the others. We used Matsumoto's seven Motives as the targeted basic movements for recognition. These motives correspond to one or more dynamic concepts.

Three example motives, which we call "major dynamic concepts", are shown in Figure 5. The concepts are "sorrow", "enthusiastic" and "happy" and each concept is expressed by other different concepts the right side as follows,

sorrow: "slow" and "sustain"
enthusiastic: "sharp" and "sustain"
happy: "fast"

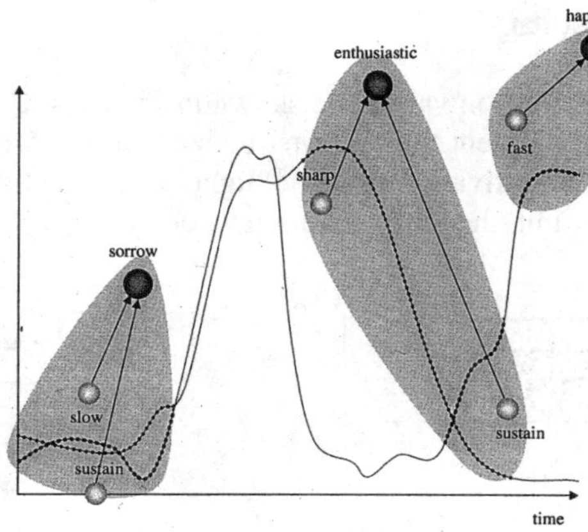

Figure 5. Composition of dynamic concept

5.2 Procedure

We used two valuables as the most primitive parameters to recognize the dance movement.

velocity: total number of blocks (4x4 pixels) moving in the rectangle including the dancer
acceleration: change in height and width of the rectangular including the dancer between successive frames (Figure 6)

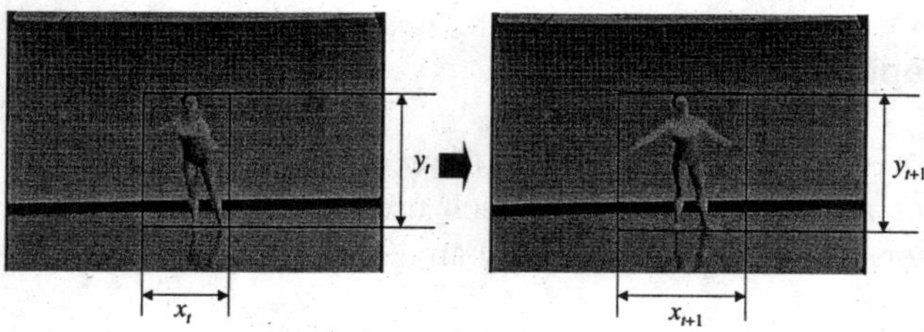

$$\Delta = |x_t - x_{t+1}| + |y_t - y_{t+1}|$$

Figure 6. Height and width of the rectangle including dancer

One or more concepts is expressed by an RNN, and the RNNs are connected, as shown in Figure 7.

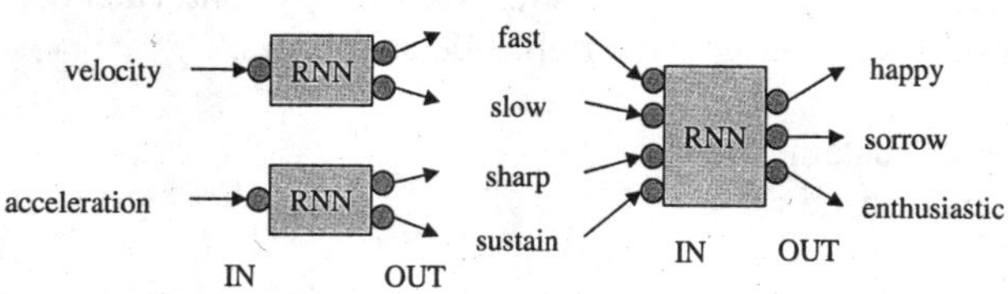

Figure 7. Connection of RNNs

5.3. Recognition results

The transitions in the activation values for the three major concepts are shown in Figure 8.

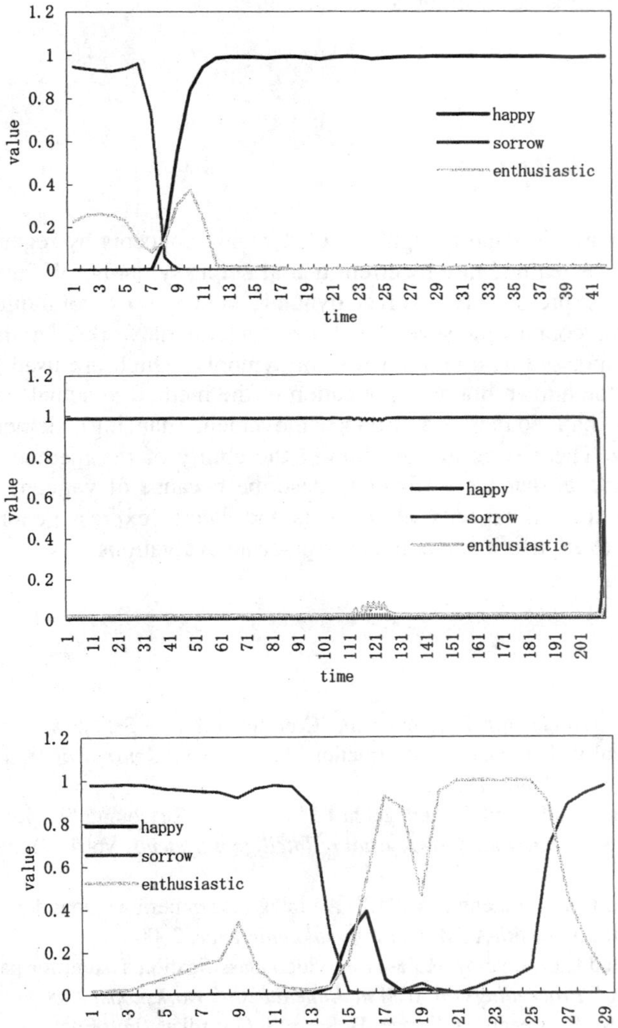

(a) Happy dance

(b) Sorrow dance

(c) Enthusiastic dance

Figure 8. Transition in activation values of three major concepts

The happy and sorrow dances were recognized correctly throughout the total dance. In the enthusiastic dances, however, only the parts with actual "enthusiastic" movements were recognized correctly, which accords with the result of human recognition. Our proposed method can thus correctly express and recognize abstract concepts that are expressed by other concepts.

6. Conclusion

We considered the expression and recognition of dynamic concepts by regarding the movement patterns learned in a recurrent neural net as symbols. We then developed a method to express more abstract dynamic concepts by combining them with symbols and connecting several recurrent neural networks. In this method, a symbol is expressed as a combination of symbols, which are used in high-level functions in the human brains. Application of the method to actual recognition cases, such as ball bouncing and dance movement (dancing), demonstrated its effectiveness. These experiments showed the ability of the method to deal with dynamic concepts that are difficult to describe because of vagueness. The current system has limited capacity of concepts and data to express general concepts. We thus need to expand it and conduct large-scale evaluations.

References

[1] T. Takagi, A. Imura, H. Ushida, and T. Yamaguchi, "Conceptual Fuzzy Sets as a Meaning Representation and their Inductive Construction",*International Journal of Intelligent Systems*, Vol.10, pp.929-945, 1995.

[2] T. Takagi, A. Imura, H. Ushida, and T. Yamaguchi, "Multilayered Reasoning by Means of Conceptual Fuzzy Sets", *International Journal of Intelligent Systems*, Vol.11, pp.97-111, 1996.

[3] R. Tusch, H. Kosch and L Boszormenyi, "VIDEX: An integrated generic video indexing approach," *Proceedings of the 8th ACM international conference*, 2000.

[4] W. Zhou, A Vellaikal and C.C. J. Kuo, "Rule-based video classification system for basketball video indexing," *Proceedings on ACM multimedia 2000 workshops*, 2000.

[5] Kuchinsky, C. Pering, M. L. Creech, D. Freeze, B. Serra, J. Gwizdka, "FotoFile: A consumer multimedia organization and retrieval system," *Proceeding of the CHI 99 conference on Human factors in computing systems*, 1999.

[6] R. J. Williams, D. Zipser, "A Learning Algorithm for Continually Fully Recurrent Neural Networks," *Neural Computation*, vol.1, pp.270-280, 1989.

[7] MIC interactive dance system: Midas, http://www.mic.atr.co.jp/~mao/midas/index.html

SORE (Self Organizable Regulating Engine) - An Example of a Possible Building Block for a "Biologizing" Control System

Paul P. Wang and Joshua Robinson
Dept. of Elec. & Comp. Engineering
Pratt School of Engineering
Duke University
Box 90291, Durham, NC 27708, USA
ppw@ee.duke.edu
Voice: (+1) 919-660-5259

Byung-Jae Choi, Professor
Dept. of Comp. & Com. Engineering
Daegu University
Kyungsan, Kyungbuk 712-714, KOREA
bjchoi@daegu.ac.kr

Abstract – The goals of this paper are threefold: (1) to introduce SORE to the biocontrol systems research community, describe how it works and explain why it could be an successful basic building block for a biocontrol system, (2) to present the basic characteristics of SORE and Boolean networks (BN) in a modern control language, with emphasis on their mathematical bases, (3) to illustrate, using some simple examples, why SORE's inherent properties enable it to realize many of the desired basic requirements for a "biologizing" control system. SORE also exhibits self-organizing, reproducing, colonization and grouping actions - essential traits of life. This paper does not report detailed research results; rather, it studies the feasibility of SORE in biocontrol systems based upon computer simulations. Rigorous research results will be presented in the future.

1. Introduction

The development of control theory has a long history due to fairly intensive research efforts for at least half a century. Some issues were thoroughly investigated and have reached a very mature status, while others were left nearly untouched. In an article entitled *"Biologizing" control theory: How to make a control system come alive* [1], John L. Casti coined the word "Biologizing" to reflect the recognition of so called "reliability and survivability" as topics of primary concern for engineers. A correct value judgment for a control system was thus finally determined. A blue-ribbon panel of 52 experts has nearly created a large set of research problems that lie ahead for control research engineers [2].

The concept of "feedback" in control theory is so far only known as a special trivial case of "homeostasis" - the function of an organism to regulate and to keep a constant "internal environment" [3]. Unfortunately, it is rather difficult to visualize that generalization based on this simplified case. These topics on biocontrol systems nevertheless bring much excitement and vision [4] – [5] to the research community. Some pioneering researchers such as James Albus have even undertaken the task of constructing a road map for the engineering of the mind [6] – [7].

There is no doubt that the system structures of biocontrol systems are very complicated hierarchical structures. But what kinds of system components are necessary to make biocontrol systems work? Without any specific examples, everything remains vague, unclear and uncertain. This paper provides such an example. SORE (Self Organizable and Regulating Engine) was first discovered as a classifier [8]. More interesting and robust properties subsequently emerged [9] – [12].

SORE is by far the most general mathematical structure of a family of automata theories listed as follows: SORE $\supset$ BN $\supset$ Cellular Automata (CA) $\supset$ Linear Automata (LA), with the exception of the condition $K < 5$ where CA could be more general than BN. In a genetic network, the total number of genes is represented by N, and K is the largest number of genes which regulates any one of the N genes in the genetic network. Based upon the theory of Stuart Kauffman, a biological genetic network usually has a much smaller K to more realistically model a biological setting. However, it is the condition of $K = N$ in which the network assumes its full strength for the best possible performance in a "biologizing" control system. This $K = N$ condition is what distinguishes SORE from the standard BN [9].

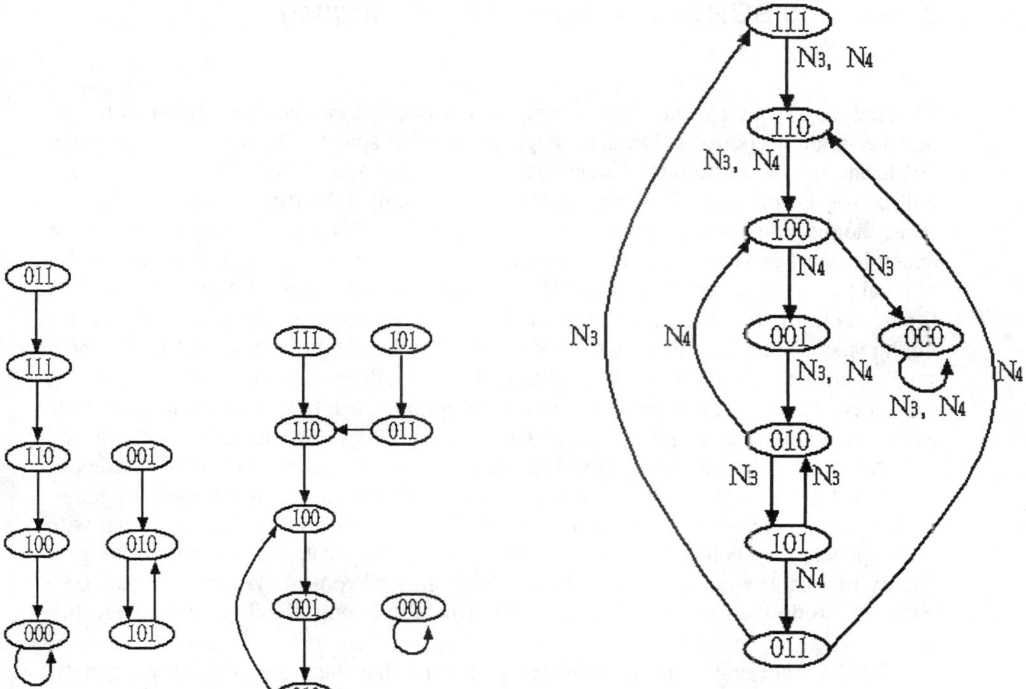

(a) A three-gene network N_3

(b) A three-gene network N_4

(c) Serial cascade of N_3 and N_4

Figure 1. Two three-gene networks and their serial cascade

2. Defining SORE in Modern Control Language

Theoretically speaking, the whole family of BN and CA can best be explained by using the modern control theoretical language of state space. The N genes' expression levels naturally constitute $N\times 1$ state vectors in a vector space. This discrete state space will consist of precisely 2^N state vectors. Unlike continuous differential dynamic systems, the discrete space can be completely displayed as long as N is not too large. For example, we can visualize a three-dimensional cube for $N = 3$. An autonomous solution subject to an initial state vector is of primary concern because it gives the sign of life in a cell. In other words, a zero-state response does not even exist, and only a zero-input response has to be dealt with. For all (N, K) BN and all $K = N$ (SORE), the whole state vector space can always be partitioned into I independent subspaces $\{S_1, S_2, ..., S_I\}$, where I is the total number of attractors or limit cycles. If α_{S_j} and α_{S_k} represent the elements of subspace S_j and S_k respectively, then $\alpha_{S_j} \cap \alpha_{S_k} = \Phi$ and $S_j \cap S_k = \Phi$ for all $j \neq k$. All isolated islands (subgraphs that describe the subspaces) are called "basins of attractors". Usually the complexity of a network is designated by the number of genes. A three-gene network is designated as N_3 and an eight-gene network as N_8. Even with a few genes, the "colonizing action" of a network can create a much more complex colony in a short amount of time. As the "biologizing" control system becomes more complex, so do the fundamental issues of reachability, controllability, and observability.

To illustrate only one example, let us assume that there are two known genetic networks and that both are made up of three genes (Fig. 1 (a) and (b)). There is only one attractor for each network where the existence of two three-gene networks would act autonomously. However, once the states of the two networks are connected (we call this a serial cascade), the behavior changes dramatically. Figure 1 (c) shows that no less than five limit cycles now exist. The existence of these limit cycles carries special meaning when considered in reference to living cells. It is precisely this constant cyclic behavior that is prevalent in many forms of life. The number of limit cycles observed is also significant. A large number of them will exhibit the biological characteristics which Stuart Kauffman called "the edge of chaos." This chaotic behavior is triggered by the time-varying rules of Boolean logic. There exists many other types of complexity, and only one of the many possibilities will be presented in this paper.

3. Concept of Colonization

As discussed in the last section, complexity in chaotic orders is due to the time-varying nature of the rules of Boolean logic. However, all of the complexity is due to autonomous activities. So far, there exists no zero-state response. The equivalence of such a response does exist, but it takes the form of dynamic systems expansion and enlargement; one may call it "colonization". Each gene is made up of a combination of the same molecules used in molecular genomics – G, A, T, and C. When a new gene is introduced into genetic networks, a larger network emerges. For example, if a new gene is added to a four-gene network N_4 and a five-gene network N_5, a single much larger ten-gene network N_{10} emerges. The new genetic network will have 2^{10} total state vectors, which is 2^6 or 64 times larger than the N_4 genetic network and 2^5 or 32 times larger than the N_5 genetic network. Why does one gene possess so much power? This single new gene may be viewed as a combination of big molecules, or it may represent a change in the chemical environment, such as temperature, pressure, or other changes [13].

The modeling of various biological phenomena within the general framework of automata has been documented for quite some time [14] (e.g. Rashevsky [15], M. Sugita [37], Rosenberg and Salomaa [16], IEEE [20]). However, the renewed interest in modeling gene regulations in molecular biology in recent years has caught the attention of a much larger set of researchers. By far, SORE is the most generalized mathematical structure among all automata theory used in modeling gene regulations. Since a generalized example of ten genes demonstrating colonization requires much more space to document, its salient features will simply be mentioned without the presentation of its computer simulations.

4. Reliability and Survivability

The "Biologizing" control theory states that reliability and survivability are the top priorities for a biocontrol system. How does one achieve this goal? One example is a system component like SORE which can deliver this requirement of reliability and survivability. SORE exhibits redundancy characteristics as described in reference [11]. Error correction code itself already exhibits redundancy characteristics. This allows SORE to have the reliability and survivability necessary for a biocontrol system. Of course, much more research is needed to have a deeper understanding of the issue.

5. Self Organizable and Regulating

What method of control should be exercised in an autonomous manner? Redundancy with a simple switching action after a decision is not a good design with respect to control and economics. Self organization, regulation, and control are basic requirements for a "biologizing" control system. Smooth control operation only be expected from a self organizable system. Self organization and regulation is a main capability of SORE [9] – [10].

6. Classification and Pattern Recognition

The main thesis of reference [12] is that SORE could be the most powerful known classifier. One of the most important issues of classification and pattern recognition theory is what classifier which yields highest percentage of correct recognition with data in any kind of structure. The emphasis of this issue is that the data points in problems will not necessarily be clustered nicely together. If this performance criterion is adopted, one can say with certainty that the best known classifiers such as Support Vector Machine (SVM) and Artificial Neural Network (ANN) will never be able to compete with SORE. In reference [10], the simplest case of SORE, the two gene network, was investigated and proven to be the best possible classifier suitable for any kind of data structure. Unfortunately, for more complicated networks, more time and research is needed because the problem becomes a synthesis problem for discovering Boolean logic functions.

Classification lies in the heart of any intelligent and autonomous control system. For a "biologizing" control system to be the most effective, some classification must be made before the best choice can be selected. SORE's abilities as a classifier may allow it to become a generic system component for a future "biologizing" control system.

7. Adaptation and Learning

With advances in MEMS and nanotechnology, all types of physical, chemical and biological sensors will soon be readily available to the public. This will allow future control systems to make more informed decisions based on their surroundings. Decision making capability under uncertainty, namely the approximation reasoning of fuzzy logic, is only a special case of SORE. The use of Boolean functions in Boolean networks allows each node or each gene to take any combination of logical connectors. Most of the methodology of fuzzy logic employed so far is a simple set of "IF ... THEN ..." rules, which is less general than SORE.

Adaptation and learning must be implemented in any "biologizing" control system. The learning aspect of SORE is quite different from the learning aspect of ANN. Due to the limitation of page space, SORE's learning algorithms will be discussed in the future.

Brain wave bio potentials have been used in experiments for mobile robot control [21]. This is just one of the many possible brain-inspired applications. Intelligent computational analysis of the human genome will drive medicine for at least the next half century [23]. Will research into bioinformatics result in useful information well beyond what is currently known? Perhaps the manipulation of brain information may be governed by the device such as SORE. Obviously these ideas are not certain, but there is a definite chance that SORE may be used in future brain-inspired applications.

8. Discussions and Conclusions

As stated from the outset, the primary objective of this paper is to introduce SORE as a generic building block for a "biologizing" control system. There has been a lot of research into control systems for many years. Everything from the relationships between intelligent systems and the fundamental rules of biology [24] [25] to system configurations and road maps for the future of control systems have been studied [6] [7] [26]. On the other hand, we present a generic system building block for control systems, because without effective system blocks, there can be no successful system.

The evolutionary processes of Boolean networks are also very important, as noted by some researchers [27] [28]. SORE demonstrates evolutionary development through its adaptation and learning capabilities. In colonization, the input of one gene (logically connected) to two genetic networks of N_4 and N_5 will produce a colony of N_{10} which is many times larger. The biological phenomena of self-reproduction can also be explained. This may very well possess scientific as well as economic value in many applications of SORE inspired technology.

Researchers are rapidly beginning to realize that problems utilizing Boolean networks can and should be viewed as control problems [29] – [31]. The stability issue has been brought up [32] and the rule capacity issue has been raised [33]. Randomizing a BN to be more realistic biologically has also been proposed [34] [37]. Imposing the assumption that BN is asynchronous is also needed [35] [36] for a more general analysis. Ultimately the difficult problem of synthesis eventually has to be raised [38].

As one can see, there is a whole spectrum of issues and problems that call for solutions using BN; SORE will be no exception. It is the condition of $K = N$ in Kauffman's (N, K) model that is the basis of SORE's ability. Finally, SORE is the most general mathematical structure in the whole automata family, making it a useful structure for many applications.

9. Acknowledgements

The authors gratefully acknowledge the financial support of Dr. Clinton W. Kelly of SAIC, McLean VA, the Pratt Fellows Program, and the Korea Science and Engineering Foundation (KOSEF) which make the publication of this paper possible.

References

[1] J. L. Casti, "Biologizing" control theory: How to make a control system come alive, Complexity 7(4), 10-12, 2002

[2] Report of the workshop Held at the University of Santa Clara on September 18-19, 1986, Challenges to control: A collective view, IEEE Trans. Automatic Control 32(4), 275-285, 1987

[3] W. B. Cannon, The wisdom of the body, W. W. Norton and Co., New York, 1939

[4] R. Lathrop, Intelligent systems in biology: Why the excitement?, IEEE Intelligent Systems, 8-13, Nov./Dec. 2001

[5] J. Leith *et al.*, Toward more intelligent annotation tools: A Prototype, IEEE Intelligent Systems, 42-51, Nov./Dec. 2001

[6] J. S. Albus, Outline for a theory of intelligence, IEEE Trans. SMC 21(3), 473-509, 1991

[7] J. S. Albus, The engineering of mind, Information Sciences 117, 1-18, 1999

[8] P. P. Wang and H. D. Cheng, "SORE: Self organizable and regulating engine" A research proposal submitted to US NSF, Feb. 20, 2003, File No.: 03248-27, Fastlane

[9] P. P. Wang and J. Robinson, What is SORE, The 7th Proceedings of the Joint Conferences on Information Sciences, Sep. 2003

[10] P. P. Wang *et al.*, A study of the Two-Gene Network - The simplest special case of SORE, The 7th Proceedings of the Joint Conferences on Information Sciences, Sep. 2003

[11] P. P. Wang and H. Tao, A Novel method of error correcting code generation based upon SORE, The 7th Proceedings of the Joint Conferences on Information Sciences, Sep. 2003

[12] P. P. Wang and J. Yu. SORE - A powerful classifier, The 7th Proceedings of the Joint Conferences on Information Sciences, Sep. 2003

[13] T. E. Ideker *et al.*, Discovery of regulatory interactions through perturbation: inference and experiment design, Pacific Symposium on Biocomputing, 5, 302-313, 2000

[14] J. Hammer, On some control problems in molecular biology, Proceedings of 33rd Conference on Decision and Control, 4098-4103, 1994

[15] N. Rashevsky, Mathematical biophysics, The University of Chicago Press, 1948

[16] G. Rozenberg and A. Salomma, L systems, Lecture Notes in Computer Science, 15, Springer Verlag, Berlin, 1975

[17] J. von Neumann, The theory of self-reproducing automata, University of Illinois Press, Urbana, 1966

[18] A. Lyndenmayer, Mathematical models for cellular interactions in development, Parts 1 and 2, J. of Theoretical Biology, 18, 280-315, 1968

[19] S. A. Kauffman, Metabolic stability and epigenesis in randomly constructed genetic nets, J. of Theoretical Biology, 22, 437-467, 1969

[20] IEEE Computer Systems Society, Proceedings of the 1974 Conference on Biologically Motivated Automata Theory, McLean VA, 1974

[21] K. H. Choi and M. Sasaki, Brain-wave bio potentials based mobile robot control: Wavelet-neural network pattern recognition approach, IEEE Int. Conference on SMC, 1, 322-328, 2001

[22] E. Mjolsness and A. Tavormina, The synergy of biology, intelligent systems, and space exploration, IEEE Intelligent Systems, 20-25, Mar./Apr. 2000

[23] R. B. Altman, Challenges for intelligent systems in biology, IEEE Intelligent Systems, 14-18, Nov./Dec. 2001

[24] J. M. Evans *et al.*, Knowledge engineering for real time intelligent control, Proceedings of the 2002 IEEE Int. Symposium on Intelligent Control, Oct. 2002

[25] H. Bolouri, Mechanisms underlying the evolution of robust nonlinear control in biology, Proceedings of the 1999 IEEE Int. Symposium on Intelligent Control/Intelligent Systems and Semiotics, Sep. 1999

[26] P. J. Fleming and R. C. Purshouse, Evolutionary algorithms in control systems engineering: A survey, Control Engineering Practice, 10, 1223-1241, 2002

[27] J. Lancharles *et al.*, Boolean networks decomposition using genetic algorithms, Microelectronics Journal, 28, 551-560, 1997

[28] L. Raeymaekers, Dynamics of Boolean networks controlled by biologically meaningful functions, J. of Theoretical Biology, 218, 331-341, 2002

[29] F. Forgelman-Soulie, Parallel and sequential computation on Boolean networks, Theoretical Computer Science, 40, 275-300, 1985

[30] S. Bilke and F. Sjunnesson, Stability of the Kauffman model, Physical Review E, 65, 016129, 2001

[31] J. A. B. Tome and J. P. Carvalho, Rule capacity in fuzzy Boolean networks, Proceedings of NAFIPS 2002, 27-29, 2002

[32] J. E. Lynch, Critical points for random Boolean networks, Physica D, 172, 49-64, 2002

[33] E. R. Dougherty and I. Shmulevich, Mappings between probabilistic Boolean networks, Signal Processing, 83, 779-809, 2003

[34] C. Gotsman *et al.*, Asynchronous dynamics of random Boolean networks, IEEE International Conference on Neural Networks, 1, 1-7, 1988

[35] M. Dorigo, Learning by probabilistic Boolean networks, IEEE International Conference on Neural Networks, 2, 887-891, 1994

[36] T. Akutsu *et al.*, Identification of genetic networks by strategic gene disruptions and gene overexpressions under a Boolean model, Theoretical Computer Science, 298, 235-251, 2003

[37] M. Sugita, Functional analysis of chemical systems in vivo using a logical circuit equivalent. II. The idea of a molecular automaton, J. of Theoretical Biology, 4, 179 – 189, 1963

Multivariate Non-Linear Feature Selection with Kernel Methods

Isabelle Guyon[(1)], Hans-Marcus Bitter[(2)], Zulfikar Ahmed[(2)], Michael Brown[(2)], and Jonathan Heller[(2)]
[(1)] Clopinet, 955 Creston Road, Berkeley, CA94708, USA Email: isabelle@clopinet.com URL: http://clopinet.com, Tel/Fax: (510) 524-6211.
Biospect, 201 Gateway Blvd, So. San Francisco, CA 94080.

Abstract: We address problems of classification in which the number of input components (variables, features) is very large compared to the number of training samples. Such problems are encountered in Internet application such as text filtering, in biomedical applications such as medical diagnosis from genomic or protemic data, and drug screening from combinatorial chemistry data. In this setting, it is often desirable to perform a feature selection to reduce the number of inputs, either for efficiency, performance, or to gain understanding of the data and the classifiers. We compare a number of methods on mass-spectrometric data of human protein sera from asymptomatic patients and prostate cancer patients. We show empirical evidence that, in spite of the high danger of overfitting, non-linear methods can outperform linear methods, both in performance and number of features selected.

1. Introduction

The problem of variable and feature selection has been tackled from many perspectives. For a review, see (Guyon and Elisseeff 2003) and references therein. This problem has recently attracted a lot of attention because new application domains produce data with huge numbers of features (10,000, 100,000, or even millions). In Internet applications, text-filtering methods (e.g. spam filters) are using "bag-of-words" representations in which texts are represented by the frequency of appearance of words, yielding sparse feature vectors of the order of 50,000 features (Drucker 1999). In bioinformatics, medical diagnosis assays on the verge of being commercialized are using DNA microarray genomic data (up to 60,000 features, see e.g. Jain 2001) or mass-spectrometry protemic data (several million features for combined techniques such as CE-MS and LC-MS, see e.g. Surman, 2002) Drug screening of candidate drugs from quantitative chemical descriptors

(QSAR) is another application actively being employed by pharmaceutical companies in which the number of features can reach 100,000 (Weston et al. 2003). A recent benchmark that we have organized presents results on a variety of datasets (http://clopinet.com/isabelle/Projects/NIPS2003/), including examples of these three tasks.

A first set of methods that are commonly used in bioinformatics and text processing consist in ranking features according to their individual predictive power. Such techniques include correlation methods, T-test, Fisher score, etc. A state-of-the art version of this generic approach has been proposed recently by (Tibshirani et al 2002). We refer to such methods as "linear univariate" because they make a convex combination of linear classifiers that are built using single variables. While it is also possible to use "non-linear univariate" methods, we do not consider these in this study because they are rarely used, particularly in our application domain. A second set of methods, which we investigate, are "linear multivariate" and make use of linear discriminant classifiers. Such classifiers are built with a subset of features. They are used to score feature subsets, according to classification performance. Finally, a third set of methods use non-linear discriminant classifiers to perform the same task. We refer to those as "non-linear multivariate". Our method categorization is summarized in Table 1.

It is common in the literature to make the distinction between "filters" and "wrappers" for feature selection. Filters are methods that select features without the direct goal of optimizing the performance of a particular classifier. The resulting features are used with any classifier. Wrappers and embedded methods are directly tied to a given classifier: they use the performance of the classifier (or a prediction of its performance) to select subsets of features, eventually conducting a search in the space of all possible feature subsets. In this study, we make use of a filter (the shrunken centroid method) that can be considered a wrapper if the input statistical independence assumptions are actually verified. Conversely, we use wrappers as filters to select features for other methods.

Table 1. Color coding for method attribute combinations.

Linear univariate	**Linear multivariate**
Non-linear univariate	**Non-linear multivariate**

In our tutorial (Guyon and Elisseff 2003), we have shown simple examples of problems that are inherently multivariate and cannot be solved with univariate techniques: the data separation lies in a subspace of dimension greater than one. We have shown cases in which a particular feature taken alone carries no class-separation power, yet it can improve classification performance when taken together with another one. Additionally, some problems have strong non-linearities. The optimum non-linear decision surface may lie in a multi-dimensional subspace: We have seen examples in which two (or more) features that individually carry no class-separation power can improve classification performance when considered simultaneously.

Even though multivariate methods lead to more universal predictors than univariate methods and non-linear methods more universal predictors than linear methods, they may turn out to provide poorer performance. This is due to the problem of overfitting. In large dimensional spaces, with the availability of a small number of training examples, being able to learn a broader class of functions is often synonymous to providing poor generalization capabilities on test data (distinct from the training data). For a theoretical treatment of this problem, see e.g. (Vapnik 1998).

The purpose of this paper is to see whether pessimistic theoretical predictions hold in practice. We show in this study that in fact they don't: non-linear multivariate methods perform better than linear multivariate methods, that, in turn perform better that linear univariate methods. These results do not completely contradict the theory, since it is known that, with proper regularization, overfitting can be overcome. The proposed algorithms have indeed regularization mechanisms that ensure their good generalization performance.

2. Material and Methods

2.1 Data and preprocessing

We analyze mass-spectrometric prostate cancer data collected at the Eastern Virginia Medical School using SELDI time-of-flight mass spectrometry. The raw data is downloadable from:
http://www.evms.edu/vpc/seldi/.
In this study, the data includes 326 spectra (the duplicates are not included) corresponding to 159 controls (77 benign prostate hyperplasia, and 82 age-matched normals) and 167 cancer spectra (84 state 1 and 2; 83 stage 3 and 4). The 60-sample test set used by the EVMS is not available for this study. The total original number of features is 48538, representing the number of ions measured at regular time intervals. It is desirable to reduce the number of features for computational reasons. Also, selecting useful peaks is a first step in identifying proteins that are useful for diagnosis (biomarkers). Small and inexpensive diagnosis kits using just a few biomarkers may then be designed. Biomarkers can be used in the drug discovery process.

Results from the original paper of the EVMS on that data are described in (Adam et al 2002). We cite the paper: "Surface enhanced laser desorption/ionization mass spectrometry protein profiles of serum from 167 PCA (prostate cancer) patients, 77 patients with benign prostate hyperplasia, and 82 age-matched unaffected healthy men were used to train and develop a decision tree classification algorithm that used a nine-protein mass pattern that correctly classified 96% of the samples. A blinded test set, separated from the training set by a stratified random sampling before the analysis, was used to determine the sensitivity and specificity of the classification system. A sensitivity of 83%, a specificity of 97%, and a positive predictive value of 96% for the study population and 91% for the general population were obtained when comparing the PCA *versus* non cancer (benign prostate hyperplasia/healthy men) groups."

We use a preprocessing that we have developed for other similar mass-spectrometric datasets. This preprocessing has proved to enhance performance and allows us to eliminate features with low information content.
The preprocessing consists of the following steps:

- **Limiting the mass range:** Indices in the range 1401:11100 were used. This corresponds roughly to eliminating small masses under m/z=200 and large masses over m/z=10000.
- **Removing the baseline:** We subtract in a window the median of the 20% smallest values. An example of baseline detection is shown in Figure 1.
- **Smoothing:** The spectra were slightly smoothed with an exponential kernel in a window of size 9.
- **Re-scaling/Normalization:** The spectra were divided by the median of the 5% top values.
- **Taking the square root:** The square root of the all values was taken to stabilize variances.
- **Limiting more the mass range:** To eliminate border effects, of the remaining variables, the index range 101:9600 was selected.

The resulting data set has 326 patterns from 2 classes and 9500 features.

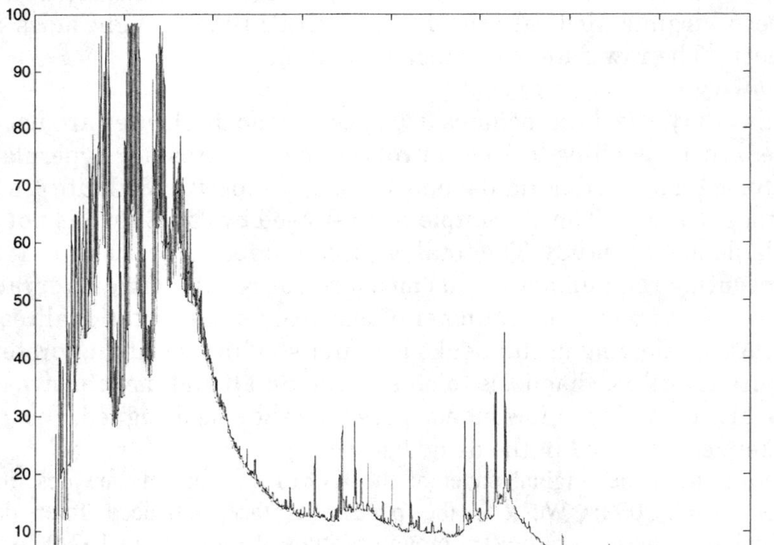

Fig. 1. Example spectrum. We show one spectrum from our data set (in blue). The estimated baseline is shown in red. The horizontal axis corresponds to time of arrival of proteins in the SELDI TOF mass-spectrometer. We plot the intensity of the signal detected as a function of time. Each intensity value, after preprocessing, is an input feature.

2.2 Performance assessment

The patterns were divided into three folds. Three non-overlapping test subsets of 108 spectra were drawn randomly from the 326 spectra. The complementary subsets of 218 spectra were used as training sets.

For performance assessment, we group together feature selection and classification, i.e. a separate feature selection is performed on each of the three folds. We refer to a system performing feature selection and classification as a "classifier". For each pair of classifiers we wish to compare, we compute an index to assess the statistical significance of the difference in performance. For each fold, we perform a McNemar test (see e.g. Guyon et al 1998). To do so, we need to keep track of the errors made by the classifiers and compute for each pair of classifiers the number of errors that one makes and the other does not, ν_1 and ν_2. We use the fact that, if the null hypothesis is true, $z=(\nu_2-\nu_1)/\mathrm{sqrt}(\nu_1+\nu_2)$ obeys approximately the Normal law. We compute p-values according to: 0.5*(1-erf(z/sqrt(2))). This allows us to make a decision: 1 for significantly better (pvalue<0.05), 0 for not significantly different (0.05≤pvalue≤0.95), -1 for significantly worse (pvalue>0.95). We then sum the decisions for the three folds to obtain an overall score.

Note that unlike other tests performed with cross-validation methods that blend the results of all the folds, our test does not violate independence assumptions of the test examples because we perform separate tests on each of the 3 folds.

When we need to adjust hyperparameters, we use another internal cross-validation loop that assesses classification performance using training data only, on each of the three folds.

2.3 Non-linear feature selection algorithms

Non-linear feature selection methods are often complex to implement and/or computationally expensive. We explore two non-linear feature selection algorithms that are simple and fast:

- Non-linear kernel multiplicative updates.
- A variant of Relief.

The non-linear multiplicative updates algorithm extends the linear version (Weston et al, 2003) by using the same idea used in non-linear backward elimination for SVM (Guyon et al 2002): eliminate the weights that change the cost function least. Relief is an algorithm proposed by Kira and Rendell that is quite popular (Kira and Rendell 1992). We propose two extensions of Relief: one adding some regularization, one removing correlations between features selected.

2.3.1 Non-linear multiplicative updates

The multiplicative updates method consists in iteratively training a classifier and rescaling the input features by multiplying them by scaling factors that de-emphasize the least promising features (from the point of view of classification accuracy).

We use a generalization to the non-linear case of the method described in (Weston et al 2003), with a different way of computing the input scaling factors. For the linear multiplicative updates, the inputs are rescaled iteratively by the weights of a linear discriminant classifier (e.g. an optimum margin classifier or linear SVM, see e.g. Boser et al 1992). For the non-linear multiplicative updates, we consider the case of kernel classifiers for the type: $f(\mathbf{x}) = \Sigma_k \alpha_k y_k K(\mathbf{x}, \mathbf{x}_k)$, where $\mathbf{x}$ is an input vector, $\mathbf{x}_k$ is a training pattern with target values y_k (±1). In the experiments, we use Gaussian kernels $K(\mathbf{x},\mathbf{x}')=\exp\text{-}\gamma\|\mathbf{x}\text{-}\mathbf{x}'\|^2$, and polynomial kernels $K(\mathbf{x},\mathbf{x}')\text{-}(1+\mathbf{x}.\mathbf{x}')^q$, where γ is a positive real number, and q is an positive integer. For more details on kernel methods, see (Schölkopf-Smola 2002).

The scaling factors used for the multiplicative updates method are obtained as:

$$s_i = \text{sqrt}(\Sigma_k \Sigma_l \alpha_k \alpha_l y_k y_l K(\mathbf{x}_{ki}, \mathbf{x}_{li}))$$

where the α_k are obtained by training and $K(\mathbf{x}_{ki}, \mathbf{x}_{li})$ is the kernel function evaluated on examples projected in the single dimension of the feature $\mathbf{x}_i$ considered. This is a variant that is computationally inexpensive of the criterion proposed in (Guyon et al 2002) for non-linear recursive feature elimination.

Note that in the case where the kernel is $K(\mathbf{x},\mathbf{x}')=\mathbf{x}.\mathbf{x}'$ (linear classifier), we obtain the scaling factors that are advocated by Weston at al:

$s_i = \text{abs}(w_i) = \text{sqrt}[(\Sigma_k \alpha_k y_k \mathbf{x}_{ki})(\Sigma_l \alpha_l y_l \mathbf{x}_{li})]$

since $w_i = \Sigma_k \alpha_k y_k \mathbf{x}_{ki}$

2.3.2 Relief

We use a slightly modified version of the original Relief algorithm that combines ideas developed by several authors.

The main idea of Relief is to compute a score for each feature measuring how well this feature separates neighboring examples in the original space. The nearest neighbor version seeks for every example its nearest example from the same class (nearest hit) and its nearest example from the opposite class (nearest miss), in the original feature space. The score is then the difference (or the ratio) between the average over all examples of the distance to the nearest miss and the average distance to the nearest hit, in projection on that feature. We use the ratio because it self-normalizes the scores.

We use an extension to that idea to K nearest hits and misses, in which we use averages of the distances to the K nearest hits and to the K nearest misses. In our experiments, we use K=4. A particularity of our method in application to mass-spectrometric spectral data is that we eliminate features that are close to one another in time of flight to remove some of the redundancy.

2.3.3 Gram-Schmidt Relief

We combine the Relief criterion with the Gram-Schmidt orthogonalization method (see e.g. Stoppiglia et al 2003) in the following way:

- The first feature is selected according to the Relief criterion (K-nearest neighbor version).
- All remaining feature vectors (columns of the training data matrix, the index varying over all training examples) are projected onto the subspace orthogonal to the feature selected.
- The next feature is selected by applying the Relief criterion in that subspace.

The procedure is iterated by projecting the remaining features into the subspace orthogonal to all the features previously selected and applying again the Relief criterion in such subspace. In this way, we select Relief features that are uncorrelated with one another.

2.3.4 Interval selection

We implemented a simple greedy algorithm that selects an optimum number of contiguous features in the spectrum. Let us call 'algo' the learning algorithm selected (e.g. linear SVM):

- Divide the feature range in 2m+1 overlapping intervals.
- Compute the cross-validation error of algo, e.g. 5x2CV on the training data (Dietterich 1998) on every interval.
- Replace the feature range by the interval with the smallest CV error.
- Iterate.

Note that this CV loop is an "internal" loop . It uses the training data only of each of the three folds whose tests sets are used to assess the final performance.
The parameter m defines the dividing factor. We chose m=2 in the experiments, which means that the number of features is divided by 2 at every iteration. There are 3 intervals of identical length: left half, right half, middle interval with half the features.

2.4 Acronyms of methods used

In our experiments, we compare the proposed methods with a number of baseline methods. The acronyms for the methods are explained below:

<u>No feature selection:</u>

OM: Optimum Margin classifier (linear SVM).

SVM_P2: Polynomial SVM of order 2 classifier (Boser et al 1992).
RBF: Radial Basis Function SVM, exponential kernel, gamma=0.1 (Boser et al 1992).
NN: K-nearest neighbor classifier, K=7.

Linear feature selection:

OMMU100_OM: OM multiplicative updates, 100 features, OM classifier.
OMMU7_OM: OM multiplicative updates to 100 features, keep top 7, OM classifier.
GS100_OM: Gram-Schmidt 100 features, Optimum Margin classifier.
GS7_OM: Gram-Schmidt, 7 features, OM classifier.
SC100_OM: Shrunken centroids (Tibshirani et al 2002), 100 features, OM classifier.
SC7_OM: Shrunken centroids, 7 features, OM classifier.

Random feature selection:

rand100_OM: random feature set, 100 features, OM classifier.
rand7_OM: random feature set, 7 features, OM classifier.
rand100_RBF: random feature set, 100 features, RBF classifier.
rand7_RBF: random feature set, 7 features, RBF classifier.

Interval feature selection:

Linear:

INTERCV_OM: Interval selected by CV, stop when CV error increases, Optimum Margin classifier (both for selection and classification).
INTERCV100_OM: Interval selected by CV, go down to 100 features, OM classifier (both for selection and classification).

Non-linear:

INTER_RBF: Interval selected by CV, stop when CV error increases, RBF classifier (both for selection and classification).
INTER100_RBF: Interval selected by CV, go down to 100 features, RBF classifier (both for selection and classification).

Non-linear feature selection:

Non-linear multiplicative updates:

NLMU100_RBF: Non-linear multiplicative updates, Radial Basis Function (both for feature selection and classifier), 100 features selected.

NLMU100_OM: Non-linear multiplicative updates, RBF feature selection to 100 features, OM classifier.

NLMU7_RBF: Non-linear multiplicative updates RBF up to 100 features, keep only top 7, RBF classifier.

NLMU7_OM: Non-linear multiplicative updates RBF up to 100 features, keep only top 7, OM classifier.

Relief (nearest hit, nearest miss):

R100_NN: Relief, 100 features, nearest neighbor classifier.

R100_OM: Relief, 100 features, OM classifier.

R100_P2: Relief 100 features, polynomial SVM order 2 classifier.

R100_P4: Relief, 100 features, polynomial SVM order 4 classifier.

R100_RBF: Relief, 100 features, RBF SVM classifier.

R7_RBF: Relief, 7 features, RBF SVM classifier.

Modified Relief (4 nearest hits, 4 nearest misses).

R100_K4_P4: Relief 4 nearest, 100 features, polynomial SVM order 4 classifier.

R100_K4_RBF: Relief 4 nearest, 100 features, RBF SVM classifier.

R100_K4_OM: Relief 4 nearest, 100 features, OM classifier.

R100_K4_NN: Relief 4 nearest, 100 features, Nearest Neighbor classifier.

R7_K4_RBF: Relief 4 nearest, 7 features, RBF classifier.

Modified Relief combined with Gram-Schmidt:

GSR_K4_PV1_RBF: Gram-Schmidt Relief 4 nearest, pvalue 1%, RBF SVM.

GSR_K4_PV10_RBF: Gram-Schmidt Relief 4 nearest, pvalue 10%, RBF SVM.

GSR7_K4_RBF: Gram-Schmidt Relief 4 nearest, 7 features, RBF SVM.

<u>RBF on linear features:</u>

OMMU100_RBF: OM multiplicative updates, 100 features, RBF classifier.

OMMU7_RBF: OM mult. updates to100 features, keep top 7, RBF classifier.

GS100_ RBF: Gram-Schmidt 100 features, RBF classifier.

GS7_ RBF: Gram-Schmidt, 7 features, RBF classifier.

SC100_ RBF: Shrunken centroids, 100 features, RBF classifier.
SC7_RBF: Shrunken centroids, 7 features, RBF classifier.

Combined methods:
INTER2373_GSR7_K4_RBF: GSR7_K4 applied to the first 2373 features to select 7 features, RBF classifier.
INTER2373_GSR30_K4_RBF: GSR7_K4 applied to the first 2373 features to select 30 features, RBF classifier.

3. Experiments

The results of our experiments are shown in Tables 2 and 3. We show only the top ranking methods in these matrices.

3.1 Results for 100 features

We show in Table 2 a comparison of the results for the selection of 100 features. We restrict the set of experiments to linear SVMs (OM) and RBF SVMs (exponential kernel, gamma=0.1).

The observations include:

- The interval selection method does not allow us to select small numbers of features, it performs poorly.
- RBF (a non-linear multivariate classifier) outperforms OM (a linear multivariate classifier).
- Features selected with a non-linear method may perform poorly with a linear classifier. Here this is the case for Relief (R and R_K4), but not for the multiplicative updates.
- Features selected with a linear method may perform better with a non-linear classifier (true here for all selection methods: Gram-Schmidt (GS), Shrunken centroids (SC), and multiplicative updates (MU)).
- The best method is Relief, but it is followed closely by Gram-Schmidt and non-linear multiplicative updates.
- Overall, we see a performance trend, be it for feature selection or classification: non-linear multivariate > linear multivariate > non-linear univariate.

3.2 Result for 7 features

We show in Table 3 a comparison of the results for the selection of 7 features. We restrict the set of experiments to linear SVMs (OM) and RBF SVMs (exponential kernel, gamma=0.1). The Relief features are not tried with OM (since they performed so poorly with OM on 100 features). We also do not include the "interval" selection method that performed poorly on 100 features.

The observations include:

- RBF still outperforms OM.
- Here NLMU performs better with the RBF SVM classifier that the linear SVM (contrarily to the 100 feature case).
- The best features are Relief features, but non-linear multiplicative updates performs well too.
- The very best is obtained with the combined method: select first an interval using CV, then reduce further the feature set using Gram-Schmidt Relief.
- Overall, we see confirm the performance trend: non-linear multivariate > linear multivariate > non-linear univariate.

Table 2. Statistical significance of the difference in performance of the best ranking methods using **100 features** according to McNemard tests on the three folds. The scores shown are the sum of the scores of the three folds: 1 for significantly better, 0 for undistinguishable, -1 for significantly worse. The matrix is antisymmetric. The circles at the top are color coding multivariate non-linear methods (orange), multivariate linear methods (green), and univariate linear methods (yellow). The center codes for the feature selection method and the outside circle for the classification method.

...

	1	2	3	4	5	6	7	8	9	10	11	12	13	14
1. R100_RBF	5.2	0	0	0	2	1	1	1	1	2	2	3	3	3
2. R100_K4_RBF		5.2	0	0	1	0	1	1	2	3	2	3	3	3
3. GS100_RBF			6.2	0	0	0	0	0	1	0	0	1	3	3
4. NLMU100_OM				6.5	1	1	1	1	1	1	1	2	3	3
5. GS100_OM					7.4	0	1	1	1	1	0	1	3	3
6. NLMU100_RBF						7.7	0	0	0	0	0	1	3	3
7.OMMU100_RBF							8.0	0	0	0	1	1	3	3
8. SC100_RBF								8.6	0	0	0	0	3	3
9. OMMU100_OM									9.6	0	1	0	3	3
10. R100_OM										10.5	0	0	2	2
11. R100K4_OM											10.8	1	2	2

12. SC100_OM											12.6	1	1
13.INTER100_RBF												19.7	0
14. INTER100_OM													20.1

4. Conclusions

We presented a set of classification experiments on a particular dataset for which input features greatly outnumber the number of training examples. In spite of the risk of overfitting, our results provide empirical evidence that non-linear multivariate methods can perform better than linear multivariate methods, which in turn perform better than linear univariate methods. From this study, we cannot draw general conclusions. In fact, with other preprocessings, we found that the linear multivariate method gave better results on the same dataset. However, we have confirmed experimentally results already reported elsewhere that multivariate methods (linear and non-linear) do not necessarily overfit the data, even in very adverse cases when the number of features is very large compared to the number of examples. Therefore, multivariate methods are worth keeping in the data analysis toolkit for such problems. They may provide more compact feature subsets for identical or even better performance. These conclusions have been confirmed by the results of a recent benchmark that we organized. The top ranking methods were non-linear multivariate methods. Such methods also yielded the most compact feature subsets (Guyon et al 2004).

Table 3. Statistical significance of the difference in performance of the best ranking methods using **7 features** according to McNemard tests on the three folds. We use the same conventions as in Table 2.

	1	**2**	**3**	**4**	**5**	**6**	**7**	**8**	**9**	**10**	**11**	**12**
1. INTER2373_GSR7_ K4_RBF	8.0	0	1	1	1	3	2	3	3	3	3	3
2. GSR7_K4_RBF		8.9	0	1	1	3	2	3	3	3	3	3
3. NLMU7_RBF			12.3	0	0	0	0	3	2	2	2	3
4. R7K4_RBF				13.6	0	1	0	2	2	2	3	3
5. R7_RBF					13.9	1	0	1	2	2	3	3
6. SC7_RBF						17.3	0	0	2	2	2	3

7. GS7_RBF							17.9	0	1	0	1	3
8. NLMU7_OM								22.5	0	0	0	0
9. GS7_OM									23.5	0	0	1
10. OMMU7_RBF										23.5	0	1
11. OMMU7_OM											25.6	1
12. SC7_OM												29.0

Acknowledgements

We would like to thank Masoud Nikravesh for challenging us with the idea that non-linear multivariate feature selection can provide more compact and better feature subsets.

References

B.-L Adam, et al, Serum Protein Fingerprinting Coupled with a Pattern-matching Algorithm Distinguishes Prostate Cancer from Benign Prostate Hyperplasia and Healthy Men, Cancer Research 62, 3609–3614, July 1, 2002.

B. Boser, I. Guyon, and V. Vapnik, An training algorithm for optimal margin classifiers. In Fifth Annual Workshop on Computational Learning Theory, pages 144--152, Pittsburgh, ACM. 1992.

T. G. Dietterich, Approximate Statistical Tests for Comparing Supervised Classification Learning Algorithms. Neural Computation, *10* (7) 1895-1924.

H. Drucker, D. Wu and V. Vapnik. Support Vector Machines for Spam Categorization. IEEE Trans. on Neural Networks , vol 10, number 5, pp. 1048-1054. 1999.

I. Guyon, J. Makhoul, R. Schwartz, and V. Vapnik, What size test set gives good error rate estimates?. PAMI, 20 (1), pages 52--64, IEEE. 1998.

I. Guyon, J. Weston, S. Barnhill, and V. Vapnik. Gene selection for cancer classification using support vector machines. Machine Learning, 46 (1-3), pages 389--422, 2002.

I. Guyon, A. Elisseeff, An Introduction to Variable and Feature Selection. JMLR, 3(Mar):1157-1182, 2003.

I. Guyon, S. Gunn, M. Nikravesh, L. Zadeh. Feature extraction: foundations and applications. Book in preparation http://clopinet.com/isabelle/Projects/NIPS2003/call-for-papers.html.

K. K. Jain. Biochips for Gene Spotting. Science, vol. 294, pages 621-625, Oct. 2001

K. Kira, and L. Rendell, A practical approach to feature selection. In D. Sleeman and P. Edwards (Eds.), Proceedings of the Ninth International Workshop on Machine Learning (ML92) (pp. 249-256). San Mateo, California: Morgan Kaufmann.

B. Schölkopf and A. Smola. Learning with Kernels. MIT Press, Cambridge, MA, 2002.

H. Stoppiglia, G. Dreyfus, R. Dubois, Y. Oussar. Ranking a Random Feature for Variable and Feature Selection. JMLR, 3(Mar):1399-1414, 2003.

C. M. Surman, The Use of Capillary Electrophoresis in Proteomics. GE Global Research Technical Report 2002GCRC138, June 2002.

R. Tibshirani, T. Hastie, B. Narasimhan, and G. Chu, Diagnosis of multiple cancer types by shrunken centroids of gene expression. R. Tibshirani, T. Hastie, B. Narasimhan, and G. Chu. PNAS, 99(10):6567--6572, 2002.

V. Vapnik, Statistical Learning Theory. V. Vapnik. John Wiley & Sons, N.Y., 1998.

J. Weston, F. Perez-Cruz, O. Bousquet, O. Chapelle, A. Elisseeff and B. Schoelkopf. "Feature Selection and Transduction for Prediction of Molecular Bioactivity for Drug Design". Bioinformatics, vol. 19 no. 6, pages 764-771, 2003.

J. Weston, A. Elisseeff, B. Schölkopf, Use of the Zero-Norm with Linear Models and Kernel Methods, Mike Tipping; JMLR, 3(Mar):1439-1461, 2003.

A New Fuzzy Spectral Approach to Information Integration in a Search Engine

Galina Korotkikh

Faculty of Informatics and Communication
Central Queensland University
Mackay, Queensland, 4740
Australia
email: g.korotkich@cqu.edu.au

Abstract. The problem of information integration is important for upgrading a search engine to a question-answering system. In the paper we consider a new fuzzy spectral approach to information integration in a search engine. The approach employs a series of variance-covariances matrices and suggests the eigenvalue spectra of the matrices as important characteristics of information integration. We show that the characteristics can be described in terms of eigenvalue dynamics. Through computational experiments we have identified an eigenvalue dynamics that can be efficiently computed by using the quadratic trace of the variance-covariance matrix. Moreover, this dynamics shows a property that can be interpreted as the eigenvalue integration. This suggests that the spectral characteristics are connected with an integration mechanism. The fuzzyfication of eigenvalues plays a key role in the observation of the dynamics. This may support the idea that fuzziness is an integral part of information integration.

1 Introduction

Recently, Lotfi Zadeh has suggested that search engines should be upgraded to question-answering systems with the ability to integrate an answer to a query by using a number of information parts [1]. To realize this integration function a search engine needs the capacity to identify the dependencies that may arise between the information parts.

In general, the problem of information integration is very challenging and remains unresolved. In the Internet search context it may be further complicated by a large number of information parts involved. Therefore, even identification of possible characteristics of information integration may be useful for upgrading a search engine to a question-answering system.

In the paper we propose a new fuzzy spectral approach to information integration in a search engine. The approach employs a series of variance-covariances matrices [2] and suggests the eigenvalue spectra of the matrices as important characteristics of information integration. The approach is hierarchical and classifies the dependencies between the parts by using different scale levels. Namely, on each scale level the eigenvalue spectrum of the variance-covariance matrix serves to characterize the dependencies arising in the information integration of the parts.

Specifically, we focus on the first scale level of the approach and use the eigenvalue spectrum of the variance-covariance matrix as a characteristic of the dependencies between the information parts. In particular, to investigate the spectral characteristic of information integration we address the following question: is it possible to describe the increase of the dependencies in terms of eigenvalue dynamics. Through computational experiments we have identified an eigenvalue dynamics that suggests an answer to the question [2]. Remarkably, the eigenvalue dynamics can be efficiently computed by using the quadratic trace of the variance-covariance matrix. Moreover, this dynamics shows a property that can be interpreted as the eigenvalue integration. This suggests that the spectral characteristic is connected with an integration mechanism.

It turns out that the fuzzyfication of eigenvalues plays a key role in the observation of the dynamics. This may support the idea that fuzziness is an integral part of information integration.

2 A Fuzzy Spectral Approach to Information Integration

Let a search engine be considered to integrate information from $N \geq 2$ separate parts. To realize this integration function a search engine needs the capacity to identify the dependencies that may arise between the information parts.

It is assumed that part i can be represented by a binary sequence $\bar{s}_i = s_{i1}...s_{in}$, $s_{ij} \in \{-1, 1\}$, $i = 1, ..., N$, $j = 1, ..., n$ of length $n \geq 2$. Then the task of a search engine is to find out the dependencies between the parts, which, when considered together, can be described by an $N \times n$ matrix $S = \{s_{ij}\}_{i=1,...,N,\ j=1,...,n}$. Let $\mathbf{S}$ be the set of such matrices.

We propose a new fuzzy spectral approach to information integration in a search engine. The approach employs a series of variance-covariances matrices [2] and suggests the eigenvalue spectra of the matrices as important characteristics of information integration. The approach is hierarchical and classifies the dependencies between the parts by using different scale levels. Namely, on each scale level the eigenvalue spectrum of the variance-covariance matrix serves to characterize the dependencies arising in the information integration of the parts.

In particular, we represent a sequence $\bar{s}_i = s_{i1}...s_{in}$, $i = 1, ..., N$ as a piecewise constant function φ_i defined on a real interval $[0, n]$, where s_{ij} is the value of the function on an interval $[j-1, j]$, $j = 1, ...n$ of length 1.

Let

$$\varphi_i^{[k]}(t) = \int_0^t \varphi_i^{[k-1]}(\tau) d\tau, \quad 0 \leq t \leq n,$$

$$\varphi_i^{[0]} = \varphi_i, \quad \varphi_i^{[k]}(0) = 0, \quad k \geq 1, \quad i = 1, ..., N.$$

We consider a series of the following variance-covariance matrices to characterize the dependencies arising in the information integration of the parts $\bar{s}_i, i = 1, ..., N$

$$V(S^{[k]}) = \{V(\varphi_i^{[k]}, \varphi_j^{[k]})\}_{i,j=1,...,N}, \quad k \geq 0,$$

$$V(S^{[0]}) = V(S), \quad S = \{s_{ij}\}_{i=1,...,N,\ j=1,...,n},$$

where the linear correlation coefficient $V(\varphi_i^{[k]}, \varphi_j^{[k]})$ is given by

$$V(\varphi_i^{[k]}, \varphi_j^{[k]}) = \frac{Cov(\varphi_i^{[k]}, \varphi_j^{[k]})}{\sigma(\varphi_i^{[k]})\sigma(\varphi_j^{[k]})}$$

$$= \frac{\frac{1}{n}\int_0^n \varphi_i^{[k]}(t)\varphi_j^{[k]}(t)dt - \frac{1}{n}\int_0^n \varphi_i^{[k]}(t)dt\frac{1}{n}\int_0^n \varphi_j^{[k]}(t)dt}{\sigma(\varphi_i^{[k]})\sigma(\varphi_j^{[k]})}, \quad i, j = 1, ..., N,$$

and

$$\sigma^2(\varphi_i^{[k]}) = \frac{1}{n}\int_0^n \varphi_i^{[k]}(t)\varphi_i^{[k]}(t)dt - (\frac{1}{n}\int_0^n \varphi_i^{[k]}(t)dt)^2,$$

$$\sigma^2(\varphi_j^{[k]}) = \frac{1}{n}\int_0^n \varphi_j^{[k]}(t)\varphi_j^{[k]}(t)dt - (\frac{1}{n}\int_0^n \varphi_j^{[k]}(t)dt)^2.$$

The eigenvalue spectra of this series of variance-covariance matrices are used to partition the matrices into classes with similar dependencies between the information parts. The classification is hierarchical and specifies the dependencies between the parts by using different scale levels. In particular, on each scale level a class of matrices may be further partitioned using the set of eigenvalue spectra of the variance-covariance matrices of the class. The partition depends on a fuzzy parameter $\varepsilon > 0$. In particular, the range of eigenvalue is divided into adjoining but not overlapping intervals of length ε. This length is a measure of uncertainty in the consideration of the eigenvalues.

First, let

$$\mathbf{S}_\varepsilon(\bar{\lambda}_0, ..., \bar{\lambda}_k) \subset \mathbf{S}, \quad \bar{\lambda}_i = (\lambda_{i1}, ..., \lambda_{iN}),\ i = 0, ..., k,\ k \geq 0$$

be a class of matrices on level $(k+1)$ of the classification such that if a matrix S belongs to the class $S \in \mathbf{S}_\varepsilon(\bar{\lambda}_0, ..., \bar{\lambda}_k)$ then the eigenvalue spectrum

$$Spec(V(S^{[i]})) = (\lambda'_{i1}, ..., \lambda'_{iN}),\ i = 0, ..., k$$

of the variance-covariance matrix $V(S^{[i]}),\ i = 0, ..., k$ satisfies

$$\lambda_{ij} \leq \lambda'_{ij} \leq \lambda_{ij} + \varepsilon,$$

where $\lambda_{ij},\ i = 0, ..., k,\ j = 1, ..., N$ are multiples of a fuzzy parameter ε.

Second, a class of matrices $S \in \mathbf{S}_\varepsilon(\bar\lambda_0, ..., \bar\lambda_k)$ of level $(k+1)$ may be further partitioned into classes of level $(k+2)$

$$\mathbf{S}_\varepsilon(\bar\lambda_0, ..., \bar\lambda_k) = \bigcup_{\bar\lambda_{k+1} \in Spec(V(\mathbf{S}_\varepsilon^{[k]}(\bar\lambda_0, ..., \bar\lambda_k)))} \mathbf{S}_\varepsilon(\bar\lambda_0, ..., \bar\lambda_k, \bar\lambda_{k+1}).$$

using the set of eigenvalue spectra

$$Spec(V(\mathbf{S}_\varepsilon^{[k]}(\bar\lambda_0, ..., \bar\lambda_k)))$$

of the variance-covariance matrices $V(S^{[k]})$ and the fuzzy parameter ε, where $S \in \mathbf{S}_\varepsilon(\bar\lambda_0, ..., \bar\lambda_k)$.

Using this classification we suggest to compare the matrices of $\mathbf{S}$ in terms of dependencies between the parts as follows:
1. The set of all matrices $\mathbf{S}$ forms one class at the zero $k = 0$ level. In this class the matrices are not distinguished in terms of dependencies between the information parts.
2. The matrices of a class $\mathbf{S}_\varepsilon(\bar\lambda_0, ..., \bar\lambda_{k-1})$ at level $k \geq 1$ are considered to have the same dependencies between the information parts.
3. Matrices $S, S' \in \mathbf{S}_\varepsilon(\bar\lambda_0, ..., \bar\lambda_{k-1})$ of a same class at level $k \geq 1$ may be compared in terms of dependencies between the information parts if they belong to different classes

$$S \in \mathbf{S}_\varepsilon(\bar\lambda_0, ..., \bar\lambda_{k-1}, \bar\lambda_k), \quad S' \in \mathbf{S}_\varepsilon(\bar\lambda_0, ..., \bar\lambda_{k-1}, \bar\lambda_k'), \quad \bar\lambda_k \neq \bar\lambda_k',$$

at level $(k+1)$. Matrices $S, S' \in \mathbf{S}$ may be compared in this sense if they belong to different classes

$$S \in \mathbf{S}_\varepsilon(\bar\lambda_0), \quad S' \in \mathbf{S}_\varepsilon(\bar\lambda_0'), \quad \bar\lambda_0 \neq \bar\lambda_0$$

at the first scale level.

To compare matrices within a class we identify two extreme cases: the minimal and maximal ones. They are suggested to give the lower and upper bounds for the dependencies existing between the parts of a matrix.

The minimal case is specified using the variance-covariance matrix, whose linear correlation coefficients are all 1

$$V_{min} = \begin{pmatrix} 1\,1\,...\,1\,1 \\ 1\,1\,...\,1\,1 \\ .\,\,.\,\,...\,.\,\,. \\ 1\,1\,...\,1\,1 \\ 1\,1\,...\,1\,1 \end{pmatrix}$$

The maximal case is determined using the variance-covariance matrix, whose non-diagonal linear correlation coefficients are all 0

$$V_{max} = \begin{pmatrix} 1\,0\,...\,0\,0 \\ 0\,1\,...\,0\,0 \\ .\,\,.\,\,...\,.\,\,. \\ 0\,0\,...\,1\,0 \\ 0\,0\,...\,0\,1 \end{pmatrix}$$

The following example illustrates the classification. Consider three matrices

$$S_1 = \begin{pmatrix} +1 & -1 & +1 & -1 \\ -1 & +1 & -1 & +1 \\ +1 & -1 & -1 & +1 \\ -1 & +1 & +1 & -1 \end{pmatrix} \quad S_2 = \begin{pmatrix} +1 & -1 & -1 & +1 \\ -1 & +1 & +1 & -1 \\ +1 & +1 & -1 & -1 \\ -1 & -1 & +1 & +1 \end{pmatrix} \quad S_3 = \begin{pmatrix} +1 & +1 & +1 & +1 \\ -1 & -1 & -1 & -1 \\ +1 & -1 & -1 & +1 \\ -1 & +1 & +1 & -1 \end{pmatrix}$$

At the first scale level of the classification, these three matrices belong to a same class, because their variance-covariance matrices are the same

$$V(S_1) = V(S_2) = V(S_3) = \begin{pmatrix} 1 & -1 & 0 & 0 \\ -1 & 1 & 0 & 0 \\ 0 & 0 & 1 & -1 \\ 0 & 0 & -1 & 1 \end{pmatrix}$$

$$Spec(V(S_1)) = Spec(V(S_2)) = Spec(V(S_3)) = (\lambda_1, ..., \lambda_4) = (0, 0, 2, 2).$$

However, at the second scale level of the classification, while matrices S_1 and S_2 are in a same class, since

$$V(S_1^{[1]}) = V(S_2^{[1]}) = \begin{pmatrix} 1 & -1 & 0 & 0 \\ -1 & 1 & 0 & 0 \\ 0 & 0 & 1 & -1 \\ 0 & 0 & -1 & 1 \end{pmatrix}$$

$$Spec(V(S_1^{[1]})) = Spec(V(S_2^{[1]})) = (\lambda_1, ..., \lambda_4) = (0, 0, 2, 2),$$

the matrix S_3 belongs to a different class when $0 \leq \varepsilon \leq 1$.

Furthermore, the matrices S_1 and S_2, being in the same classes at the first and second scale levels of the classification, at the third scale level belong to different classes when $0 \leq \varepsilon \leq 1$.

3 Fuzzy Spectral Characteristic of Information Integration

In this section we focus only on the first scale level of the classification. For a matrix

$$S = \{s_{ij}\}_{i=1,\dots,N,\ j=1,\dots,n} \in \mathbf{S}$$

we consider the variance-covariance matrix

$$V(S) = \{V(\bar{s}_i, \bar{s}_j)\}_{i,j=1,\dots,N},$$

where $V(\bar{s}_i, \bar{s}_j)$ is the linear correlation coefficient between sequences $\bar{s}_i$ and $\bar{s}_j$. In particular, on the first scale level of the classification the eigenvalue spectrum

$$Spec(V(S)) = (\lambda_1, ..., \lambda_N)$$

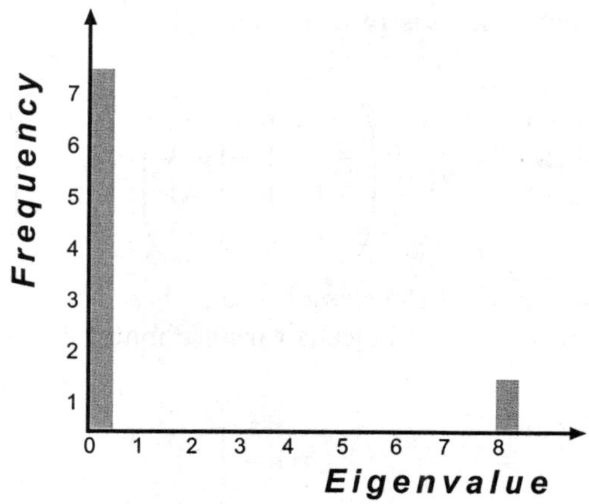

Fig. 1. The eigenvalue spectrum when all information parts are the same.

of the variance-covariance matrix $V(S)$ is used as a characteristic of the dependencies between the information parts of the matrix S. For this purpose, we suggest to consider two extreme cases:

1. The minimal case. The variance-covariance matrix is

$$V_{min} = \begin{pmatrix} 1\,1\,...\,1\,1 \\ 1\,1\,...\,1\,1 \\ .\;.\;...\;.\;. \\ 1\,1\,...\,1\,1 \\ 1\,1\,...\,1\,1 \end{pmatrix}, \; Spec(V_{min}) = (0, ..., 0, N),$$

$$tr(V_{min}^2) = \lambda_1^2 + ... + \lambda_N^2 = N^2.$$

As an example, a frequency distribution of the eigenvalue spectrum of V_{min} when $N = 8$ is shown in Figure 1. In this case all information parts of the matrix S are the same, because the linear correlation coefficient between any two parts is 1. Therefore, there are no dependencies between the information parts, since each part is a replica of one and the same data. There is no need for a search engine to integrate information in this case.

2. The maximal case. The variance-covariance matrix is

$$V_{max} = \begin{pmatrix} 1\,0\,...\,0\,0 \\ 0\,1\,...\,0\,0 \\ .\;.\;...\;.\;. \\ 0\,0\,...\,1\,0 \\ 0\,0\,...\,0\,1 \end{pmatrix}, \; Spec(V_{max}) = (1, ..., 1),$$

$$tr(V_{max}^2) = \lambda_1^2 + ... + \lambda_N^2 = N.$$

As an illustration, a frequency distribution of the eigenvalue spectrum of V_{max} when $N = 8$ is shown in Figure 2. In this case, the linear correlation coefficient between any two information parts is 0. But zero linear

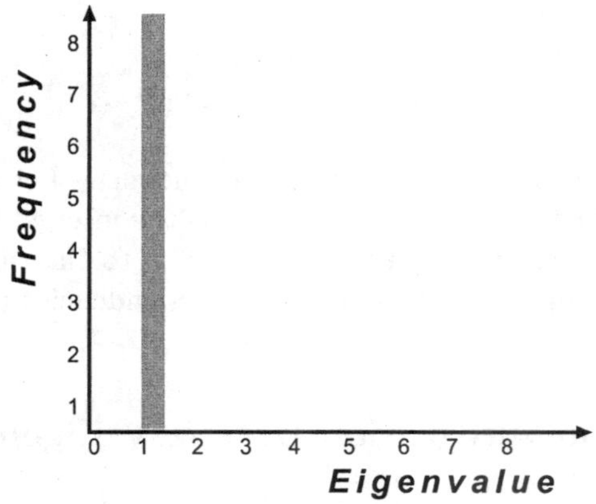

Fig. 2. The eigenvalue spectrum is suggested to describe the upper bound of the dependencies between the information parts.

correlation does not, in general, imply independence. Only in the case of the multivariate normal distribution, it is possible to interpret uncorrelatedness as implying independence [3]. Furthermore, we suggest that this case gives us a way describe the upper bound of *nonlinear* dependencies existing between the information parts of a matrix [2]. We view that it is important for a search engine to have the capacity to identify such dependencies in order to integrate the information for the answer.

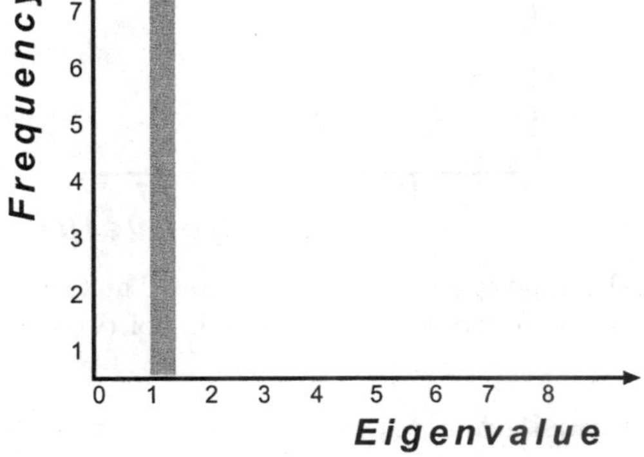

Fig. 3. The two eigenvalues following the spectral composition property start to approach each other in order to meet and integrate in the middle.

It is proposed that for a matrix $S \in \mathbf{S}$ we have [4]

$$1 \leq \frac{N^2}{tr(V^2(S))} \leq N. \tag{1}$$

In order to describe and measure the dependencies between the information parts of a matrix $S \in \mathbf{S}$, we suggest to use the interval (1). In particular, the closer the point specifying a matrix $S \in \mathbf{S}$ is to the right end of the interval, then, we assume, the greater are the dependencies between the parts of the matrix.

4 Information Integration by Fuzzy Eigenvalue Dynamics

In this section we investigate the spectral characteristic of information integration by addressing the following question: is it possible to describe the increase of the dependencies, i.e., under our assumption, the movements from the left to the right in the interval (1), in terms of eigenvalue dynamics. Through computational experiments we have identified an eigenvalue dynamics that suggests an answer to the question [2]. Remarkably, this dynamics shows a property that can be interpreted as the eigenvalue integration. This suggests that the spectral characteristic is connected with an integration mechanism. The eigenvalue dynamics can be described by using a spectral composition property [2].

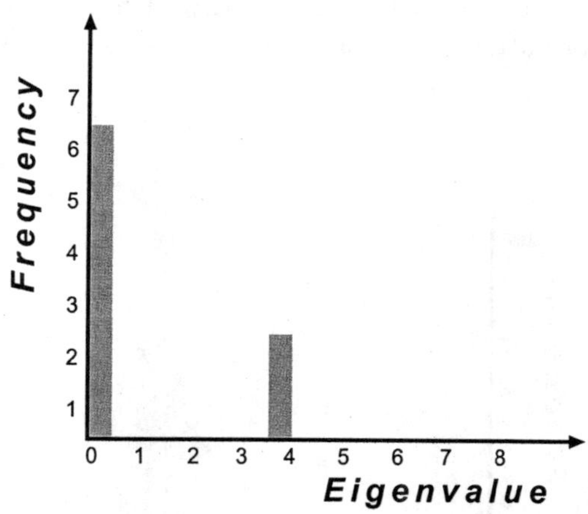

Fig. 4. The spectral composition property in action. The two eigenvalues of Figure 3 have integrated into a composite object consisting of two eigenvalues.

Spectral Composition Property. *Let $S, S' \in \mathbf{S}$ be two $N \times n$ matrices such that*

$$Spec(V(S)) = (\lambda_1, ..., \lambda_N), \quad Spec(V(S')) = (\lambda'_1, ..., \lambda'_N).$$

If there exist an integer $q = 1, ..., \lfloor N/2 \rfloor$, integers l, r with $l+1 < r$ and a fuzzy parameter $\varepsilon > 0$ such that $\lambda_l = l\varepsilon$, $\lambda_r = r\varepsilon$

$$0 \leq \lambda_i - \lambda_l < \varepsilon, \quad 0 \leq \lambda_{q+i} - \lambda_r < \varepsilon, \quad i = 1, ..., q,$$

$$0 \leq \lambda_i' - \frac{\lambda_l + \lambda_r}{2} < \varepsilon, \quad i = 1, ..., 2q \quad \textit{if } l + r \textit{ is even,}$$

$$0 \leq |\lambda_i' - (\lfloor \frac{l+r}{2} \rfloor + 1)\varepsilon| < \frac{\varepsilon}{2}, \quad i = 1, ..., 2q \quad \textit{if } l + r \textit{ is odd}$$

and

$$|\lambda_i - \lambda_i'| \leq \varepsilon, \quad i = 1, ..., N - 2q, \ k \neq \lfloor N/2 \rfloor, \tag{2}$$

where $\lfloor x \rfloor$ is the integer part of x, then we say that the matrices S and S' have the spectral composition property, denoted $S \prec S'$, for the fuzzy parameter ε.

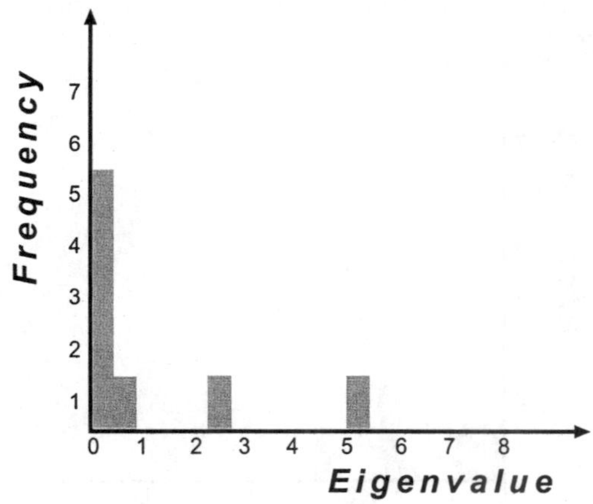

Fig. 5. The composite object of Figure 4 has collapsed and the two eigenvalues split. The third nonzero eigenvalue has come to produce new composite eigenvalue objects.

The spectral composition property (2) admits the following interpretation. Given the eigenvalues are viewed as some objects at positions $\lambda_i, i = 1, ..., N$, the spectral composition property means that q of the objects located at λ_l, and q of the objects located at λ_r, integrate into a composite object. The position of the composite object is approximately in the middle of the interval $[\lambda_l, \lambda_r]$ and the uncertainty is specified by the fuzzy parameter ε.

The existence of such distinctive dynamics provides us with a "map" that is useful to locate the results of information integration. The eigenvalue dynamics, illustrated in Figures 3-7, specifically presents the spectral composition property. Remarkably, it has been found that this property cannot be observed for information matrices based on random sequences (see Figure 8).

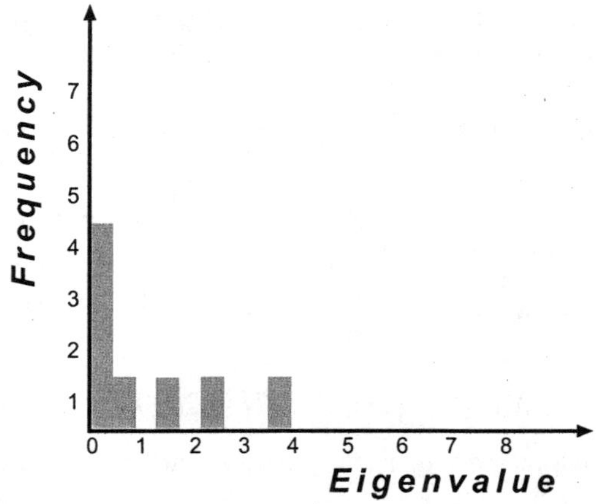

Fig. 6. There are four nonzero eigenvalues. The second and third nonzero eigenvalues move to each other in order to integrate into a composite object.

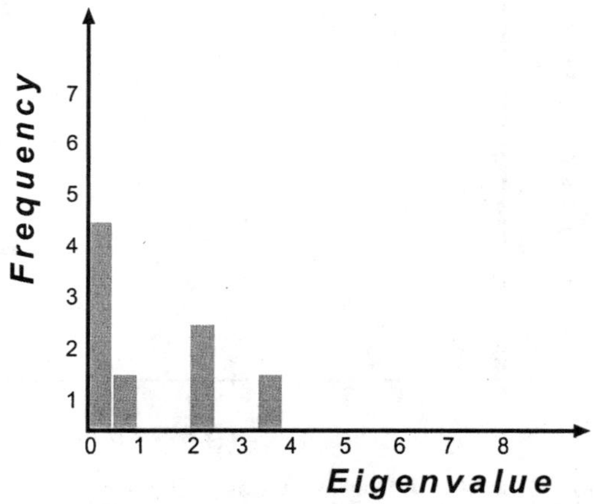

Fig. 7. The composite eigenvalue object has been created.

The spectral composition property can be efficiently expressed in computations using the following.

Proposition 1. *Let matrices* S *and* S' *such that*

$$Spec(V(S)) = (\lambda_1, ..., \lambda_N), \quad Spec(V(S')) = (\lambda'_1, ..., \lambda'_N)$$

have the spectral composition property $S \prec S'$ *for a fuzzy parameter* ε, *then the quadratic traces of their variance-covariance matrices* $V(S)$ *and* $V(S')$ *satisfy the condition*

$$tr(V^2(S)) > tr(V^2(S')). \tag{3}$$

Proof. It is known that if the eigenvalue spectrum of a matrix V is $Spec(V) = (\lambda_1, ..., \lambda_N)$ then the trace of a matrix $V^2 = V * V$ or the quadratic trace of matrix V equals

$$tr(V^2) = \sum_{i=1}^{N} \lambda_i^2 = \lambda_1^2 + ... + \lambda_N^2.$$

By using the spectral composition property (2) we have

$$tr(V^2(S)) = \sum_{i=1}^{N} \lambda_i^2 = \sum_{i=1}^{q} \lambda_i^2 + \sum_{i=q+1}^{2q} \lambda_i^2 + \sum_{i=2q+1}^{N} \lambda_i^2$$

$$= \sum_{i=1}^{q} \lambda_l^2 + \sum_{i=q+1}^{2q} \lambda_r^2 + \sum_{i=2q+1}^{N} \lambda_i'^2 = q(\lambda_l^2 + \lambda_r^2) + \sum_{i=2q+1}^{N} \lambda_i'^2. \qquad (4)$$

Since

$$\lambda_l^2 + \lambda_r^2 > 2(\frac{\lambda_l + \lambda_r}{2})^2,$$

then (4) can be written as

$$tr(V^2(S)) = q(\lambda_l^2 + \lambda_r^2) + \sum_{i=2q+1}^{N} \lambda_i'^2 > 2q(\frac{\lambda_l + \lambda_r}{2})^2 + \sum_{i=2q+1}^{N} \lambda_i'^2$$

$$= 2q\lambda_m^2 + \sum_{i=2q+1}^{N} \lambda_i'^2 = \sum_{i=1}^{2q} \lambda_i'^2 + \sum_{i=2q+1}^{N} \lambda_i'^2$$

$$= \sum_{i=1}^{N} \lambda_i'^2 = tr(V^2(S')).$$

Thus we arrive at condition (3)

$$tr(V^2(S) > tr(V^2(S'))$$

and the proposition follows.

From Proposition 1 for matrices $S, S' \in \mathbf{S}$ such that $S \prec S'$, we obtain a condition

$$1 \leq \frac{N^2}{tr(V^2(S))} < \frac{N^2}{tr(V^2(S'))} \leq N.$$

This condition describes the spectral composition property of the matrices S, S' in terms of the interval (1). Therefore, if two matrices $S, S' \in \mathbf{S}$ have the spectral composition property $S \prec S'$, then under our assumption the dependencies between the information parts of the matrix S are less then the dependencies between the information parts of the matrix S'.

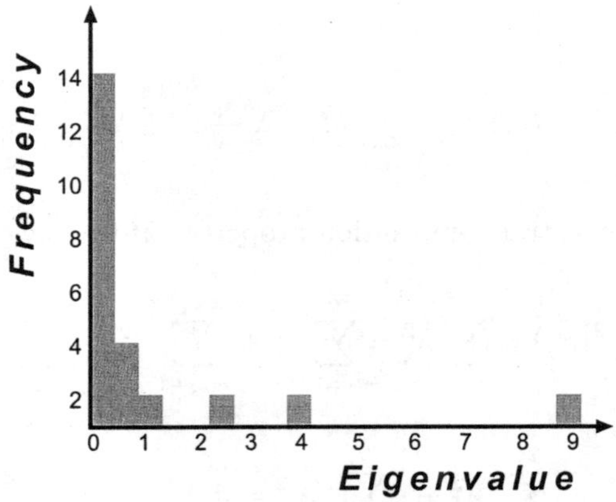

Fig. 8. A typical eigenvalue spectrum of random information matrix.

References

1. L. A. Zadeh, *From Search Engines to Question-Answering Systems*, 2003 BISC FLINT-CIBI International Joint Workshop on Soft Computing for Internet and Bioinformatics, M. Nikravesh, S. S. Bensafi and L. A. Zadeh (Eds.), Memorandum No. UCB/ERL M03/47, 2003, pp. 33-34.
2. G. Korotkikh, *A Computational Approach in Dealing with Uncertainty of Financial Markets*, PhD Thesis, Central Queensland University, 2003.
3. P. Embrechts, A. McNeil and D. Straumann, *Correlation and Dependence in Risk Management: Properties and Pitfalls*, Department Mathematik, ETHZ, July 1999.
4. V. Korotkikh and G. Korotkikh, *On a New Quantization in Complex Systems*, in Quantitative Neuroscience, P. Pardalos, C. Sackellares, P. Carney and L. Iasemidis (Eds.), Kluwer Academic Publishers, Dordrecht/Boston/London, 2004, pp. 71-91.

Towards Irreducible Modeling of Structures and Functions of Protein Sequences

Victor Korotkikh

Faculty of Informatics and Communication
Central Queensland University
Mackay, Queensland, 4740
Australia
email: v.korotkich@cqu.edu.au

Abstract. A major aim of bioinformatics is to contribute to our understanding of the relationship between protein sequence and its structure and function. In the paper we present an approach that allows us to derive a new type of hierarchical structures and formation processes from sequences. These structures and formation processes are irreducible, because they are based only on the integers and develop within existing rules of arithmetic. Therefore, a key feature of the approach is that it may model structures and functions of protein sequences in an irreducible way.

1 Introduction

A major aim of bioinformatics is to contribute to our understanding of the relationship between protein sequence and its structure and function. In this context approaches that can produce structures and processes from sequences attract our special attention. Furthermore, if such an approach works consistently with observations, then naturally we may ask: why the approach is able to describe the nature of living things. We may even question whether the approach is fundamental or whether it may be explained in terms of deeper concepts. Therefore, when we look for an approach to model structures and functions of protein sequences, we should aim to find it in an irreducible form. In this case there will be no deeper level of understanding possible and thus nothing further exists that could explain the approach.

In the paper we present an approach that allows us to derive a new type of hierarchical structures and formation processes from sequences [1]. Such a formation process begins with integers acting as ultimate building blocks and produce integer relations of one level from the integer relations of the previous one. These integer relations in turn may form integer relations of the next level. Thus, this is a continual process, where the integer relations of a level already reached form the integer relations of the following level and so on. Notably however, the formation process is only maintained until eventually rules of arithmetic prevent it from producing more integer relations.

The formation processes are irreducible, because they are based only on the integers and develop within existing rules of arithmetic. This eliminates the need for further explanations of these formation processes and any possible questions arising: why they occur in the way they do [1]. Therefore, a key feature of the approach is that it may model structures and functions of protein sequences in an irreducible way.

Although the hierarchical structures and the formation processes are rigorously defined mathematically, in computations, however, they resist to be easily described and processed. It is shown that fuzzy logic can propose a solution to this problem [2].

2 Nonlocal Correlation Leads from Sequences to Structures and Processes

In this section we present a notion of nonlocal correlation [3]. This notion gives us a convenient way to derive structures and formation processes from sequences [1]. The correlation is assumed to exist between the parts of a sequence and act upon them in a specific manner. In particular, the correlation controls the changes of the parts of a sequence so that some of its global characteristics stay the same.

Now our objects of consideration are sequences. Let I be an integer alphabet and

$$I_n = \{s = s_1 ... s_n, \ s_i \in I, \ i = 1, ..., n\}$$

be the set of all sequences of length $n \geq 2$ with symbols in I. If $I = \{-1, +1\}$ then I_n is the set of all binary sequences of length n.

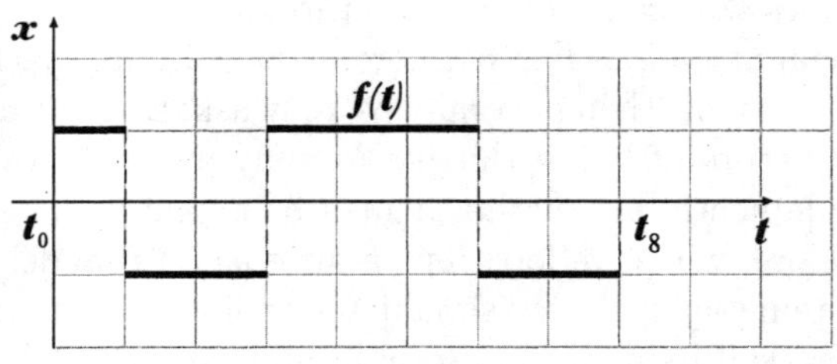

Fig. 1. The graph of a function $f = \rho_{011}(s)$, where $s = +1 -1 -1 +1 +1 +1 -1 -1$ and $m = 0, \delta = 1, \varepsilon = 1$.

We consider a geometric representation of sequences in 1+1 space-time by using piecewise constant functions. Let $\delta > 0$ and $\varepsilon > 0$ be respective spacings of a space-time lattice (δ, ε) in $1+1$ dimensions. Let $W_{\delta\varepsilon}([t_m, t_{m+n}])$ be a class of piecewise constant functions such that a function f of the class is constant on $(t_{i-1}, t_i]$, $i = m+1, ..., m+n$ and equals

$$f(t_m) = s_1 \delta, \quad f(t) = s_i \delta, \quad t \in (t_{m+i-1}, t_{m+i}], \ i = 1, ..., n,$$

$$t_i = i\varepsilon, \ i = m, ..., m+n,$$

where m is an integer and s_i, $i = 1, ..., n$ are real numbers. The sequence $s = s_1 ... s_n$ is called a code of the function f and is denoted by $s = c(f)$.

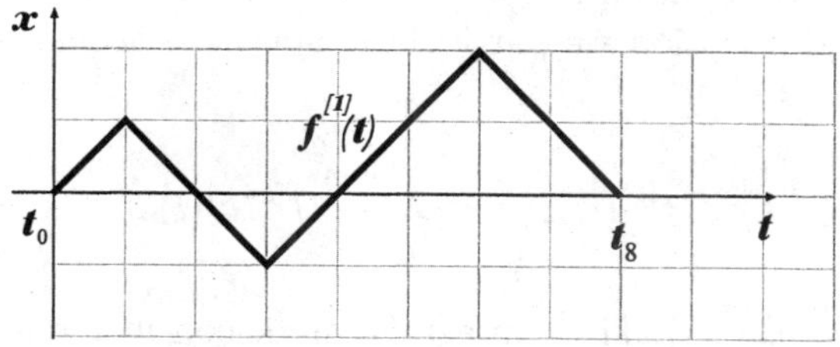

Fig. 2. The graph of the first integral $f^{[1]}$ of the function f in Figure 1.

Let $\rho_{m\delta\varepsilon} : s \to f$ be a mapping that associates a sequence $s \in I_n$ with a function $f \in W_{\delta\varepsilon}[t_m, t_{m+n}]$, denoted by $f = \rho_{m\delta\varepsilon}(s)$, such that $s = c(f)$ and whose kth integral satisfies the condition $f^{[k]}(t_m) = 0$, $k = 1, 2, \ldots$. For example, Figure 1 shows a function $f \in W_{11}([t_0, t_8])$ such that $f = \rho_{011}(s)$ and

$$s = +1 - 1 - 1 + 1 + 1 + 1 - 1 - 1. \tag{1}$$

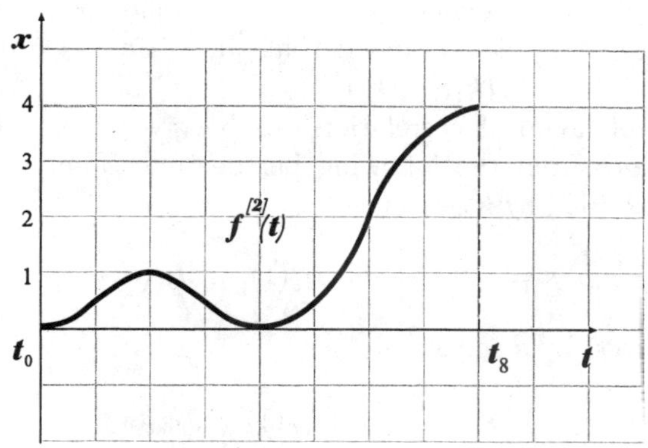

Fig. 3. The graph of the second integral $f^{[2]}$ of the function f in Figure 1.

To describe a sequence $s \in I_n$ we consider successive integrals

$$s \Longrightarrow f^{[1]}(t), f^{[2]}(t), ..., f^{[k]}(t), \ldots ,$$

$$f^{[k]}(t) = \int_{t_m}^{t} f^{[k-1]}(t')dt', \quad f^{[0]} = f, \quad k \geq 1, \quad t \in (t_m, t_{m+n}]$$

of a function

$$f = \rho_{m\delta\varepsilon}(s) \in W_{\delta\varepsilon}[t_m, t_{m+n}],$$

which gives the geometric representation of the sequence s. Figures 2 and 3 illustrate the description. They show the first integral $f^{[1]}$ and the second integral $f^{[2]}$ of the function f depicted in Figure 1. The function f provides the geometric representation of the sequence (1).

Within this description we particularly specify a sequence $s \in I_n$ using definite integrals

$$J_{\delta\varepsilon}(s,k) = f^{[k]}(t_{m+n}) = \int_{t_m}^{t_{m+n}} f^{[k-1]}(t)dt, \quad k \geq 1$$

of a function $f = \rho_{m\delta\varepsilon}(s)$. These integrals can be considered as global characteristics of the sequence s, because they use information about it as a whole object.

We are interested to partition and classify sequences in terms of such global characteristics. For this purpose let

$$I_n(J_{\delta\varepsilon}(1), ..., J_{\delta\varepsilon}(k)) = \{s \in I_n : J_{\delta\varepsilon}(s,1) = J_{\delta\varepsilon}(1), ..., J_{\delta\varepsilon}(s,k) = J_{\delta\varepsilon}(k)\}$$

be a set of sequences, whose $k \geq 1$ global characteristics are the same.

We present the notion of nonlocal correlation assuming that we have the following complex system. It consists of n parts and values of a component s_i can specify the states of part i. Moreover, a state of the complex system can be described by a sequence $s = s_1...s_n$, where the components take particular values. It is also assumed that the state space of the complex system is $I_n(J_{\delta\varepsilon}(1), ..., J_{\delta\varepsilon}(k)),\ k \geq 1$.

The notion of nonlocal correlation can be given when we look at the system's dynamics from the following perspective. Namely, the transition from a state described by a sequence

$$s = s_1...s_n \in I_n(J_{\delta\varepsilon}(1), ..., J_{\delta\varepsilon}(k))$$

to a state described by a sequence

$$s' = s'_1...s'_n \in I_n(J_{\delta\varepsilon}(1), ..., J_{\delta\varepsilon}(k))$$

is made under a correlation that controls the behaviors of the parts of the complex system in order to preserve k of its global characteristics. In other words, the correlation admits only those interdependent changes of the parts that preserve k global characteristics of the complex system. Therefore, in the first place the dynamics of a part of the complex system is specified by its interdependent role in the system as a whole, rather than by something that can be derived from the part as a separate entity.

It turns out that the mechanism of the correlation does not have its origin in space-time. Namely, it does not involve interactions and signals between the parts of a complex system. Moreover, the correlation is nonlocal in the sense that it acts on the parts, irrespective of their distances, at once [2]. It is shown that integer relations appear to be the "glue", which not requiring accommodation in space-time, nevertheless connects the parts of a complex system together and allows to control their behavior in such a correlated manner [2].

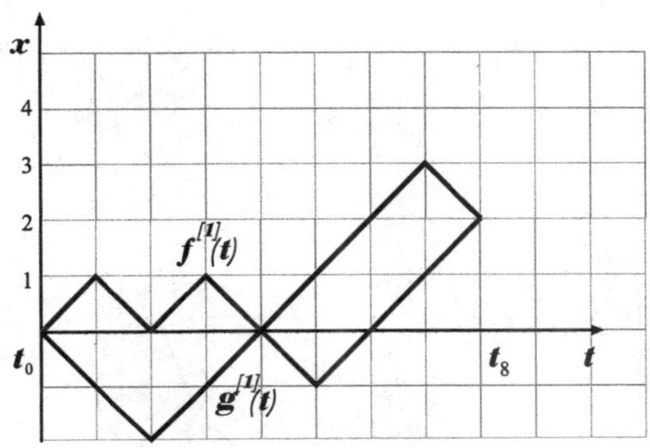

Fig. 4. The figure shows that for the sequences (4) we have $f^{[1]}(t_8) = g^{[1]}(t_8)$.

It is useful to consider the following condition between $C(s,s') \geq 1$ global characteristics of two different sequences $s,\ s' \in I_n$, i.e., successive integrals

$$f^{[k]}(t_{m+n}) = g^{[k]}(t_{m+n}), \quad k = 1, ..., C(s,s') \tag{2}$$

$$f^{[C(s,s')+1]}(t_{m+n}) \neq g^{[C(s,s')+1]}(t_{m+n}) \tag{3}$$

of functions $f = \rho_{m\delta\varepsilon}(s),\ g = \rho_{m\delta\varepsilon}(s')$. Recall, by definition

$$f^{[k]}(t_m) = g^{[k]}(t_m) = 0, \quad k = 1, ..., C(s,s').$$

If for sequences $s,\ s'$ we have $f^{[1]}(t_{m+n}) \neq g^{[1]}(t_{m+n})$, then $C(s,s') = 0$.

The conditions (2) and (3) appear as a proper form to investigate the nature of the optimization realized by the correlation. Namely, we may ask the question: why the correlation can control the changes of the parts of a sequence s to preserve at most $C(s,s')$ global characteristics in a resulting sequence s'. In other words, why the correlation cannot maximize for sequences s and s' the number of the same global characteristics to be greater than $C(s,s')$.

We illustrate the conditions (2) and (3) for binary sequences

$$s = +1 - 1 + 1 - 1 - 1 + 1 + 1 + 1,$$

$$s' = -1 - 1 + 1 + 1 + 1 + 1 + 1 - 1. \tag{4}$$

In this case we have $C(s,s') = 2$, because

$$f^{[1]}(t_8) = g^{[1]}(t_8), \quad f^{[2]}(t_8) = g^{[2]}(t_8),$$

but $f^{[3]}(t_8) \neq g^{[3]}(t_8)$, where $f = \rho_{011}(s)$, $g = \rho_{011}(s')$.

Figures 4 and 5 present this situation. In particular, from Figure 5 we can conclude

$$f^{[3]}(t_8) \neq g^{[3]}(t_8),$$

because this figure shows us that

$$f^{[2]}(t) > g^{[2]}(t),\ t \in (t_0, t_8), \quad f^{[2]}(t_8) = g^{[2]}(t_8)$$

and by definition $f^{[2]}(t_0) = g^{[2]}(t_0) = 0, \quad f^{[3]}(t_0) = g^{[3]}(t_0) = 0.$

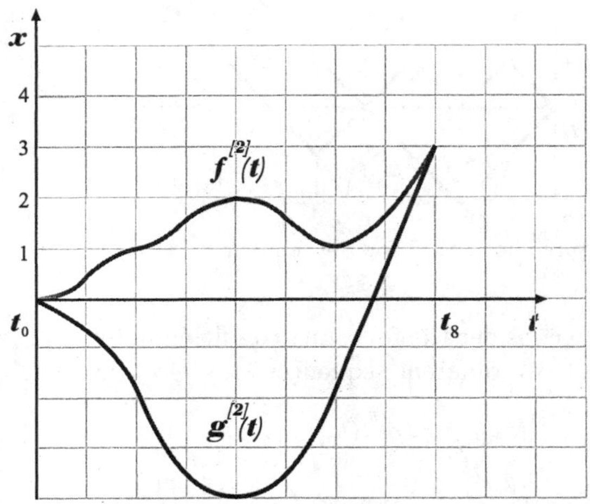

Fig. 5. The figure shows that for the sequences (4) we have $f^{[2]}(t_8) = g^{[2]}(t_8)$ and $f^{[2]}(t) > g^{[2]}(t),\ t \in (t_0, t_8)$.

The sequences (4) do not seem to have regularities and even may be initial segments of random sequences. However, there is an interesting pattern in how the parts of the sequence

$$s = s_1 ... s_n = +1 - 1 + 1 - 1 - 1 + 1 + 1 + 1$$

change to preserve two of its global characteristics in the resulting sequence

$$s' = s'_1 ... s'_n = -1 - 1 + 1 + 1 + 1 + 1 + 1 - 1.$$

In particular, these changes can be specified by the sequence $s'' = s - s'$, where $s'' = s''_1 ... s''_n$ and $s''_i = s_i - s'_i,\ i = 1, ..., n$. Thus, we have

$$s'' = +2 \ \ 0 \ \ 0 - 2 - 2 \ \ 0 \ \ 0 + 2.$$

Figure 6 shows us a function $h = \rho_{011}(s'')$ and reveals that the sequence s'' has a certain global pattern. We may say that this pattern serves the

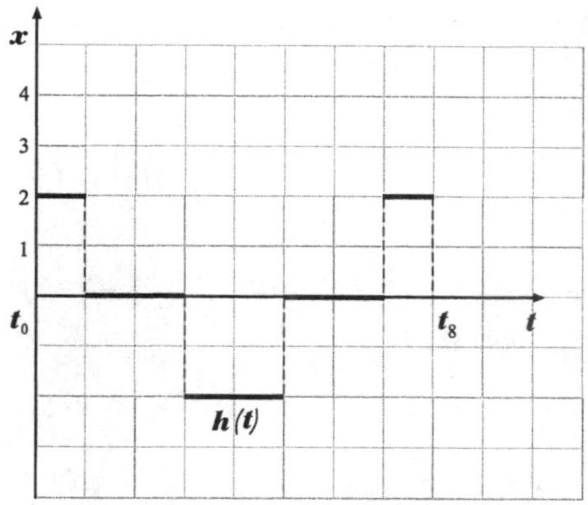

Fig. 6. The figure shows a function $h = \rho_{011}(s'')$, where $s'' = +2\ 0\ 0{-}2{-}2\ 0\ 0{+}2$. We can see that the sequence s'' has a certain global pattern.

correlation to control the changes of the parts of the sequence s in order to preserve two of its global characteristics in the resulting sequence s'. The pattern is global because it belongs to the sequence as a whole. Uncorrelated local changes of the sequence s'' destroy the pattern.

Using an integer code series [4] the nonlocal correlation can be expressed as a system of linear equations, where the coefficients are the powers of integers. Furthermore, it is possible to arrange these coefficients into a number of integer relations and identify their structure [1].

The integer code series can express an integral of a piecewise constant function $f \in W_{\delta\varepsilon}([t_m, t_{m+n}])$ in terms of the code $c(f)$ of the function f, powers of integers and combinatorial coefficients [4].

Integer Code Series. *Let $f \in W_{\delta\varepsilon}([t_m, t_{m+n}])$ be a piecewise constant function such that $c(f) = s_1 ... s_n$. Then the kth $k \geq 1$ integral $f^{[k]}$ of the function f at a point $t_{m+l}, l = 1, ..., n$ can be given by*

$$f^{[k]}(t_{m+l}) = \sum_{i=0}^{k-1} \alpha_{kmi}((m+l)^i s_1 + ... + (m+1)^i s_l)\varepsilon^k \delta + \sum_{i=1}^{k} \beta_{kli} f^{[i]}(t_m)\varepsilon^{k-i}, \tag{5}$$

where $\alpha_{kmi}, \beta_{kli}$, $i = 1, ..., k$ are combinatorial coefficients [4].

By using (5) it is proved that if $s = s_1 ... s_n$, $s' = s'_1 ... s'_n \in I_n$ then $C(s, s') \leq n$ and the condition (2) reduces to a system of $C(s, s')$ linear equations [1]

$$\begin{aligned} (m+n)^0(s_1 - s'_1) + ... + (m+1)^0(s_n - s'_n) = 0 \\ . \quad . \quad . \quad . \quad . \quad . \\ (m+n)^{C(s,s')-1}(s_1 - s'_1) + ... + (m+1)^{C(s,s')-1}(s_n - s'_n) = 0 \end{aligned} \tag{6}$$

and the condition (3) results in an inequality

$$(m+n)^{C(s,s')}(s_1 - s'_1) + ... + (m+1)^{C(s,s')}(s_n - s'_n) \neq 0. \tag{7}$$

For example, for the Prouhet-Thue-Morse sequences of length 8, starting with +1 and −1, we have $C(s,s') = 3$. In this case the system of linear equations (6) becomes

$$+5^0 - 4^0 - 3^0 + 2^0 - 1^0 + 0^0 + (-1)^0 - (-2)^0 = 0$$
$$+5^1 - 4^1 - 3^1 + 2^1 - 1^1 + 0^1 + (-1)^1 - (-2)^1 = 0$$
$$+5^2 - 4^2 - 3^2 + 2^2 - 1^2 + 0^2 + (-1)^2 - (-2)^2 = 0 \quad (8)$$

and the inequality (7) becomes

$$+5^3 - 4^3 - 3^3 + 2^3 - 1^3 + 0^3 + (-1)^3 - (-2)^3 \neq 0. \quad (9)$$

when $n = 8, m = -3$ and factor 2 is ignored for clarity.

3 Revealing Hierarchical Structures and Formation Processes of Integer Relations

The system of linear equations (6) does not immediately show us a structure of integer relations. However, if we focus on the system's coefficients rather than the sequences' components, we can recognize that these coefficients can be arranged into a number of integer relations. Moreover, these integer relations can be defined as the elements of a hierarchical structure, where the relationships are specified by a single organizing principle. The relationship between an element of a level and some of the elements of the previous level is set up if the element can be formed from the elements according to the organizing principle.

In particular, if in the system (6) first we pay attention to the coefficients, which in our case are not set free but are consecutive powers $k = 0, ..., C(s,s') - 1$ of integers $m+n, m+n-1, ..., m+1$, and second consider the sequences' components in the form of $s_i - s'_i, i = 1, ..., n$ as multiples of these powers, then we can see that (6) is a specific system of integer relations. This can be demonstrated clearly for particular cases, because then the coefficients appear as powers of concrete integers and the sequences' components are expressed explicitly.

For example, consider sequences

$$s = +1 - 1 - 1 + 1 - 1 + 1 + 1 - 1 - 1 + 1 + 1 - 1 + 1 - 1 - 1 + 1,$$

$$s' = -1 + 1 + 1 - 1 + 1 - 1 - 1 + 1 + 1 - 1 - 1 + 1 - 1 + 1 + 1 - 1,$$

when $n = 16, m = 0$. It turns out that $C(s,s') = 4$ and the system of linear equations (6) becomes a system of specific integer relations between integers $16, 15, ..., 1$

$$+16^0 - 15^0 - 14^0 + 13^0 - 12^0 + 11^0 + 10^0 - 9^0 - 8^0 + 7^0 + 6^0 - 5^0 + 4^0 - 3^0 - 2^0 + 1^0 = 0$$

$$+16^1 - 15^1 - 14^1 + 13^1 - 12^1 + 11^1 + 10^1 - 9^1 - 8^1 + 7^1 + 6^1 - 5^1 + 4^1 - 3^1 - 2^1 + 1^1 = 0$$

$$+16^2-15^2-14^2+13^2-12^2+11^2+10^2-9^2-8^2+7^2+6^2-5^2+4^2-3^2-2^2+1^2=0$$

$$+16^3-15^3-14^3+13^3-12^3+11^3+10^3-9^3-8^3+7^3+6^3-5^3+4^3-3^3-2^3+1^3=0 \quad (10)$$

and the inequality (7) becomes

$$+16^4-15^4-14^4+13^4-12^4+11^4+10^4-9^4-8^4+7^4+6^4-5^4+4^4-3^4-2^4+1^4 \neq 0, \quad (11)$$

where for clarity common factor 2 originated from the sequences' components is taken of site in (10) and (11).

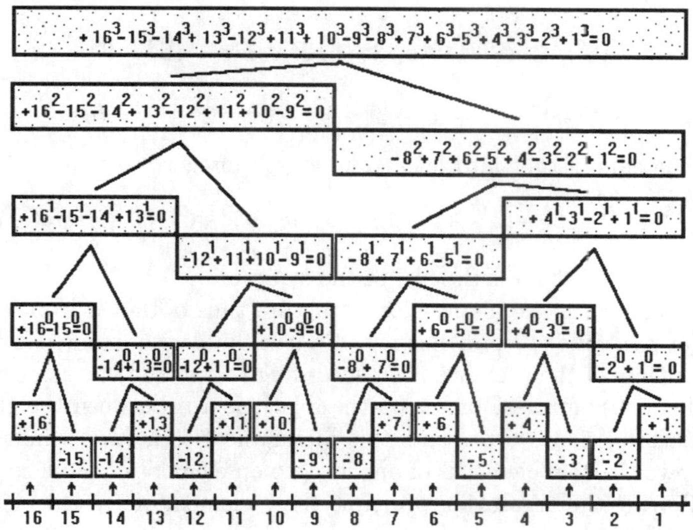

Fig. 7. The hierarchical structure of integer relations underlying the system of integer relations (10) and inequality (11). The structure may be interpreted in terms of a formation process of integer relations with integers $16, ..., 1$ as initial building blocks. Integers $16, 13, 11, 10, 7, 6, 4, 1$ are viewed to be in the positive state while integers $15, 14, 12, 9, 8, 5, 3, 2$ are viewed to be in the negative state. In the formation process all integer relations follow the same organizing principle.

Next, we need to realize that there may be more integer relations involved with the system of linear equations (6) than it is able to show us directly. Such an integer relation may exist between a set of integers selected from $m+n, m+n-1, ..., m+1$ in a certain manner. Moreover, the integer relations can be seen as the elements of a hierarchical structure and their relationships can be described by an organizing principle.

For example, from the system of integer relations (10) we can identify integer relations as the elements of a hierarchical structure. The first integer relation of (10) gives integer relations

$$+16^0-15^0=0, \quad -14^0+13^0=0, \quad -12^0+11^0=0, \quad +10^0-9^0=0,$$

$$-8^0 + 7^0 = 0, \quad +6^0 - 5^0 = 0, \quad +4^0 - 3^0 = 0, \quad -2^0 + 1^0 = 0$$

as the elements of the first level. The second integer relation of (10) - integer relations

$$+16^1 - 15^1 - 14^1 + 13^1 = 0, \quad -12^1 + 11^1 = +10^1 - 9^1 = 0,$$

$$-8^1 + 7^1 + 6^1 - 5^1 = 0, \quad +4^1 - 3^1 - 2^1 + 1^1 = 0$$

as the elements of the second level. The third integer relation of (10) - integer relations

$$+16^2 - 15^2 - 14^2 + 13^2 - 12^2 + 11^2 + 10^2 - 9^2 = 0,$$

$$-8^2 + 7^2 + 6^2 - 5^2 + 4^2 - 3^2 - 2^2 + 1^2 = 0$$

as the elements of the third level. The fourth integer relation of (10)

$$+16^3 - 15^3 - 14^3 + 13^3 - 12^3 + 11^3 + 10^3 - 9^3 - 8^3 + 7^3 + 6^3 - 5^3 + 4^3 - 3^3 - 2^3 + 1^3 = 0$$

is the element of the fourth level of the structure.

Remarkably, in the hierarchical structure thus obtained from the system of integer relations (10), an element of a level higher than one can be formed from elements of the previous level according to one and the same organizing principle. These formations set up the relationships between the elements of the structure. Figure 6 presents this hierarchical structure, where following edges upwards from elements of one level to an element of the next one reads that following the organizing principle these elements form the element.

In general, it is shown that the system of linear equations (6) can be associated with a hierarchical structure $WR(s, s', n, m, I_n)$ of integer relations [1]. The structure has $C(s, s')$ levels. In particular, the structure can be defined using some integers from $m + n, m + n - 1, ..., m + 1$ as its initial building blocks. This may be said because the integer relations of the structure can be made from these integers by the organizing principle. It is convenient to think that there is as a "source" of integers that can generate integers in the positive and negative states. Namely, a number of $|\ (s_i - s'_i)\ |$ integers $(m+n-i)$ is generated in the state specified by the sign of $(s_i - s'_i)$, $i = 1, ..., n$ at the zero level of the structure $WR(s, s', n, m, I_n)$.

The elements of the zero level combine and form the elements of the first level of the structure $WR(s, s', n, m, I_n)$. In the operations the state of an element of the zero level is converted into the arithmetic sign of the integer in the integer relation, where the integer is raised to the zero power. The organizing principle starts working on the first level. Namely, following the organizing principle the elements of the first level form the elements of the second level, and then the elements of the second level form the elements of the third level. In this manner this formation process continues up to the level as high as $C(s, s')$.

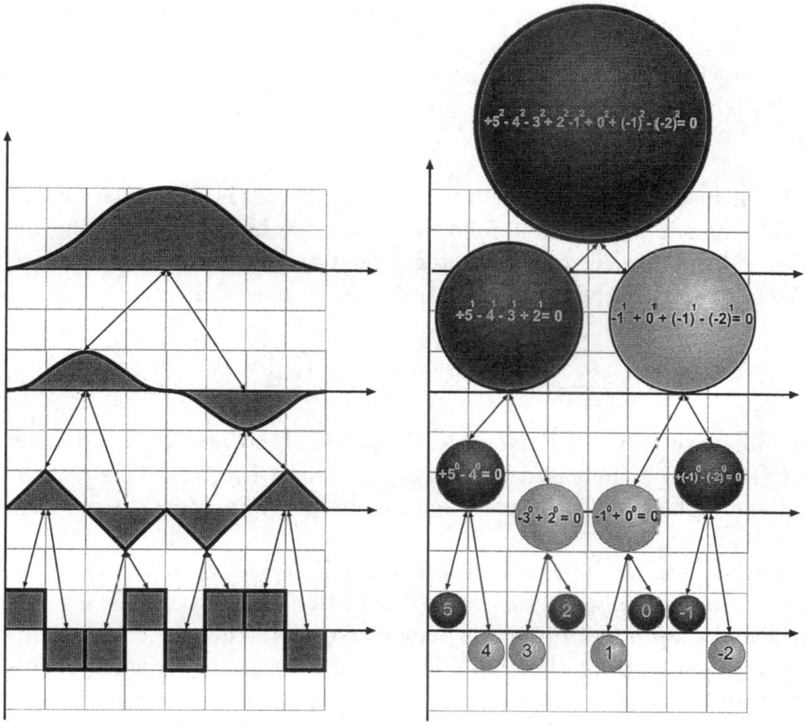

Fig. 8. The right side of the figure shows a formation process of integer relations, where the integers and integers relations are shown as some sort of particles. Following the organizing principle integer relation $+5^1 - 4^1 - 3^1 + 2^1 = 0$ and integer relation $-1^1 + 0^1 + (-1)^1 - (-2)^1 = 0$ form a new integer relation, because $+5^2 - 4^2 - 3^2 + 2^2$ and $-1^2 + 0^2 + (-1)^2 - (-2)^2$, when added together equal zero In the left part of the figure a corresponding formation process of two-dimensional geometric patterns is depicted [1].

The elements of the structure $WR(s, s', n, m, I_n)$ of level $k = 1, ..., C(s, s')$ are integer relations of the form

$$A_1 d_1^{k-1} + ... + A_l d_l^{k-1} = 0,$$

where A_i, $i = 1, ..., l$ are integers, d_i, $i = 1, ..., l$ are integers such that $d_i > d_{i+1}$, $i = 1, ..., l-1$ and k is the power of d_i, $i = 1, ..., l$. It is possible to make general statements on their nature.

The elements of level $k = 2, ..., C(s, s')$ of the structure $WR(s, s', n, m, I_n)$ can be formed from the elements of level $(k-1)$ according to the organizing principle. It is the same for all levels and can be described as follows. If $r \geq 1$ integer relations

$$A_{i1} d_{i1}^{k-1} + ... + A_{il(i)} d_{il(i)}^{k-1} = 0 \tag{12}$$

of level $k = 1, ..., C(s, s') - 1$, where relation i, $i = 1, ..., r$ contains $l(i)$ terms and $d_{il(i)} > d_{i+1,l(i+1)}$, $i = 1, ..., r$, satisfy

$$\sum_{i=1}^{r} A_{i1} d_{i1}^{k} + ... + A_{il(i)} d_{il(i)}^{k} = 0, \tag{13}$$

and the inclusion of each of the integer relations (12) is a necessary condition for (13), then it is said that following the organizing principle the integer relation (13) is formed from the integer relations (12).

We can see that the integer relation (13) is more than the simple sum

$$\sum_{i=1}^{r} A_{i1} d_{i1}^{k-1} + ... + A_{il(i)} d_{il(i)}^{k-1} = 0. \tag{14}$$

of the integer relations (12). Notably, the power of integers d_{ij}, $i = 1, ..., r$, $j = 1, ..., l(i)$ in (13) is increased by 1 in comparison with (14). This means that the integer relations (12) have a special property in order to produce the integer relation (13).

It makes sense to interpret the structure $WR(s, s', n, m, I_n)$ in terms of a formation process. Indeed, the above description of the structure seems like a description of a process. From this perspective, the structure $WR(s, s', n, m, I_n)$ is a convenient form to show all stages of the formation process. For example, Figure 8 shows the structure of integer relations underlying the system of integer relations (8). Notably, what is depicted in the figure may be interpreted as a formation process. Indeed, we can see that integers are generated on the zero level in the positive and negative states, and then form the integer relations of the first level. These integer relations in turn form the integer relations of the second level. The formation process proceeds to the third level, where the integer relation by itself can not form an element of the next level due to (9).

References

1. V. Korotkikh, *A Mathematical Structure for Emergent Computation*, Kluwer Academic Publishers, Dordrecht/Boston/London, 1999.
2. V. Korotkikh, *An Approach to the Mathematical Theory of Perception-Based Information*, in Fuzzy Partial Differential Equations and Relational Equations: Reservoir Characterization and Modeling, M. Nikravesh, L. Zadeh and V. Korotkikh (Eds.), Springer, Berlin/Heidelberg/New York, 2004, pp. 80-115.
3. G. Korotkikh and V. Korotkikh, *On the Role of Nonlocal Correlations in Optimization*, in Optimization and Industry: New Frontiers, P. Pardalos and V. Korotkikh (Eds.), Series in Applied Optimization, Kluwer Academic Publishers, Dordrecht/Boston/London, 2003, pp. 181-220.
4. V. Korotkikh, *Integer Code Series with Some Applications in Dynamic Systems and Complexity*, Communications on Applied Mathematics, Computing Center of the Russian Academy of Sciences, Moscow, pp. 1-65, 1993.

Mining Fuzzy Association Rules: An Overview

M. Delgado[1], N. Marín, M.J. Martín-Bautista, D. Sánchez, M.-A. Vila
Department of Computer Science and A.I., University of Granada,
Granada, Spain

Abstract

The main aim of this paper is to present a revision of the most relevant results about the use of Fuzzy Sets in Data Mining, specifically in relation with the discovery of Association Rules. Fuzzy Sets Theory has been shown to be a very useful tool in Data Mining in order to represent the so-called Association Rules in a natural and human-understandable way.

First of all we will introduce the basic concepts of Data Mining to justify the need of using Fuzzy Sets. A historical revision on developments in this field is made too.

Next we will present our researches about Fuzzy Association Rules, starting with the formulation of a general model to discover association rules among items in a (crisp) set of fuzzy transactions. This general model can be particularized in several ways so that each particular instance allows to represent and mine a different kind of pattern on some kind of data. We describe some applications of this scheme, paying special attention to its application in Text Mining.

The paper finishes with some suggestions about future researches and problems to be solved.

Keywords: Data mining, association rules, fuzzy transactions, quantified sentences.

1 Introduction

In the last years the development of Information Technologies has motivated a parallel growing of the facilities to store and manage data in databases. The larger the amount of stored data, the more important the demand of extracting the implicit information they contain to aid decision-making in business, health care services, research, etc. and thus to obtain useful knowledge from data in large repositories, i. e. the "Knowledge Discovery", is recognized as a basic necessity in many areas.

Since the nineties the research area named "Data Mining" has become a central topic in Databases and Artificial Intelligence. Although Data Mining is sometimes considered as synonymous of Knowledge Discovery, it seems to be more accurate to see it as a particular task within the general Knowledge Discovery process.

A "classical" characterization usually accepted establishes that "Data Mining is the process of nontrivial extraction of implicit, previously unknown and potentially useful patterns (rules constraints, regularities, etc) from data in databases".

Some authors consider Knowledge Discovery, and more particularly Data Mining, are within the Machine Learning paradigm. However others consider that the characteristics of the explored basic data structures (very large databases) and the request for "implicit, previously unknown and potentially useful information" introduces special features to differentiate Data Mining as a very broad field where different problems and methods may be distinguished. According to [78], the most important problems in Data Mining are:

[1] Corresponding author: M. Delgado (mdelgado@ugr.es)

- Dependence modelling and link analysis.- The idea is to describe significant dependencies between the variables included in the database.
- Mining Association Rules.- The basic idea is to discover meaningful associations between different pairs of sets of attribute values, in such a way that the presence of any value of some set in a database element (tuple, record, object etc.) implies the presence of other value belonging to another set.
- Multilevel Data Generalization, Summarization and Characterization.- Data and objects in databases often contains detailed information at primitive levels. The summarization (generalization) idea is to provide compact descriptions for subsets of data such that the concepts they represent are in a higher conceptual level than those existing in the database.
- Pattern identification and description.- One of the most important task in Knowledge Discovery is to look for interesting patterns and to describe them in a concise and meaningful manner. Two clear phases appear in this task, pattern identification and pattern description. The first basically consists of a clustering process which groups the items in the database according to natural classes, based on similarity metrics or probability density models. The second attempts to describe the classes by using relevant attribute values; this is usually named the classification process.

In all the above mentioned problems, Soft-Computing and more specifically Fuzzy Sets, play a significant role as they provide tools being particularly well-suited to cope with imprecise knowledge in the setting of complex systems. It is widely recognized that many real world relations are intrinsically fuzzy. Many techniques used in crisp data mining models have their corresponding "fuzzy version" and some problems treated in a "pure crisp" data mining context have a more natural formulation by using a fuzzy knowledge and fuzzy data representation (see [78]). For instance , it has been shown that fuzzy clustering generally provides a more suitable partition of a set of objects than crisp clustering.

This paper is devoted to analyze the use of Fuzzy Sets tools in the task of Mining Association Rules. In the following section we introduce the basic concepts and algorithms for (crisp) Association Rules. We also analyze the drawbacks of using a crisp approach to motivate the "natural" introduction of fuzzy tools. After that, section 3.2 will be devoted to study Fuzzy Association Rules. We will describe current approaches paying especial attention to our developments. Last section contains a revision of some current applications as well as future interesting ones. The paper finish with a selected Bibliography about this topic.

2 MINING (CRISP) ASSOCIATION RULES

Starting from the seminal paper by Agrawal et al in 1993, [2], the first works on the topic were devoted to disclose patterns in transactional databases from the retail industry and business and thus, some usual present terminology of this field preserves its origin. Sometimes looking for Association Rules is also named "market basket analysis" in allusion to looking for associations among the items that a purchaser in a retail shop selects to his/her purchase. From the point of view of the company any purchase of a client constitute a transaction being represented by a set of items.

A classical example is to analyze the connections among different types of goods in a sale database, that is to check if those customers which buy a kind of goods (e.g. bread) usually also buy another kind (e.g. milk). Let us suppose Table 1 reflects some information collected by a retailer.

This table describes the composition of six purchases i.e. six transactions (trans.id), in terms of four items: bread, butter, biscuits and milk. The value 1 (value 0) means that the corresponding item was included (not included). It is very easy to detect an association rule *Bread&Butter* $\Rightarrow$ *Milk* that represents the fact: when a client bought bread and butter then he/she also bought milk. However this rule has one exception, transaction 5 and so the retailer can not completely trust on this statement

To measure the reliability/accuraccy of a rule two values, *Support* and *Confidence*, that have been extensively used, were initially introduced. The "support" measures the reliability by the relative frequency of co-occurrence of the rule's items. The "confidence" measures the rule accuracy as the quotient between the

trans − id	Bread	butter	Biscuits	Milk
1	1	1	0	1
2	0	1	1	0
3	1	0	1	0
4	1	1	0	1
5	1	1	1	0
6	1	1	1	1

Table 1. A set of transactions

support of that rule and the relative frequency of the items belonging to the left part of the rule. It is easy to show that for our rule above the support is 3/6 whereas the confidence is 3/4.

The former approaches to Data Mining look for Association Rules by searching itemsets (sets of items) with support larger than a threshold *minsupp* which is supposed to be fixed by the users. On the basis of these so-called *frequent* itemsets, rules and their confidence are computed, and only those rules whose confidence is greater than another threshold *minconf* are given as result.

In the following section we will present the above ideas in a more formal way.

2.1 Formal model

Let I be a set of items (objects) and T a set of transactions with items in I, both assumed to be finite.

Definition 1. *An association rule is an expression of the form $A \Rightarrow C$, where $A, C \subseteq I$, $A, C \neq \emptyset$, and $A \cap C = \emptyset$.*

The rule $A \Rightarrow C$ means "every transaction of T that contains A contains C too".

As we said before, the usual measures to assess association rules are support and confidence, both based on the concept of support of an *itemset* (i.e. a subset of items).

Definition 2. *The support of an itemset $I_0 \subseteq I$ with respect to a set of transactions T is*

$$supp(I_0, T) = \frac{|\{\tau \in T \mid I_0 \subseteq \tau\}|}{|T|} \quad (1)$$

i.e., the probability that a transaction of T contains I_0.

Definition 3. *The support of the association rule $A \Rightarrow C$ in T is*

$$Supp(A \Rightarrow C, T) = supp(A \cup C, T) \quad (2)$$

and its confidence is

$$Conf(A \Rightarrow C, T) = \frac{Supp(A \Rightarrow C, T)}{supp(A, T)} = \frac{supp(A \cup C, T)}{supp(A, T)}. \quad (3)$$

It is usual to assume that T is fixed for each problem and thus, it is customary to avoid any reference to it. Then, the above introduced values are simply noted as $supp(I_0)$, $Supp(A \Rightarrow C)$ and $Conf(A \Rightarrow C)$ respectively. Notice that the names of these measures start with small letter for items, *supp*, whereas they start by capital letter *Supp*, *Conf* for rules.

Support is the percentage of transactions where the rule holds. Confidence is the conditional probability of C with respect to A or, in other words, the relative cardinality of C with respect to A.

The techniques employed to mine for association rules attempt to discover rules whose support and confidence are greater than two user-defined thresholds called *minsupp* and *minconf* respectively. Such rules are called *strong rules*.

2.2 The certainty framework to measure accuracy and importance

Several authors have pointed out some drawbacks of the support/confidence framework to assess association rules [19, 73, 70, 10].

To avoid some of these drawbacks and to ensure that the discovered rules are interesting and accurate, a new approach was proposed in [70, 10]. It employs certainty factors [71] and the new concept of very strong rule.

Definition 4. *The certainty factor of a fuzzy association rule $A \Rightarrow C$ is the value*

$$CF(A \Rightarrow C) = \frac{Conf(A \Rightarrow C) - supp(C)}{1 - supp(C)}$$

if $Conf(A \Rightarrow C) > supp(C)$, and

$$CF(A \Rightarrow C) = \frac{Conf(A \Rightarrow C) - supp(C)}{supp(C)}$$

if $Conf(A \Rightarrow C) \leq supp(C)$, assuming by agreement that if $supp(C) = 1$ then $CF(A \Rightarrow C) = 1$ and if $supp(C) = 0$ then $CF(A \Rightarrow C) = -1$.

The certainty factor takes values in $[-1, 1]$. It is positive when the dependence between A and C is positive, 0 when there is independence, and a negative value when the dependence is negative. The following proposition is an interesting property shown in [12]:

Proposition 1. *$Conf(A \Rightarrow C) = 1$ if and only if $CF(A \Rightarrow C) = 1$*

This property guarantees that the certainty factor of a fuzzy association rule achieves its maximum possible value, 1, if and only if the rule is totally accurate.

2.3 Algorithms to disclose Association Rules

The first algorithms to disclose association rules, were developed by Agrawal et al [2]. These algorithms obtain strong rules from *frequent* itemsets which on its turn are those itemsets with support greater than *minsupp*.

The most known algorithm, *APriori*, is based on a simple but key fundamental observation about frequent itemsets, the emph A Priori Property, that may be stated as follows: *"Every subset of a frequent itemset must be a frequent itemset too"*. From this, the algorithm is designed to proceed iteratively starting from frequent itemsets containing a single item.

Although this Data Mining model is associated to binary transactional databases, it is easy to generalize it to cover more complex data structures taking into account that the concepts of item and itemset are abstract ones that may represent general objects and sets of objects, respectively. After the first papers by Agrawal, a lot of work has been devoted to this topic and several different situations have been investigated. Many papers have been devoted to develop algorithms to mine ordinary association rules. The original algorithms have been adapted, modified or improved in multiple senses by several authors in order to cover the different situations that may appear in Data Mining. The early efficient algorithms like AIS [2], Apriori and AprioriTid [3], SETM [49], OCD [58], and DHP [63] were continued with more recent developments like DIC [19], CARMA [45], TBAR [11], and FP-Growth [44]. See [46] [8] for recent surveys about the topic.

Most of the existing algorithms work in two steps:

Step P.1. Find the frequent itemsets. In this step it is usual to consider transactions one by one, updating the support of the itemsets each time a transaction is considered. This step is the most expensive from the computational point of view.

Step P.2. Obtain rules with accuracy greater than an user-defined threshold, from the frequent itemsets obtained in step P.1. Specifically, if the itemsets A and $A \cup C$ are frequent, we can obtain the rule $A \Rightarrow C$. The support of that rule will be high enough, since it is equal to the support of the itemset $A \cup C$. However, we must verify the accuracy of the rule, in order to determine whether it is strong.

2.4 Association Rules in Relational Databases

Since the nineties relational databases are the usual store for data and it is supposed that they contains today a very important amount of hidden valuable and useful knowledge. Thus, having methods to disclose patterns in relational databases becomes a need.

Data in relational databases are stored in tables, where each row (tuple) describes an object and each column is one characteristic/attribute of the object. For each tuple t, $t[X]$ stands for the value of attribute (column) X. Algorithms to mine for association rules have been applied to represent patterns in relational databases by defining items as pairs $\langle attribute, value\rangle$ and transactions as tuples. The following formalization is described in [33]. Let $RE = \{X_1, \ldots, X_m\}$ be a set of attributes. We denote I^{RE} to the set of items associated to RE, i.e.

$$I^{RE} = \{\langle X_j, x\rangle \text{ such that } x \in Dom(X_j), j \in \{1, \ldots, m\}\}$$

Every instance r of RE is associated to a T-set, denoted T^r, with items in I^{RE}. Each tuple $t \in r$ is associated to an unique transaction $\tau^t \in T^r$ in the following way:

$$\tau^t = \{\langle X_j, t[X_j]\rangle \mid j \in \{1, \ldots, m\}\}$$

No pair of items in one transaction shares the same attribute, because of the First Normal Form constraint. Any other itemset must also hold this special feature.

The earliest researches dealt with categorical attributes although the topic of discovering association rules involving quantitative values, called *quantitative association rules* [75], arose very early too. Two difficulties arise at once when numerical values (the finest possible granularity) are directly used: the mining task is very expensive [75, 80] and, the support and the semantic content of rules are quite poor [33]. One solution is to split into intervals the domain of the quantitative attributes , and to take this set of clusters as the new domain of the attribute (i.e. to take a coarser granularity). Several approaches based on this idea have been proposed, either performing the clustering during the mining process [75, 61] or before it [83, 87]. This solution has two drawbacks: it is difficult for clusters to fit a meaningful (for users) concept [70], and the importance and accuracy of rules can be very sensitive to (even small) variations of boundaries [75, 54].

A soft alternative allows to solve these drawbacks. The very idea is to define a set of linguistic labels with semantics given by fuzzy sets on the domain of the quantitative attributes, and to use them as a new domain. Now, the meaning of the new values is clear, and the rules are not sensitive to small changes of the boundaries because they are fuzzy.

3 Fuzzy Transactions and Fuzzy Association Rules

3.1 An historical overview

Data Mining in general and Association Rules Mining in particular are young topics but the number of paper devoted to them (from both theoretical and practical point of view) is quite impressive and most papers deal with mining association rules involving quantitative attributes in relational databases by using fuzzy sets/linguistic labels to diminish the granularity and to translate the problem in a more "natural" and understandable one.

To our knowledge, [56] is the first paper introducing fuzzy sets into association rules to diminish the granularity of quantitative attributes. The model uses a membership threshold to change fuzzy transactions into crisp ones before looking for ordinary association rules in the set of crisp transactions. Items keep being pairs $\langle attribute, label\rangle$.

In [4, 5, 6], a set of predefined linguistic labels is employed. The importance and accuracy of fuzzy association rules are assessed by means of two measures called *adjusted difference* and *weight of evidence*. A rule is said to be important when its adjusted difference is greater than 1.96 (the 95th percentile of $N(0,1)$). This avoids the need for a user's importance threshold, but has the drawback of making symmetric the adjusted difference and thus, when a rule $A \Rightarrow C$ is found to be interesting, then $C \Rightarrow A$ will be too. The weight of evidence is a measure of information gain that is provided to the user as an estimation of how interesting a rule is.

In [54], the usefulness of itemsets and rules is measured by means of a *significance factor*, defined as a generalization of support based on sigma-counts (to count the percentage of transactions where the item is) and the product (for the intersection in the case of k-itemsets with $k > 1$). The accuracy is based on a kind of *certainty factor* (with different formulation and semantics of our measure). In fact, two different formulations of the certainty factor are proposed in this work: the first one is based on the significance factor, in the same way that confidence is based on support. This provides a generalization of the ordinary support/confidence framework for association rules. The second proposal is based on correlation and it is not a generalization of confidence.

In [48], only one item per attribute is considered: the pair $\langle attribute, label \rangle$ with greater support among those items based on the same attribute. The model is the usual generalization of support and confidence based on sigma-counts.The proposed mining algorithm first transform each quantitative value into a fuzzy sets in linguistic terms. The algorithm then calculates the scalar cardinalities of all linguistic terms in the transaction data. Now the linguistic term with maximal cardinality is used for each attribute and thus the number of items keeps. The algorithm is therefore focused on the most important linguistic terms, so reducing its time complexity. The mining process is then performed by using fuzzy counts.

In [86] an extension of the Equi-depth (EDP) algorithm [75] for mining fuzzy association rules involving quantitative attributes is presented. The approach combines the obtained partitions with predefined linguistic labels.

To cope with the task of diminishing the granularity in quantitative attribute representations to obtain useful and natural association rules, several authors opted for using crisp grid partition or clustering based approaches/algorithms like Partial Completeness [75], Optimized Association Rules [42] or CLIQUE [1]. Hu et al. [50] have extended these ideas allowing non empty intersections between neighborhood sets in partitions and describing that by fuzzy sets. In this way an effective algorithm named "Fuzzy Grid Based Rules Mining Algorithm" (FGBRMA) is constructed. This algorithm deals with both quantitative and qualitative algorithm in a similar manner. The concepts of large fuzzy grid and effective fuzzy association rule are introduced by using specifically fuzzy support and fuzzy confidence measures. FGBRMA generates large fuzzy grids and the fuzzy association rules by using boolean operations on suitable table data structures.

With a similar methodology in [84] a method for inductive Machine Learning Problems to disclose classification rules from a set of examples is developed. Very related with the above mentioned approaches is the methodology in [41] that finds the fuzzy sets to represent suitable linguistic labels for data (in the sense that they allow to obtain rules with good support/accuracy) by using fuzzy clustering techniques. This way, the user does not need to define them, and that can be an advantage in certain cases, but the obtained fuzzy sets could be hard to fit to meaningful labels. Another methodology that follows this line is proposed in [77].

In all the above mentioned papers, the items in transactions (on which the search is made to detect associations) are considered to be defined with a single granularity (conceptual) level. However items in real world applications are usually organized in some hierarchies and thus mining multiple concept level fuzzy rules may produce very useful knowledge. A very interesting research field (initiated by Srikant et al. in [74] in the crisp case) deals with mining association rules by identifying relationships between transactions with quantitative values and items being organized into a hierarchical structure. Hong et al. [47] have investigated the construction of a fuzzy data algorithm to deal with quantitative data under a given taxonomy. They have developed an algorithm modifying that of Srikant et al. in [74] under the approach developed in [48] to transform quantitative values into linguistic terms. In this algorithm each item uses only that linguistic terms with maximal cardinality thus keeping the number of fuzzy regions equal to the number of original items.

Chen et al. [22],[21] have considered the case in which there are certain fuzzy taxonomic structures reflecting partial belonging of one item to another in the hierarchy. To deal with these situations, association rules are requested to be of the form $X \Rightarrow Y$ were either X or Y is a collection of Fuzzy sets. The model is based on a generalization of support and confidence by means of sigma-counts, and the algorithms are again extensions of the classical Srikant and Agrawal's ones [2, 74].

An alternative/complementary research line is devoted to mine association rules taking into account some "relevance measure" for items and/or rules themselves. In [55] the concept of "ordinal fuzzy set" is introduced as an alternative interpretation of the membership degrees of values to labels. This carries out an alternative interpretation of fuzzy rules. Paper [72] studies fuzzy association rules with weighted items, i.e., an importance degree is given to each item. Weighted support and confidence are defined. Also, numerical values of attributes are mapped into linguistic terms by using Kohonen's self-organized maps. A generalization of the Apriori algorithm is proposed to discover fuzzy rules among weighted items.

The definition of fuzzy association rule introduced in [7] is different from most of the existing in the literature. Fuzzy degrees are associated to items, and their meaning is the relative importance of items in rules. The model is different from [72], because linguistic labels are not considered. An item i with associated degree α is said to be in a fuzzy transaction $\tilde{\tau}$ when $\tilde{\tau}(i) \geq \alpha$. This seems to be a generalization of the model in [56] which uses the degree associated to an item as the threshold to turn fuzzy transactions into crisp ones, instead of using the same thresholds for all the items. In summary, the support of a "fuzzy itemset" $\tilde{I}$ (a set of items with associated degrees) is the percentage of fuzzy transactions $\tilde{\tau}$ such that $\tilde{I} \subseteq \tilde{\tau}$. Ordinary support and confidence are employed. A very interesting algorithm is proposed, which has the valuable feature that performs only one pass over the database in the mining process. Within this research line papers [43], [79], [51] are to be remarked also. In [43] and [51], weights are associated to the whole fuzzy rules which are obtained from the support/confidence assessment. In its turn in [79], it is supposed that each item in a transaction has a weight that reflects the interest/intensity of such item within the transaction. That is translated into a weight parameter for each item in each resulting association rule.

Let us point out that the mining algorithms for fuzzy association rules are mainly inspired on crisp ones, but some papers extending inductive learning methods may be found also.

3.2 Fuzzy Transactions and Fuzzy Association Rules: Our Approach

This section contains a summary of our researches on this topic. First, we introduce some definitions and after that some ideas about the corresponding search algorithms. In [28] a detailed description of them can be found.

Definition 5. *A fuzzy transaction is a nonempty fuzzy subset $\tilde{\tau} \subseteq I$.*

For every $i \in I$, we note $\tilde{\tau}(i)$ the membership degree of i in a fuzzy transaction $\tilde{\tau}$. We note $\tilde{\tau}(I_0)$ the degree of inclusion of an itemset $I_0 \subseteq I$ in a fuzzy transaction $\tilde{\tau}$, defined as

$$\tilde{\tau}(I_0) = \min_{i \in I_0} \tilde{\tau}(i)$$

According to definition 5 a transaction is a special case of fuzzy transaction. We represent a set of fuzzy transactions by means of a table. Columns and rows are labelled with identifiers of items and transactions respectively. The cell for item i_k and transaction $\tilde{\tau}_j$ contains a $[0,1]$-value: the membership degree of i_k in $\tilde{\tau}_j$, $\tilde{\tau}_j(i_k)$.

Example 1. [28] Let $I = \{i_1, i_2, i_3, i_4\}$ be a set of items. Table 2 shows six fuzzy transactions defined on I.

Here, $\tilde{\tau}_1 = 0.6/i_2 + 0.7/i_3 + 0.9/i_4$, $\tilde{\tau}_2 = 1/i_2 + 1/i_4$, and so on. In particular, $\tilde{\tau}_2$ is a crisp transaction, $\tilde{\tau}_2 = \{i_2, i_4\}$.

Some inclusion degrees are:

$\tilde{\tau}_1(\{i_3, i_4\}) = 0.7$, $\tilde{\tau}_1(\{i_2, i_3, i_4\}) = 0.6$, $\tilde{\tau}_4(\{i_1, i_4\}) = 1$.

We call *T-set* a set of ordinary transactions, and *FT-set* a set of fuzzy transactions. Example 1 shows the FT-set $\{\tilde{\tau}_1, \ldots, \tilde{\tau}_6\}$ that contains six transactions. Let us remark that a FT-set is a crisp set.

	i_1	i_2	i_3	i_4
$\tilde{\tau}_1$	0	0.6	0.7	0.9
$\tilde{\tau}_2$	0	1	0	1
$\tilde{\tau}_3$	1	0.5	0.75	1
$\tilde{\tau}_4$	1	0	0.1	1
$\tilde{\tau}_5$	0.5	1	0	1
$\tilde{\tau}_6$	1	0	0.75	1

Table 2. The set T_6 of fuzzy transactions

Definition 6. *Let I be a set of items, T a FT-set, and $A, C \subseteq I$ two crisp subsets, with $A, C \neq \emptyset$, and $A \cap C = \emptyset$. A fuzzy association rule $A \Rightarrow C$ holds in T iff*

$$\tilde{\tau}(A) \leq \tilde{\tau}(C) \;\; \forall \tilde{\tau} \in T$$

i.e., the inclusion degree of C is greater than that of A for every fuzzy transaction $\tilde{\tau}$.

This definition preserves the meaning of association rules, because if we assume $A \subseteq \tilde{\tau}$ in some sense, we must assume $C \subseteq \tilde{\tau}$ given that $\tilde{\tau}(A) \leq \tilde{\tau}(C)$. Since a transaction is a special case of fuzzy transaction, an association rule is a special case of fuzzy association rule.

Let us remark that the main characteristic feature of our approach is to model Fuzzy Transactions with crisp items. This is quite general because in the case of having actually fuzzy items: labels, fuzzy numbers, etc, finally they will produce a table as the one before.

Support and confidence of fuzzy association rules

To assess these rule values, we employ a semantic approach based on the evaluation of quantified sentences [85]. A quantified sentence is an expression of the form "Q of F are G", where F and G are two fuzzy subsets of a finite set X, and Q is a relative fuzzy quantifier. Relative quantifiers are linguistic labels for fuzzy percentages that can be represented by means of fuzzy sets on $[0, 1]$, such as "most", "almost all", or "many".

A family of relative quantifiers, called coherent quantifiers [24], is specially relevant for us. Coherent quantifiers are those that verify the following properties:

- $Q(0) = 0$ and $Q(1) = 1$
- If $x < y$ then $Q(x) \leq Q(y)$ (monotonicity)

An example is "many young people are tall", where $Q = many$, and F and G are possibility distributions induced in the set $X = people$ by the imprecise terms "young" and "tall" respectively. A special case of quantified sentence appears when $F = X$, as in "most of the terms in the profile are relevant". The evaluation of a quantified sentence yields a $[0, 1]$-value, that assesses the accomplishment degree of the sentence.

Definition 7. *Let $I_0 \subseteq I$. The support of I_0 in T is the evaluation of the quantified sentence*

$$Q \text{ of } T \text{ are } \widetilde{\Gamma}_{I_0}$$

where $\widetilde{\Gamma}_{I_0}$ is a fuzzy set on T defined as

$$\widetilde{\Gamma}_{I_0}(\tilde{\tau}) = \tilde{\tau}(I_0)$$

Definition 8. *The support of the fuzzy association rule $A \Rightarrow C$ in the set of fuzzy transactions T is supp$(A \cup C)$, i.e., the evaluation of the quantified sentence*

$$Q \text{ of } T \text{ are } \widetilde{\Gamma}_{A \cup C} = Q \text{ of } T \text{ are } \left(\widetilde{\Gamma}_A \cap \widetilde{\Gamma}_C\right)$$

Definition 9. *The confidence of the fuzzy association rule $A \Rightarrow C$ in the set of fuzzy transactions T is the evaluation of the quantified sentence*

$$Q \text{ of } \widetilde{\Gamma}_A \text{ are } \widetilde{\Gamma}_C$$

Let us remark that these definitions establish families of support and confidence measures, depending on the choice of both, the evaluation method and the quantifier. The only constraint we consider to do that is looking for the following four intuitive properties of the measures for ordinary association rules to hold:

1. If $\widetilde{\Gamma}_A \subseteq \widetilde{\Gamma}_C$ then $Conf(A \Rightarrow C) = 1$.
2. If $\widetilde{\Gamma}_A \cap \widetilde{\Gamma}_C = \emptyset$ then $Supp(A \Rightarrow C) = 0$ and $Conf(A \Rightarrow C) = 0$.
3. If $\widetilde{\Gamma}_A \subseteq \widetilde{\Gamma}_{A'}$ (particularly when $A' \subseteq A$) then $Conf(A' \Rightarrow C) \leq Conf(A \Rightarrow C)$.
4. If $\widetilde{\Gamma}_C \subseteq \widetilde{\Gamma}_{C'}$ (particularly when $C' \subseteq C$) then $Conf(A \Rightarrow C) \leq Conf(A \Rightarrow C')$.

We have selected the method GD to evaluate the sentences [35], which has been shown to verify good properties with better performance than others. The evaluation of "Q of F are G" by means of GD is defined as

$$GD_Q(G/F) = \sum_{\alpha_i \in \Delta(G/F)} (\alpha_i - \alpha_{i+1})\, Q\left(\frac{|(G \cap F)_{\alpha_i}|}{|F_{\alpha_i}|}\right) \tag{4}$$

where $\Delta(G/F) = \Lambda(G \cap F) \cup \Lambda(F)$, $\Lambda(F)$ being the level set of F, and $\Delta(G/F) = \{\alpha_1, \ldots, \alpha_p\}$ with $\alpha_i > \alpha_{i+1}$ for every $i \in \{1, \ldots, p\}$. The set F is assumed to be normalized. If not, F is normalized and the normalization factor is applied to $G \cap F$.

The evaluation of a quantified sentence "Q of F are G" by means of method GD can be interpreted as:

- the evidence that the percentage of objects in F that are also in G (relative cardinality of G with respect to F) is Q [70], or
- a quantifier-guided aggregation [52, 82] of the relative cardinalities of G with respect to F for each α-cut of the same level of both sets.

Thus $Supp(A \Rightarrow C)$ can be interpreted as the evidence that the percentage of transactions in $\widetilde{\Gamma}_{A \cup C}$ is Q, and $Conf(A \Rightarrow C)$ can be seen as the evidence that the percentage of transactions in $\widetilde{\Gamma}_A$ that are also in $\widetilde{\Gamma}_C$ is Q. In both cases, the quantifier is a linguistic parameter that determines the final semantics of the measures. Since the properties of the evaluation method GD [35], it is easy to show that any coherent quantifier yields support and confidence measures that verify the four aforementioned properties.

On the other hand, we have proposed to choose the quantifier Q_M defined by $Q_M(x) = x$, since it is coherent and the measures obtained by using it in definitions 7, 8 and 9 are the ordinary measures in the crisp case, (see [35], [28]).A possible interpretation of the values of the measures for crisp association rules is the evidence that the support (resp. confidence) of the rule is Q_M.

Let us finally remark that the choice of the quantifier allows us to change the semantics of the values in a linguistic way. This flexibility is very useful when using this general model to fit different types of patterns, as we shall see in the applications section. Unless specific references were made, from now on we shall consider support and confidence based on Q_M and GD. The study of the support/confidence framework with other choices is left to future research.

Example 2. [28] Let's illustrate the concepts introduced in this subsection. According to definition 7, the support of several itemsets in table 2 is shown in table 3.

Table 4 shows the support and confidence of several fuzzy association rules in T_6.

We remark that

$$Conf(\{i_1, i_3\} \Rightarrow \{i_4\}) = GD_{Q_M}(\widetilde{\Gamma}_{\{i_4\}} / \widetilde{\Gamma}_{\{i_1,i_3\}}) = 1$$

since $\widetilde{\Gamma}_{\{i_1,i_3\}} \subseteq \widetilde{\Gamma}_{\{i_4\}}$.

Itemset	Support
$\{i_1\}$	0.583
$\{i_4\}$	0.983
$\{i_2, i_3\}$	0.183
$\{i_1, i_3, i_4\}$	0.266

Table 3. Support in T_6 of four itemsets

Rule	Support	Confidence
$\{i_2\} \Rightarrow \{i_3\}$	0.183	0.283
$\{i_1, i_3\} \Rightarrow \{i_4\}$	0.266	1
$\{i_1, i_4\} \Rightarrow \{i_3\}$	0.266	0.441

Table 4. Support and confidence in T_6 of three fuzzy rules

A different framework to measure accuracy and importance

As we have previously pointed out (see subsection 2.2) the certainty factors and the associated idea of very strong rules was introduced in the setting of crisp Data Mining to avoid some of these drawbacks of using support and confidence measures. This section will be devoted to present the extension of these ideas to the fuzzy case.

Definition 10. *We call certainty factor of a fuzzy association rule $A \Rightarrow C$ to the value*

$$CF(A \Rightarrow C) = \frac{Conf(A \Rightarrow C) - supp(C)}{1 - supp(C)}$$

if $Conf(A \Rightarrow C) > supp(C)$, and

$$CF(A \Rightarrow C) = \frac{Conf(A \Rightarrow C) - supp(C)}{supp(C)}$$

if $Conf(A \Rightarrow C) \leq supp(C)$, assuming by agreement that if $supp(C) = 1$ then $CF(A \Rightarrow C) = 1$ and if $supp(C) = 0$ then $CF(A \Rightarrow C) = -1$.

This definition straightforward extends the crisp one. The only difference is how to assess the used confidence and support values.

Definition 11. *We say that a fuzzy association rule is strong when its certainty factor and support are greater than two user-defined thresholds* minCF *and* minsupp *respectively. A fuzzy association rule $A \Rightarrow C$ is said to be very strong if both $A \Rightarrow C$ and $\neg C \Rightarrow \neg A$ are strong.*

The itemsets $\neg A$ and $\neg C$, whose meaning is "absence of A" (resp. C) in a transaction, are defined in the usual way: $\widetilde{\Gamma}_{\neg A}(\tilde{\tau}) = 1 - \widetilde{\Gamma}_A(\tilde{\tau})$ and $\widetilde{\Gamma}_{\neg C}(\tilde{\tau}) = 1 - \widetilde{\Gamma}_C(\tilde{\tau})$. The logical basis of this definition is that the rules $A \Rightarrow C$ and $\neg C \Rightarrow \neg A$ represent the same knowledge. Therefore, if both rules are strong we can be more certain about the presence of that knowledge in a set of transactions.

Several experiments described in [70, 10] have shown that, by using certainty factors and very strong rules, we avoid to report a large amount of false, or at least, doubtful rules. In some experiments, the number of rules was diminished by a factor of 20 and even more. This is important not only because the discovered rules are reliable, but also because the set of rules is smaller and more manageable. Hence, from now on, we propose the use of certainty factors to measure the accuracy of the fuzzy association rules. Anyway, let us point out that the assessment of the quality of (fuzzy) association rules is an open problem that is still receiving considerable attention (see [37], [38])

Algorithms

Most of the existing association rule mining algorithms can be adapted in order to discover fuzzy association rules. The generalization of crisp algorithms to mine fuzzy rules is being worked now, and we have already extended some of them. Let us point out that an important objective for us is to keep (in the worst case) the complexity of the existing algorithms when they are modified in order to find fuzzy rules (see [28] for details).

Roughly speaking we adapt step P.1 of the general algorithm (see subsection 2.3) by obtaining the support from the evaluation of the corresponding quantified sentences as proposed before.

To adapt step P.2. of the general procedure (see subsection 2.3) is rather straightforward. We only modify this step in the sense that we obtain the certainty factor of the rules from the confidence and the support of the consequent, both available.

Let us note that it is easy to decide whether a rule is strong, because its support and certainty factor are available in this step.

There are several possible solutions to deal with very strong rules. They are described in [12], and it is to remark that all the algorithms can be easily adapted (as the basic algorithm) to find strong rules.

To finish this section let us remember again that most algorithms (crisp and fuzzy) lie on the user determination of certain thresholds for support, confidence, etc. Although these thresholds are supposed to depend only on the user, the practice in real world applications shows that mining different databases requires different assessments of those thresholds, but it seems almost impossible for the user to do that without any heuristic knowledge guidance. Several authors have worked on this problem in order to develop that kind of heuristics (see [88]).

4 Applications

Let us point out that "an item" and "a transaction" are abstract concepts that may be seen as representing some kind of "an object" and "a subset of objects", respectively. Depending on the particular instantiation one makes, association rules can provide different kinds of patterns. In this section we shall describe briefly some different instances of carrying out this simple idea

4.1 Fuzzy Association Rules in Relational Databases

This section contains a summary of the results in [33].

Let $Lab(X_j) = \{L_1^{X_j}, \ldots, L_{c_j}^{X_j}\}$ be a set of linguistic labels for attribute X_j. We shall use the labels to name the corresponding fuzzy set, i.e.

$$L_k^{X_j} : Dom(X_j) \to [0, 1]$$

Let $L = \bigcup_{j\in\{1,\ldots,m\}} Lab(X_j)$. Then, the set of items with labels in L associated to RE is

$$I_L^{RE} = \left\{ \langle X_j, L_k^{X_j} \rangle \mid X_j \in RE \text{ and } \begin{array}{l} k \in \{1, \ldots, c_j\} \\ j \in \{1, \ldots, m\} \end{array} \right\}$$

Every instance r of RE is associated to a FT-set, denoted T_L^r, with items in I_L^{RE}. Each tuple $t \in r$ is associated to a single fuzzy transaction $\tilde{\tau}_L^t \in T_L^r$

$$\tilde{\tau}_L^t : I_L^{RE} \to [0, 1]$$

such that

$$\tilde{\tau}_L^t \left(\langle X_j, L_k^{X_j} \rangle \right) = L_k^{X_j} \left(t[X_j] \right)$$

In this case, a fuzzy transaction can contain more than one item corresponding to different labels of the same attribute, because it is possible for a single value in the table to fit more than one label to a certain degree. However, itemsets keep restricted to contain one item per attribute at most, because otherwise fuzzy rules wouldn't make sense.

The following example is taken also from [33].

Example 3. Let r be the relation of table 5, containing the age and hour of birth of six people. The relation r is an instance of $ER = \{Age, Hour\}$.

	Age	Hour
t_1	60	20:15
t_2	80	23:45
t_3	22	15:30
t_4	55	01:00
t_5	3	19:30
t_6	18	06:51

Table 5. Age and hour of birth of six people

We shall use for age the set of labels *Lab*(*Age*)={Baby, Kid, Very young, Young, Middle age, Old, Very old} of figure 1. Figure 2 shows a possible definition of the set of labels for hour *Lab*(*Hour*) = {Early morning, Morning, Noon, Afternoon, Night}.

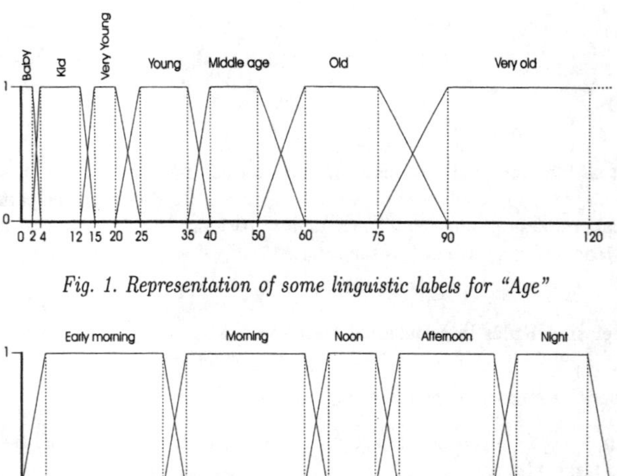

Fig. 1. Representation of some linguistic labels for "Age"

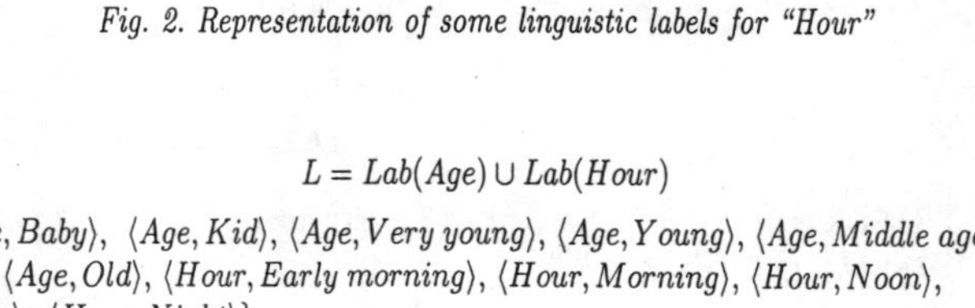

Fig. 2. Representation of some linguistic labels for "Hour"

Then

$$L = Lab(Age) \cup Lab(Hour)$$

and $I_L^{ER} = \{\langle Age, Baby\rangle, \langle Age, Kid\rangle, \langle Age, Very\ young\rangle, \langle Age, Young\rangle, \langle Age, Middle\ age\rangle,$ $\langle Age, Very\ Old\rangle, \langle Age, Old\rangle, \langle Hour, Early\ morning\rangle, \langle Hour, Morning\rangle, \langle Hour, Noon\rangle,$ $\langle Hour, Afternoon\rangle, \langle Hour, Night\rangle\}$.

The FT-set T^r on I_L^{ER} is

$$T_L^r = \{\tilde{\tau}_L^{t_1}, \tilde{\tau}_L^{t_2}, \tilde{\tau}_L^{t_3}, \tilde{\tau}_L^{t_4}, \tilde{\tau}_L^{t_5}, \tilde{\tau}_L^{t_6}\}$$

The columns of table 6 define the fuzzy transactions of T_L^r as fuzzy subsets of I_L^{ER} (we have interchanged rows and columns from the usual representation of fuzzy transactions, for the sake of space). For instance

$$\tilde{\tau}_L^{t_1} = \{1/_{\langle Age,Old\rangle} + 0.75/_{\langle Hour,Afternoon\rangle} + \\ +0.25/_{\langle Hour,Night\rangle}\}$$
$$\tilde{\tau}_L^{t_3} = \{0.6/_{\langle Age,Very\ young\rangle} + 0.4/_{\langle Age,Young\rangle} + \\ +0.5/_{\langle Hour,Noon\rangle} + 0.5/_{\langle Hour,Afternoon\rangle}\}$$

	$\tilde{\tau}_L^{t_1}$	$\tilde{\tau}_L^{t_2}$	$\tilde{\tau}_L^{t_3}$	$\tilde{\tau}_L^{t_4}$	$\tilde{\tau}_L^{t_5}$	$\tilde{\tau}_L^{t_6}$
$\langle Age, Baby\rangle$	0	0	0	0	**0.5**	0
$\langle Age, Kid\rangle$	0	0	0	0	**0.5**	0
$\langle Age, Very\ young\rangle$	0	0	**0.6**	0	0	**1**
$\langle Age, Young\rangle$	0	0	**0.4**	0	0	0
$\langle Age, Middle\ age\rangle$	0	0	0	**0.5**	0	0
$\langle Age, Old\rangle$	**1**	**0.67**	0	**0.5**	0	0
$\langle Age, Very\ old\rangle$	0	**0.33**	0	0	0	0
$\langle Hour, Early\ morning\rangle$	0	0	0	**1**	0	**0.85**
$\langle Hour, Morning\rangle$	0	0	0	0	0	**0.15**
$\langle Hour, Noon\rangle$	0	0	**0.5**	0	0	0
$\langle Hour, Afternoon\rangle$	**0.75**	0	**0.5**	0	**1**	0
$\langle Hour, Night\rangle$	**0.25**	**1**	0	0	0	0

Table 6. Fuzzy transactions with items in I_L^{ER} for the relation of table 5

In table 6 the row for item i_L contains the fuzzy set $\tilde{\Gamma}_{\{i_L\}}^r$. For instance

$$\tilde{\Gamma}_{\{\langle Age,Old\rangle\}}^r = \{1/\tilde{\tau}_L^{t_1} + 0.67/\tilde{\tau}_L^{t_2} + 0.5/\tilde{\tau}_L^{t_4}\}$$
$$\tilde{\Gamma}_{\{\langle Hour,Night\rangle\}}^r = \{0.25/\tilde{\tau}_L^{t_1} + 1/\tilde{\tau}_L^{t_2}\}$$

Descriptions of itemsets with more than one fuzzy item are, for instance

$$\tilde{\Gamma}_{\{\langle Age,Old\rangle,\langle Hour,Night\rangle\}}^r = \{0.25/\tilde{\tau}_L^{t_1} + 0.67/\tilde{\tau}_L^{t_2}\}$$
$$\tilde{\Gamma}_{\{\langle Age,Kid\rangle,\langle Hour,Afternoon\rangle\}}^r = \{0.5/\tilde{\tau}_L^{t_5}\}$$

Some rules involving fuzzy items in I_L^{ER} are:

$$\langle Age, Old\rangle \Rightarrow \langle Hour, Afternoon\rangle$$
$$\langle Hour, Afternoon\rangle \Rightarrow \langle Age, Baby\rangle$$

This general approach has been tested to find fuzzy association rules in several relational database instances. Algorithms, implementations and some experimental results are detailed in [33, 70]. In [28] an experiment to obtain fuzzy association rules from CENSUS database may be found.

Let us finally remark that only crisp databases have been considered. The linguistic labels are defined by fuzzy sets on the domains of crisp quantitative attributes. However it is quite possible to have data being intrinsically fuzzy to be represented and stored by means of one of the existing fuzzy relational database models. In these cases, our approach keeps suitable by working like in the examples of section 3.2.

4.2 Fuzzy and Approximate Functional Dependencies

By using a suitable definition of items and transactions, fuzzy association rules are able to characterize pattern structures different from the described in the previous sections. In the following we summarize a methodology to mine Functional Dependencies in relational databases by mining association rules with an appropriated representation of items and transactions.

Let RE be a set of attributes and r an instance of RE. A functional dependence $X \to Y$, $X, Y \subset RE$, holds in r if the value $t[X]$ determines $t[Y]$ for every tuple $t \in r$. Formally, such a dependence is a rule of the form

$$\forall t, s \in r \quad \text{if } t[X] = s[X] \text{ then } t[Y] = s[Y] \tag{5}$$

Any dependence is said to hold in RE if it holds in every instance of RE.

To disclose such eventually hidden knowledge is very interesting but it is difficult to find "perfect" dependencies, mainly because of usually there exist exceptions. To cope with this, two main approaches (both introducing a kind of smoothed dependence) have been proposed: fuzzy functional dependencies and approximate dependencies. The former introduce some fuzzy components into (5) (e.g. the equality can be replaced by a similarity relation) while the latter establishes the functional dependencies with exceptions (i.e., with some uncertainty). Approximate dependencies can be interpreted as a relaxation of the universal quantifier in rule (5). A detailed study of different definitions of fuzzy and approximate dependencies can be found in [70, 16, 13].

We have used association rules to represent approximate dependencies. For this purpose,now transactions and items are to be associated to pairs of tuples and attributes respectively. We consider that the item associated to the attribute X, I_X, is in the transaction τ_{ts} associated to the pair of tuples $\langle t, s\rangle$ when $t[X] = s[X]$. The set of transactions associated to an instance r of RE is denoted T_r, and contains $|r|^2$ transactions. We define an approximate dependence in r to be an association rule in T_r [30, 70, 13].

Obviously the support and certainty factor of an association rule $I_X \Rightarrow I_W$ in T_r measure the importance and accuracy of the corresponding approximate dependence $X \to W$. The main drawback of this approach is its computational complexity , because $|T_r| = |r|^2$ and the underlying algorithm have linear complexity on the number of transactions. We have solved the problem by analyzing several transactions at a time. The algorithm, (see [70, 13]) stores the support of every item of the form $\langle X, x\rangle$ with $x \in Dom(X)$ in order to obtain the support of I_X. Its complexity is linear on the number of tuples in r. An additional feature is that it finds dependencies and the associated models at the same time. We have shown that our definitions and algorithms provide a reasonable and manageable set of dependencies [70, 13].

Example 4. [13]: Let consider the relation r_3 of table 7. It is an instance of $RE = \{ID, Year, Course, Lastname\}$.

ID	Year	Course	Lastname
1	1991	3	Smith
2	1991	4	Smith
3	1991	4	Smith

Table 7. Table r_3 with data about three students

Table 8 shows the T-set T_{r_3} and table 9 contains some association rules that hold in T_{r_3}. They define approximate dependencies that hold in r_3. Confidence and support of the association rules in table 9 measure the accuracy and support of the corresponding dependencies.

Fuzzy association rules are needed again in this context when quantitative attributes are involved. Our algorithms provide not only an approximate dependence, but also a model that consists of a set of association rules (in the usual sense in relational databases) relating values of the antecedent with values of the consequent of the dependencies. The support and certainty factor of the dependencies have been shown to be related to the same measures of the rules in the model [13]. But when attributes are quantitative, this model suffers from the same problem discussed in the previous subsection. To cope with this, we propose to use a set of linguistic labels. A set of labels $Lab(X_j)$ induces a fuzzy similarity relation $S_{Lab(X_j)}$ in the domain of X in the following way:

Pair	it_{ID}	it_{Year}	it_{Course}	$it_{Lastname}$
$\langle 1,1\rangle$	1	1	1	1
$\langle 1,2\rangle$	0	1	0	1
$\langle 1,3\rangle$	0	1	0	1
$\langle 2,1\rangle$	0	1	0	1
$\langle 2,2\rangle$	1	1	1	1
$\langle 2,3\rangle$	0	1	1	1
$\langle 3,1\rangle$	0	1	0	1
$\langle 3,2\rangle$	0	1	1	1
$\langle 3,3\rangle$	1	1	1	1

Table 8. The set T_{r_3} of transactions for r_3

Ass. Rule	Confidence	Support	App. dependence
$\{it_{ID}\} \Rightarrow \{it_{Year}\}$	1	1/3	$ID \rightarrow Year$
$\{it_{Year}\} \Rightarrow \{it_{Course}\}$	5/9	5/9	$Year \rightarrow Course$
$\{it_{Year}, it_{Course}\} \Rightarrow \{it_{ID}\}$	3/5	1/3	$Year, Course \rightarrow ID$

Table 9. Some association rules in T_{r_3} that define approximate dependencies in r_3

$$S_{Lab(X_j)}(x_1, x_2) = \max_{L_k^{X_j} \in Lab(X_j)} \min\left(L_k^{X_j}(x_1), L_k^{X_j}(x_2)\right)$$

for all x_1,$x_2 \in Dom(X_j)$, assuming that for every $x \in Dom(X_j)$ there is one $L_k^{X_j} \in Lab(X_j)$ such that $L_k^{X_j}(x) = 1$.

Then, the item I_X is in the transaction τ_{ts} with degree $S_{Lab(X_j)}(t[X], s[X])$. Now, the transactions for the table r are fuzzy, and we denote $T_{S_L}^{\tau}$ this FT-set. In this new situation, we can find approximate dependencies in r by looking for fuzzy association rules in $T_{S_L}^{\tau}$. The model of such approximate dependencies will be a set of association rules in T_L^{τ}. These dependencies can be used to summarize data in a relation.

Functional dependencies may be smoothed into fuzzy functional dependencies in several ways alternative of the afore described one [16]. We have shown that most of them can be obtained by replacing the equality and the universal quantifier in the rule 5 by a similarity relation S and a fuzzy quantifier Q respectively [34].

For instance, let S_Y be a resemblance relation [25] and $\varphi \in (0, 1]$ such that

$$S(y_1, y_2) = \begin{cases} 1 & S_Y(y_1, y_2) \geq \varphi \\ 0 & \text{otherwise} \end{cases}$$

Also let $Q = \forall$, with

$$\forall(x) = \begin{cases} 1 & x = 1 \\ 0 & \text{otherwise} \end{cases}$$

The fuzzy functional dependence $X \rightarrow W$ defined by [26]

$$\forall t, s \in r \quad \text{if } t[X] = s[X] \text{ then } S_Y(t[Y], s[Y]) \geq \varphi$$

can be modelled in r by an association rule in T_S^{τ}. Here, T_S^{τ} stands for the FT-set of fuzzy similarities given by S between pairs of tuples of r.

We have also faced a more general problem: the integration of fuzzy and approximate dependencies in what we have called *fuzzy quantified dependencies* [34] (i.e. fuzzy functional dependencies with exceptions). Let us remark that our semantic approach, based on the evaluation of quantified sentences, allows to assess rules in a more flexible way. Hence, to deal with certain kinds of patterns is possible, as we have seen before.

4.3 Gradual Rules

Gradual rules are expressions of the form "The more X is L_i^X, the more Y is L_j^Y", like "the more *Age* is *Young*, the more *Salary* is *Low*". Roughly speaking the semantics of that rules is "the greater the membership degree of the value of X in L_i^X, the greater the membership degree of the value of Y in L_j^Y" [36, 17]. There are several possibilities to formulate this idea.

In [16] the authors propose a direct interpretation of the above idea:

$$\forall t \in r \quad L_i^X(t[X]) \leq L_j^Y(t[Y]) \tag{6}$$

In this case, the items are pairs $\langle Attribute, Label \rangle$ and the transactions are associated to tuples. The item $\langle X, L_i^X \rangle$ is in the transaction τ_t associated to the tuple t when $L_i^X(t[X]) \leq L_j^Y(t[Y])$, and the set of transactions, denoted $T^r_{G_1(L)}$, is a T-set. This way, ordinary association rules in $T^r_{G_1(L)}$ are gradual rules in r.

A more general formulation for gradual rules is

$$\forall t \in r \quad L_i^X(t[X]) \rightarrow_* L_j^Y(t[Y]) \tag{7}$$

where $\rightarrow_*$ is a fuzzy implication. The expression (6) is a particular case where Rescher-Gaines implication ($\alpha \rightarrow_{R-G} \beta = 1$ when $\alpha \leq \beta$ and 0 otherwise) is employed.

One interesting alternative is to use the FT-set T_L^r described in subsection 4.1 and the quantifier $Q = \forall$. From the properties of GD, the evaluation of "$\forall$ of F are G" provides an inclusion degree of F in G that can be interpreted as a kind of implication.

In our opinion, the meaning of the rules above is closer to "the membership degree of the value of Y to L_j^Y is greater than the membership degree of the value of X to L_i^X". But expression (7) is not the only possible general semantics for a gradual rule. Another possibility is, $\forall t, s \in r$

$$\text{if } L_i^X(t[X]) \leq L_i^X(s[X]) \text{ then } L_j^Y(t[Y]) \leq L_j^Y(s[Y]) \tag{8}$$

where it is not assumed that the degrees in Y are greater than those of X. Items keep being pairs $\langle Attribute, Label \rangle$ but transactions are associated to pairs of tuples. The item $\langle X, L_i^X \rangle$ is in the transaction τ_{ts} when $L_i^X(t[X]) \leq L_i^X(s[X])$, and the set of transactions, denoted $T^r_{G_2(L)}$, is a T-set. Now, $\left|T^r_{G_2(L)}\right| = |r|^2$. This alternative can be extended with fuzzy implications in a similar way that (7) extends (6).

4.4 Fuzzy association rules in text mining

It is well known that searching the web is not always so successful as users expect. Most of the retrieved sets of documents in a web search meet the search criteria but do not satisfy the user needs. One of the reasons for this is the lack of specificity in the formulation of the queries, that in its turn is mainly due to the user does not know the vocabulary of the topic or query terms do not come to user's mind at the query moment.

One solution to this problem is to carry out the process known as *query expansion* or *query reformulation.* After retrieving a first set of documents new terms are added and/or removed to the corresponding query in order to improve the results, i.e., to discard uninteresting retrieved documents and/or to retrieve interesting documents that were not initially obtained. A good review of the topic in the Information Retrieval field can be found in [39].

Our proposal is to use mining technologies to build systems with queries reformulation ability . In the following we summarize the main results contained in [60]. Let us mention here that data mining techniques have been already applied to solve some classical Information Retrieval problems such as document classification [57] and query optimization [76].

Text Items

In the context of Text-Mining the items may be built either at term-level or at document-level depending on relations among terms or among documents are looked for. [53]. Here term-level items are considered.

Different representations of text for association rules extraction at term-level can be found in the literature: bag of words, indexing keywords, term taxonomy and multi-term text phrases [31]. We have decided to use automatic indexing techniques coming from Information Retrieval [69] to obtain *word-items*, that is, items will be associated to single words appearing in a document where stop-list and/or stemming processes may be carried out.

Text Fuzzy Transactions

In a text framework, each transaction is associated with the representation of a document.

According to our term-level item choice, each document will be represented by a set of pairs ¡term,weight¿ where the weight assesses the presence of that term in the document. Several weighting schemes can be used depending on the assessment criteria [68] and among them we have explore the use of:

Boolean scheme: The weights values are in {0,1} indicating the absence or presence of the word in the document, respectively.

Frequency scheme: It weights each term by the relative frequency of that term in the document. In a fuzzy framework, the normalization of this frequency can be carried out by dividing the number of occurrences of a term in a document by the number of occurrences of the most frequent term in that document [15].

TFIDF scheme: It is a combination of the within-document word frequency (*TF*) and the inverse document frequency (*IDF*). We use normalized weights in the interval [0, 1] according to [14]. Under this scheme a high weight is associated to any term that occurs frequently in a document but infrequently in the collection.

From a collection of documents $D = \{d_1, \ldots, d_n\}$ we can obtain a set of terms $I = \{t_1, \ldots, t_m\}$ which is the union of the keywords for all the documents in the collection. The weights associated to these terms in a document d_i are represented by $W = \{w_{i1}, \ldots, w_{im}\}$. For each document d_i, we consider an extended representation where a weight of 0 will be assigned to every term appearing in some of the documents of the collection but not in d_i.

Considering these elements, we can define a *text transaction* $\tau_i \in T$ as the extended representation of document d_i. Without loosing generality, we can write the set of transactions associated to the collection of document D as $T_D = \{d_1, \ldots, d_n\}$.

When the boolean weighting scheme is used, the transactions can be called boolean or crisp, since the values of the tuples are 1 or 0 meaning that the attribute is present in the transaction or not, respectively.

When we consider a normalized weighting scheme in the unit interval we may speak about fuzzy text transactions. Concretely, we have considered the frequency weighting scheme and the TDIDF weighting scheme, both normalized, and therefore, analogously to the former definition of text transactions, we can define a set of *fuzzy text transactions* $FT_D = \{d_1, \ldots, d_n\}$, where each document d_i corresponds to a fuzzy transaction $\tilde{\tau}_i \in FT$, and where the weights $W = \{w_{i1}, \ldots, w_{im}\}$ of the keyword set $I = \{t_1, \ldots, t_m\}$ are fuzzy values from a fuzzy weighting scheme.

Association Rules and Fuzzy Association Rules for Query Refinement

The objective now is the application of disclosing association rules and fuzzy association rules to the problem of query reformulation.

In our proposal we start from a set of documents obtained from an initial query. The representation of the documents is obtained following one of the weighting schemes as we have commented before. The document representation building process is shown in the Algorithm 2. Given a query and a set of retrieved documents, the query representation is matched to each document representation in order to obtain a relevance value for every document. If a document term does not appear in the query, its value will be assumed to be 0. In the crisp case, the considered model is the Boolean one [69], while in the fuzzy case the considered model is the generalized Boolean model with fuzzy logic [20].

The user's initial query generates a set of ranked documents. If the top-ranked documents do not satisfy user's needs, the query improvement process starts, (i.e. is a *local feedback analysis* technique).

Because we consider each document to be a transaction, it is clear that:

- $T_D = \{d_1, \ldots d_n\}$ is the whole set of possible transactions (from the collection of documents D) ,
- $I = \{t_1, \ldots, t_m\}$ is the set of (text) items obtained as representation of each $d_i \in D$ with their membership to the transaction assessed by the weights $W = \{w_{i1}, \ldots, w_{im}\}$.

On this set of transactions we apply the Algorithm 3 to extract the association rules. Let us note that this algorithm does not distinguish between crisp and fuzzy approach, because that only depends on the considered item weighting scheme.

The whole process is summarized in the Algorithm 1. The process may be said to perform a *semi-automatic process* (see [60])as it only suggest a list of terms to refine query.

Algorithm 1 Semi-automatic query refinement process using association rules

1. The user queries the system.
2. A first set of documents is retrieved.
3. From this set, the representation of documents is extracted following Algorithm 1 and association rules are generated following Algorithm 2 and the extraction rule procedure.
4. Terms that appear in certain rules are shown to the user(subsection ??).
5. The user selects those terms more related to her/his needs.
6. The selected terms are added to the query, which is used to query the system again.

Algorithm 2 Basic algorithm to obtain the representation of documents in a collection

Input: a set of documents $D = \{d_1, \ldots d_n\}$.
Output: a representation for all documents in D.

1. Let $D = \{d_1, \ldots d_n\}$ be a collection of documents
2. Extract an initial set of terms S from each document $d_i \in D$
3. Remove stop words
4. Apply stemming (*via* Porter's algorithm [66])
5. The representation of d_i obtained is a set of keywords $S = \{t_1, \ldots, t_m\}$ with their associated weights $\{w_{i1}, \ldots, w_{im}\}$

Algorithm 3 Basic algorithm to obtain the association rules from text

Input: a set of transactions $T_D = \{d_1, \ldots d_n\}$
a set of term items $I = \{t_1, \ldots, t_m\}$ with their associated weights $W = \{w_{i1}, \ldots, w_{im}\}$.
Output: a set of association rules.

1. Construct the itemsets from the set of transactions T.
2. Establish the threshold values of minimum support *minsupp* and minimum confidence *minconf*
3. Find all the itemsets that have a support above threshold *minsupp*, that is, the *frequent itemsets*
4. Generate the rules, discarding those rules below threshold minconf

Once the strong association rules are extracted, their utility for query refinement depends on their form i.e. how the query terms appear in antecedent and/or consequent of these rules. Let us suppose that $qterm$ is a term that appears in the query and let $term \in S$, $S_0 \subseteq S$. Some possibilities are the following:

- Rules of the form $term \Rightarrow qterm$. We could suggest the term $term$ to the user as a way to restrict the set of results.
- Rules of the form $S_0 \Rightarrow qterm$ with $S_0 \subseteq S$. We could suggest the set of terms S_0 to the user as a whole, i.e., to add S_0 to the query.
- Rules of the form $qterm \Rightarrow term$ with $term \in S$. We could suggest the user to replace $qterm$ with $term$ in order to obtain a set of documents that include the actual set (this is interesting if we're going to perform the query again in the web, since perhaps $qterm$ is more specific that the user intended).

The previous examples illustrate us how to provide the user with some alternatives in order to reformulate a query. However the problem may be more complicated because in most cases it is necessary to take into account the reciprocal of the rules. For example, if both $term \Rightarrow qterm$ and $qterm \Rightarrow term$ are strong, then that means that having $term$ and $qterm$ in a query is equivalent to some extent (depending on the $mincf$ threshold employed). Then new documents not previously retrieved and interesting for the user can be obtained by replacing $term$ with $qterm$.

In the applications of mining techniques to text, documents are usually represented by a set of keywords (see [40], [67])and thus usually a full text is not considered. This produces a lost of the discriminatory power of frequent terms (see [64]). Now then, if the terms are not good discriminators, then the reformulation of a query may not improve its result. Almost automatic discriminatory schemes can be used alternatively to a preprocessing stage for selecting the most discriminatory terms. This is the case of the *TFIDF* weighting scheme .

However, in a dynamic environment, where the response-time is important, the application of a pre-processing stage to select good discriminatory terms may not be suitable. Moreover to calculate the term weights by the *TFIDF* scheme, to know the presence of a term in the whole collection is needed just in the assessment time, which constraints its use in dynamic collections, as usually occurs in Internet. To face this situation, we can improve the rule obtaining process instead of improving the document representation. The use of alternative measures of importance and accuracy such as the ones presented before is to be considered in order to avoid the problem of non appropriate rule generation.

Besides the representation of the documents by terms, an initial categorization of the documents can be available. In that case, the categories can appear as items to be included in the transactions with value $\{0, 1\}$ based on the membership of the document to that category.

In this case, the extracted rules may provide additional information about the relation between terms and categories.

For instance, if a rule of the form $term \rightarrow category$ appears with enough accuracy, we can assert that documents where that term appears can be classified in that category.

5 Further remarks, future researches

Mining fuzzy association rules (i.e., association rules in fuzzy transactions) is a useful technique to find patterns in data in the presence of imprecision, either because data are fuzzy in nature or because we must improve their semantics.

The proposed models has been tested on some of the applications described in this paper, mainly to discover fuzzy association rules in relational databases that contain quantitative data.

We have introduced a very general model for fuzzy transactions in which the items are considered as being crisp. We have shown that this characteristic makes our model easy to formulate and use.

We have also study how our model can be employed in mining distinct types of patterns, from ordinary association rules to fuzzy and approximate functional dependencies and gradual rules. They will be used in multimedia data mining and web mining. We have paid an special attention to the problem of tex mining.

Technical issues we will study in the future, such as the analysis of measures given by quantifiers others than Q_M, have been pointed out in previous sections.

From a "practical point of view, we consider our model will be successful in dealing with other problems in the field of Data Mining and Knowledge Discovering.

We are now working on (fuzzy) mining transactional data about images given by artificial vision models and more generally to disclose pattern in multimedia databases. [59].

Knowledge Discovery in Databases (KDD) is undoubtedly recognized as a key "technology" in business and industry (see [18]). On the other hand, Fuzzy Sets and Fuzzy Logic are also considered as a need to represent the inherent non random uncertainty which lies in most part of information and decision processes. However the papers about disclosing fuzzy patterns within this field scarce. We have started a multidisciplinary research project to disclose and use Fuzzy Association Rules from financial data.

Let us also mention that our proposal for query reformulation procedure is to be implemented into a software package and the comparison with other approaches to query refinement coming from Information Retrieval is to be carried out.

Mining (crisp) association rules has been shown to be interesting in the field of medical and clinic data. Also in proteomics and genomics to detect biological patterns in the protein composition and the conditioning from a particular genoma in the presence of some disease([62],[23]). We plan to study the use of our model to cope with these problems because we consider that fuzzy tools will be very appropriated to model these associations as wagueness is present everywhere in this field.

References

1. R. Agrawal, J. Gehrke, D. Gunopoulos, P. Raghavan, "Automatic subspace clustering of high dimensional data for Datamining Applications," *Proceedings of the ACM SIGMOD International Conference on Management of Data*, pp. 94–105. June 1998.
2. R. Agrawal, T. Imielinski, and A. Swami, "Mining association rules between sets of items in large databases," in *Proc. Of the 1993 ACM SIGMOD Conference*, 1993, pp. 207–216.
3. R. Agrawal and R. Srikant, "Fast algorithms for mining association rules," in *Proc. Of the 20th VLDB Conference*, Sep. 1994, pp. 478–499.
4. W.H. Au and K.C.C. Chan, "Mining fuzzy association rules," in *Proc. Of 6th Int. Conf. On Information and Knowledge Management. Las Vegas, NV, USA*, 1997, pp. 209–215.
5. W.H. Au and K.C.C. Chan, "An effective algorithm for discovering fuzzy rules in relational databases," in *Proc. IEEE Int. Conf. On Fuzzy Systems Vol. II*, 1998, pp. 1314–1319.
6. W.H. Au and K.C.C. Chan, "FARM: A data mining system for discovering fuzzy association rules," in *Proc. FUZZ-IEEE'99, Seoul, South Korea, Vol. 3, pp. 22–25*, 1999.
7. S. Ben-Yahia and A. Jaoua, "A top-down approach for mining fuzzy association rules," in *Proc. 8th Int. Conf. on Information Processing and Management of Uncertainty in Knowledge-Based Systems, IPMU'2000*, 2000, pp. 952–959.
8. F. Berzal "ART: Un método alternativo para la construcción de árboles de decisión". Ph.D. Thesis, Department of Computer Science and Artificial Intelligence, University of Granada, September 2002.
9. F. Berzal, I. Blanco, D. Sánchez, and M.A. Vila, "A new framework to assess association rules," in *Advances in Intelligent Data Analysis. Fourth International Symposium, IDA'01. Lecture Notes in Computer Science 2189*, F. Hoffmann, Ed., pp. 95–104. Springer-Verlag, 2001.
10. F. Berzal, I. Blanco, D. Sánchez, and M.A. Vila, "Measuring the accuracy and interest of association rules: A new framework," An extension of [9]. Intelligent Data Analysis 6 (3), pp. 221–235. , 2002.
11. F. Berzal, J.C. Cubero, N. Marín, and J.M. Serrano, "TBAR: An efficient method for association rule mining in relational databases," *Data & Knowledge Engineering*, Vol. 37 (1), pp. 47–64. 2001.
12. F. Berzal, M. Delgado, D. Sánchez, and M.A. Vila, "Measuring the accuracy and importance of association rules," Tech. Rep. CCIA-00-01-16, Department of Computer Science and Artificial Intelligence, University of Granada, 2000.

13. I. Blanco, M.J. Martín-Bautista, D. Sánchez, J.M. Serrano, and M.A. Vila, "Using Association Rules to Mine for Strong Approximate Dependencies", Data Mining and Knowledge Discovery, Submitted.
14. Bordogna, G., Carrara, P. & Pasi, G. "Fuzzy Approaches to Extend Boolean Information Retrieval". In Bosc., Kacprzyk, J. *Fuzziness in Database Management Systems*, 231-274. Germany: Physica Verlag, 1995.
15. Bordogna, G. & Pasi, G. "A Fuzzy Linguistic Approach Generalizing Boolean Information Retrieval: A Model and Its Evaluation". *Journal of the American Society for Information Science 44(2)*, pp. 70–82, 1993.
16. P. Bosc and L. Lietard, "Functional dependencies revisited under graduality and imprecision," in *Annual Meeting of NAFIPS*, 1997, pp. 57–62.
17. B. Bouchon-Meunier, D. Dubois, LL. Godó, and H. Prade, *Fuzzy Sets and Possibility Theory in Approximate and Plausible Reasoning*, chapter 1, pp. 15–190, Handbooks of Fuzzy Sets. Series Editors: D. Dubois and H. Prade. Kluwer Academic Publishers, 1999, Edited by J.C. Bezdek, D. Dubois and H. Prade.
18. R.J. Brachman, T. Khazaba, W. Kloesgen, G. Piatesky-Shapiro, E. Simopudis, "Mining Business Databases" Communications of the ACM, vol. 90, n? 11, 1996
19. S. Brin, R. Motwani, J.D. Ullman, and S. Tsur, "Dynamic itemset counting and implication rules for market basket data," *SIGMOD Record*, vol. 26, no. 2, pp. 255–264, 1997.
20. Buell, D.A. & Kraft, D.H. "Performance Measurement in a Fuzzy Retrieval Environment". In *Proceedings of the Fourth International Conference on Information Storage and Retrieval, ACM/SIGIR Forum 16(1)*, pp. 56–62. Oakland, CA, USA, 1981.
21. Guoqing Chen, Quiang Wei "Fuzzy Association Rules and the extended mining algorithms," *Information Sciences*, vol 147, pp. 201–228, 2002
22. G. Chen, Q. Wei, and E. Kerre, "Fuzzy data mining: Discovery of fuzzy generalized association rules," in *Recent Issues on Fuzzy Databases*, G. Bordogna and G. Pasi, Eds. Physica-Verlag, 2000, "Studies in Fuzziness and Soft Computing" Series.
23. C. Creighton, S. Hanash, "Mining gene expression databases for association rules," *Bioinformatics*, vol 19, pp. 79-86, 2003
24. J.C. Cubero, J.M. Medina, O. Pons, and M.A. Vila, "The generalized selection: An alternative way for the quotient operations in fuzzy relational databases," in *Fuzzy Logic and Soft Computing*, B. Bouchon-Meunier, R. Yager, and L.A. Zadeh, Eds. World Scientific Press, 1995.
25. J.C. Cubero, O. Pons, and M.A. Vila, "Weak and strong resemblance in fuzzy functional dependencies," in *Proc. IEEE Int. Conf. on Fuzzy Systems, Orlando/FL, USA*, 1994, pp. 162–166.
26. J.C. Cubero and M.A. Vila, "A new definition of fuzzy functional dependence in fuzzy relational databases," *Int. Journal on Intelligent Systems*, vol. 9, no. 5, pp. 441–448, 1994.
27. A. De Luca and S. Termini, "Entropy and energy measures of a fuzzy set," in *Advances in Fuzzy Set Theory and Applications*, vol. 20, pp. 321–338. M.M.Gupta and R.K.Ragade and R.R.Yager, 1979.
28. M. Delgado,M. Marín, D.Sánchez, and M.A. Vila, "Fuzzy Association Rules: General Model and Applications," IEEE Transactions on Fuzzy Systems, vol.11, pp. 214–225, 2003.
29. M. Delgado, M.J. Martín-Bautista, D. Sánchez, and M.A. Vila, "A probabilistic definition of a nonconvex fuzzy cardinality," *Fuzzy Sets and Systems*, vol. 126, no. 2, pp. 41–54, 2002.
30. M. Delgado, M.J. Martín-Bautista, D. Sánchez, and M.A. Vila, "Mining strong approximate dependencies from relational databases," in *Proceedings of IPMU'2000*, 2000.
31. M. Delgado, M.J. Martín-Bautista, D. Sánchez, M.A. Vila, "Mining Text Data: Special Features and Patterns". In *Proc. of EPS Exploratory Workshop on Pattern Detection and Discovery in Data Mining*, London, September 2002. Lecture Notes in Computer Science 2447, D. Hand et.al., Eds., pp. 140–153. Springer-Verlag.
32. M. Delgado, D. Sánchez, J.M. Serrano, and M.A. Vila, "A survey of methods to evaluate quantified sentences," *Mathware and soft computing*, vol. VII, no. 2-3, pp. 149–158, 2000.
33. M. Delgado, D. Sánchez, and M.A. Vila, "Acquisition of fuzzy association rules from medical data," in *Fuzzy Logic in Medicine*, S. Barro and R. Marín, Eds. pp. 286–310, Physica Verlag, 2002
34. M. Delgado, D. Sánchez, and M.A. Vila, "Fuzzy quantified dependencies in relational databases," in *Proc. of EUFIT'99*, 1999.
35. M. Delgado, D. Sánchez, and M.A. Vila, "Fuzzy cardinality based evaluation of quantified sentences," *International Journal of Approximate Reasoning*, vol. 23, pp. 23–66, 2000.
36. D. Dubois and H. Prade, "Fuzzy rules in knowledge-based systems. modelling gradedness, uncertainty and preference," in *An introduction to fuzzy logic applications in intelligent systems*, R.R. Yager and L.A. Zadeh, Eds., pp. 45–68. Kluwer, Dordrecht, 1992.
37. D. Dubois, H. Prade, T. Sudkamp, "A Discussion of Indices for the Evaluation of Fuzzy Associnations in Relational Databases," *T Bilgic et al. (eds.): IFSA 2003, Lectures Notes on Artificial Intelligence 2715*, pp. 111-118, Springer Verlag Berlin Heidelberg 2003.
38. D. Dubois, E. Hüllermeier, H. Prade "A Note on Quality Measures for Fuzzy Association Rules," *T Bilgic et al. (eds.): IFSA 2003, Lectures Notes on Artificial Intelligence 2715*, pp. 346-353, Springer Verlag Berlin Heidelberg 2003.

39. Efthimiadis, E. "Query Expansion". *Annual Review of Information Systems and Technology 31*:121-187, 1996.
40. Feldman, R., Fresko, M., Kinar, Y., Lindell, Y., Liphstat, O., Rajman, M., Schler, Y. & Zamir, O. "Text Mining at the Term Level". In *Proc. of the 2nd European Symposium of Principles of Data Mining and Knowledge Discovery*, 65-73, 1998.
41. A.W.C. Fu, M.H. Wong, S.C. Sze, W.C. Wong, W.L. Wong, and W.K. Yu, "Finding Fuzzy Sets For The Mining Of Fuzzy Association Rules For Numerical Attributes," In *Proc. Int. Symp. On Intelligent Data Engineering And Learning (Ideal'98), Hong Kong*, 1998, Pp. 263–268.
42. T. Fukuda, Y. Morimoto, S. Morishita, T. Tokuyama, "Mining optimized association rules for numeric attributes," *Proceedings of the ACM SIGMOD International Conference on Management of Data* pp. 182-191, June 1996
43. A. Gyenesei, "Mining Weighted Association Rules for Fuzzy Quantitative Items," *Turku Center for Computer Science Technical Report No 346*, May 2000
44. J. Han, J. Pei, and Y. Yin, "Mining frequent patterns without candidate generation," in *Proc. 2000 ACM SIGMOD Int. Conf. On Management of Data, Dallas, TX, USA*, 2000, pp. 1–12.
45. C. Hidber, "Online association rule mining," in *Proc. 1999 ACM SIGMOD Int. Conf. On Management of Data*, 1999, pp. 145–156.
46. J. Hipp, U. Güntzer, and G. Nakhaeizadeh, "Algorithms for association rule mining - a general survey and comparison," *SIGKDD Explorations*, vol. 2, no. 1, pp. 58–64, 2000.
47. Tzung-Pei Hong, Kuei-Ying Ling, Syue-Liang Wang "Fuzzy Data Mining for Interesting Generalized Association Rules," *Fuzzy Sets and Systems*, vol 138, pp. 255–269,
48. T.P. Hong, C.S. Kuo, and S.C. Chi, "Mining association rules from quantitative data," *Intelligent Data Analysis*, vol. 3, pp. 363–376, 1999.
49. M. Houtsma and A. Swami, "Set-oriented mining for association rules in relational databases," in *Proc. Of the 11th International Conference on Data Engineering*, 1995, pp. 25–33.
50. Yi-Chung Hu, Ruey-Shun Chen, Gwo-Hshiung Tzeng "Discovering fuzzy Association Rules using Fuzzy Partition Methods," *Knowledge Based Systems*,vol. 16 , pp. 137–147. 2003.
51. H. Ishibuchi, T. Yamamoto, T. Nakashima, "Determination of Rule Weights of Fuzzy Association Rules," *IEEE International Fuzzy Systems Conference*, pp. 1555–1558, 2001.
52. J. Kacprzyk, "Fuzzy logic with linguistic quantifiers: A tool for better modeling of human evidence aggregation processes?," in *Fuzzy Sets in Psychology*, T. Zétényi, Ed., pp. 233–263. North-Holland, 1988.
53. Kraft, D.H., Martín-Bautista, M.J., Chen, J. & Vila, M.A., "Rules and fuzzy rules in text: concept, extraction and usage". *International Journal of Approximate Reasoning* 34, pp. 145–161, 2003.
54. C.-M. Kuok, A. Fu, and M. H. Wong, "Mining fuzzy association rules in databases," *SIGMOD Record*, vol. 27, no. 1, pp. 41–46, 1998.
55. J.W.T. Lee, "An ordinal framework for data mining of fuzzy rules," in *Proc. FUZZ-IEEE 2000, San Antonio, TX, USA*, 2000.
56. J.H. Lee and H.L. Kwang, "An extension of association rules using fuzzy sets," in *Proc. of IFSA'97*, 1997.
57. Lin, S.H., Shih, C.S., Chen, M.C., Ho, J.M., Ko, M.T., Huang, Y.M. "Extracting Classification Knowledge of Internet Documents with Mining Term Associations: A Semantic Approach". In *Proc. of ACM/SIGIR'98*, 241-249. Melbourne, Australia, 1998.
58. H. Mannila, H. Toivonen, and I. Verkamo, "Efficient algorithms for discovering association rules," in *Procd AAAI Workshop on Knowledge Discovery in Databases*, 1994, pp. 181–192.
59. M.J. Martín-Bautista, *Modelos de Computación Flexible para la Recuperación de Información (in spanish)*, Ph.D. thesis, Department of Computer Science and Artificial Intelligence, University of Granada, September 2000.
60. M.J. Martín-Bautista, D. Sánchez, J, Chamorro-Martínez, J.M. Serrano, M.A. Vila, "Mining web documents to find additional query terms using fuzzy association rules," Fuzzy Sets and Systems, submitted.
61. R.J. Miller and Y. Yang, "Association rules over interval data," in *Proc. of the ACM-SIGMOD Int. Conf. Management of Data*, 1997, pp. 452–461.
62. T. Oyama, K. Kitano, K. Satou, T. Ito "Extraction of knowledge on protein-protein interaction by association rule discovery," *Bioinformatics*, vol 18, pp. 705-714, 2002
63. J.-S. Park, M.-S. Chen, and P.S. Yu, "An effective hash based algorithm for mining association rules," *SIGMOD Record*, vol. 24, no. 2, pp. 175–186, 1995.
64. Peat, H.P. & Willet, P. "The limitations of term co-occurrence data for query expansion in document retrieval systems". *Journal of the American Society for Information Science 42(5)*,378-383, 1991.
65. W. Pedrycz, "Fuzzy set technology in knowledge discovery," *Fuzzy Sets and Systems*, vol. 98, pp. 279–290, 1998.
66. Porter, M.F. "An algorithm for suffix stripping". *Program 14(3)*:130-137, 1980.
67. Rajman, M. & Besançon, R. "Text Mining: Natural Language Techniques and Text Mining Applications". In *Proc. of the 3rd International Conference on Database Semantics (DS-7)*. Chapam & Hall IFIP Proceedings serie, 1997.
68. Salton, G. & Buckley, C. "Term weighting approaches in automatic text retrieval". *Information Processing and Management 24(5)*, 513-523, 1988.

69. Salton, G. & McGill, M.J. *Introduction to Modern Information Retrieval.* McGraw-Hill, 1983.
70. D. Sánchez, *Adquisición de Relaciones Entre Atributos En Bases de Datos Relacionales (Translates to: Acquisition of Relationships Between Attributes in Relational Databases) (in Spanish)*, Ph.D. thesis, Department of Computer Science and Artificial Intelligence, University of Granada, December 1999.
71. E. Shortliffe and B. Buchanan, "A model of inexact reasoning in medicine," *Mathematical Biosciences*, vol. 23, pp. 351–379, 1975.
72. Shu-Yue-J, Tsang-E, Yenng-D, and Daming-Shi, "Mining fuzzy association rules with weighted items," in *Proc. IEEE Int. Conf. On Systems, Man and Cybernetics*, 2000.
73. C. Silverstein, S. Brin, and R. Motwani, "Beyond market baskets: Generalizing association rules to dependence rules," *Data Mining and Knowledge Discovery*, vol. 2, pp. 39–68, 1998.
74. R. Srikant and R. Agrawal, "Mining generalized association rules," in *Proc 21th Int'l Conf. Very Large Data Bases*, September 1995, pp. 407–419.
75. R. Srikant and R. Agrawal, "Mining quantitative association rules in large relational tables," in *Proc. 1996 ACM SIGMOD Int'l. Conf. Management Data*, 1996, pp. 1–12.
76. Srinivasan, P., Ruiz, M.E., Kraft, D.H. & Chen, J. "Vocabulary mining for information retrieval: rough sets and fuzzy sets". *Information Processing and Management 37*:15-38, 2001.
77. M. Vazirgiannis, "A Classification And Relationship Extraction Scheme For Relational Databases Based On Fuzzy Logic," In *Research And Development In Knowledge Discovery And Data Mining. Pakdd-98, Melbourne, Australia*, 1998, Pp. 414–416.
78. M.A. Vila, J.C. Cubero, J.M. Medina, O. Pons, *Soft Computing: A new perspective for Some Data Mining Problems*, Vistas in Astronomy, vol 41, pp. 379–386, 1997.
79. Wei Wang, Jiong Yang, Philip S. Yu,, "Efficient Mining of Weighted Association Rules (WAR)," *IBM Research Report RC21692 (97734)* , March 2000
80. J. Wijsen and R. Meersman, "On the complexity of mining quantitative association rules," *Data Mining and Knowledge Discovery*, vol. 2, pp. 263–281, 1998.
81. M. Wygralak, *Vaguely Defined Objects. Representations, Fuzzy Sets and Nonclassical Cardinality Theory*, Kluwer Academic Press, Dordrecht, Boston, London, 1996.
82. R.R. Yager, "Quantifier guided aggregation using OWA operators," *International Journal of Intelligent Systems*, vol. 11, pp. 49–73, 1996.
83. S.-J. Yen and A. L.P. Chen, "The analysis of relationships in databases for rule derivation," *Journal of Intelligent Information System*, vol. 7, pp. 235–259, 1996.
84. Yi-Chung Hu, Gwo-Hshiuhg Tzeng, "Elicitation of classification rules by fuzzy data mining," Engineering Applications of Artificial Intelligence, 16, 2003, pp. 709-716
85. L. A. Zadeh, "A computational approach to fuzzy quantifiers in natural languages," *Computing and Mathematics with Applications*, vol. 9, no. 1, pp. 149–184, 1983.
86. W. Zhang, "Mining fuzzy quantitative association rules," in *Proc. 11th Int. Conf. On Tools with A.I., Chicago, IL, USA, Pp. 99-102*, 1999.
87. Z. Zhang, Y. Lu, and B. Zhang, "An effective partitioning-combining algorithm for discovering quantitative association rules," *KDD: Techniques and Applications*, H.Lu,H. Motoda, H. Liu, eds, pp. 241-251, World Scientific, 1997
88. S. Zhang, J. Lu, and C. Zhang, "A Fuzzy Logic based method to acquire user threshold of minimum-support for mining association rules," Will appears in *Information Sciences*

A Foundation for Computing with Words: Meta-Linguistic Axioms

I. BURHAN TÜRKŞEN, Fellow, IFSA, IEEE
Director, Knowledge / Intelligence Systems Laboratory
Mechanical and Industrial Engineering, University of Toronto
Toronto, Ontario, M5S 3G8 CANADA
Tel: (416) 978-1298; Fax: (416) 978-3453
turksen@mie.utoronto.ca
http://www.mie.utoronto.ca/staff/profiles/turksen.htm

Abstract: As a foundation for Computing With Words, meta-linguistic axioms are proposed in analogy to the axioms of classical theory. Consequences of these meta-linguistic expressions are explored in the light of Interval-valued Type 2 Fuzzy Sets. This once again demonstrates that fuzzy set theories and hence CWW have a richer and more expressive power that classical theory.

1. Introduction

Meta-Linguistic axioms are proposed as a foundation for Computing With Words, CWW. Zadeh (1991, 2001) proposed CWW as an extension of fuzzy sets and logic theory. Over the last 40 plus years, we have discussed and made considerable progress on the foundations of fuzzy set and logic theory and their applications in domains of mainly fuzzy control and partially fuzzy decision support systems. But in all these works, we generally have started out with the classical axioms of classical set and logic theory which are expressed in set notation and then relaxed some of these axioms, such as distributivity absorption, idempotency, etc., in order to come up with the application of t-norms and t-conorms in various domains.

In this paper, we propose that a unique foundation for CWW can be established by re-stating the original classical axioms in terms of meta-linguistic expressions where "AND", "OR" are expressed linguistically as opposed to their set theoretic symbols "$\cap$", "$\cup$", respectively.

Next these meta-linguistic expressions can be interpreted in terms of their Fuzzy Disjunctive and Conjunctive Canonical Forms, i.e., FDCF and FCCF.

We explore the consequences of this proposal when these meta-linguistic expressions are interpreted with their Fuzzy Disjunctive and Conjunctive Canonical Forms, FDCF, FCCF.

In our previous writings Türkşen (1986-2004), we have explored various aspects of FDCF and FCCF, including their generation, their non-equivalence, i.e., $FDCF_i(.)TFCCF_i(.)$, i=1,...,16, for the sixteen well known linguistic expressions that form the foundation of any set and logic theory. We have also explored that, for specific cases of t-norms and t-conorms, that are strict and nilpotent Archimedean, we get:

$FDCF_i(.) \subseteq FCCF_i(.)$ (Taner, 1995).

In this paper, we explore in detail, the consequences of re-stating axioms of the classical theory as meta-linguistic expression in the development of a foundation for CWW proposed by Zadeh (1999, 2001).

In turn, we show that new formulas are generated in fuzzy set and logic theory as a new foundation for CWW. This demonstrates the richness and expressive power of fuzzy set and logic theories and CWW that collapse into the classical theory under restricted assumptions of reductionism. That is we obtain the equivalence of the Disjunctive and Conjunctive Normal Forms, $DNF_i(.) \equiv CNF_i(.)$, i=1,...,16, in classical theory as well as the classical set and logic theory axioms.

In our opinion, the break down of these classical equivalences are important in establishing the foundations of fuzzy set theories and the basic formulations of Computing With Words. This break-down and generation of additional formulas expose part of the uncertainty expressed in the combination of concepts that are generated by linguistic operators, "AND", "OR".

In the rest of this paper, we first state the meta-linguistic expression of the axioms for CWW in comparison to their set theoretic forms in section 2.

In Section 3, we explore the consequences of expressing the proposed meta-linguistic axioms in FDCF and FCCF, i.e., Fuzzy Disjunctive and Fuzzy Conjunctive Canonical Forms. In Section 4, we state our conclusions.

2. Meta-Linguistic Axioms

In order to form a sound foundation for the research to be conducted in Computing With Words, CWW, we believe, it is rather necessary that we begin with a statement of the basic axioms with their meta-linguistic expressions to form a sound foundation for CWW proposed by Zadeh (1999, 2001).

In particular, we propose that we need to re-state the classical axioms shown in Table 1 in terms of their meta-linguistic expressions shown in Table 2.

Table 1. Axioms of Classical & Set & Logic Theory, where A, B, C are crisp, two valued sets and c(.) is the complement, "$\cap$", "$\cup$" are set notations and they stand for "AND", "OR" in a one-to-one correspond once with "$\cap$", "$\cup$", respectively. X is the universal set and ϕ is the empty set.

Involution	$c(c(A)) = A$
Commutativity	$A \cup B = B \cup A$
	$A \cap B = B \cap A$
Associativity	$(A \cup B) \cup C = A \cup (B \cup C)$
	$(A \cap B) \cap C = A \cap (B \cap C)$
Distributivity	$A \cup (B \cap C) = (A \cup B) \cap (A \cup C)$
	$A \cap (B \cup C) = (A \cap B) \cup (A \cap C)$
Idempotence	$A \cup A = A$
	$A \cap A = A$
Absorption	$A \cup (A \cap B) = A$
	$A \cap (A \cup B) = A$
Absorption by X and ϕ	$A \cup X = X$
	$A \cap \phi = \phi$
Identity	$A \cup \phi = A$
	$A \cap X = A$
Law of contradiction	$A \cap c(A) = \phi$
Law of excluded middle	$A \cup c(A) = X$
De Morgan Laws	$c(A \cap B) = c(A) \cup c(B)$
	$c(A \cup B) = c(A) \cap c(B)$

It is to be noted that in Table 1, in the Axioms of Classical Set and Logic Theory, A, B, C stand for classical sets such that, for example, $\mu_A(X)=a \in \{0,1\}$, where $\mu_A(X)=a$ is the crisp membership value of $x \in X$, the universe of discourse X, c(.) is involutive complementation operator in the set domain which is the standard negation, n(a)=1-a, where n(.) is the involutive negation operator in the membership domain. Furthermore, "$\cap$", "$\cup$", are set theoretic "intersection", "union" operators and are taken in one-to-one correspondence to the linguistic operators "AND", "OR", respectively, in the classical reductionist perspective.

Whereas in Table 2, in the Meta-Linguistic expression of the proposed Axioms for CWW, A, B, C stand for fuzzy sets which are linguistic terms of linguistic variables, such that, for example, $\mu_A(X)=a \in [0,1]$ where $\mu_A(X)=a$ is the fuzzy membership value of $x \in X$, NOT(.) is a linguistic negation operator, which will be taken to be equivalent to the involutive negation of the classical theory, i.e., for the purposes of this paper, NOT(.)=c(.) and hence n(a)=1-a. However, the linguistic "AND", "OR" operators will be taken as linguistic connectives which do not map in a one-to-one correspondence to "$\cap$", "$\cup$", respectively, which are symbols of the classical set theory that map to t-norms and t-conorms, respectively, within the perspective of fuzzy theory.

Table 2. Meta-Linguistic Expression of the Axioms for CWW where A, B, C are fuzzy sets and stand for linguistic terms of linguistic variables, NOT(.) is the complementation operator,"AND", "OR" are linguistic connectives that are not in a one-to-one correspondence with "∩", "∪", respectively.

Involution:	NOT(NOT(A))=A
Commutativity:	A AND B = B AND A
	A OR B = B OR A
Associativity:	(A AND B) AND C = A AND (B AND C)
	(A OR B) OR C = A OR (B OR C)
Distributivity:	A AND (B OR C)=(A AND B) OR (A AND C)
	A OR (B AND C)=(A OR B) AND (A OR C)
Idempotency:	A AND A = A
	A OR A = A
Absorption by X and ∅:	A OR X = X
	A AND ∅ =∅
Identity:	A OR ∅ = A
	A AND X = A
Law of contradiction:	A AND NOT(A) ⊆ ∅
Law of excluded middle:	A OR NOT(A) ⊆ X
De Morgan's Laws:	NOT(A AND B) = NOT(A) AND NOT(B)
	NOT(A OR B) = NOT(A) OR NOT(B)

This notion that linguistic "AND", "OR" do not correspond in a one-to-one mapping to "∩", "∪", respectively, is supported by Zimmermann, Zysno(1980) experiments and our investigations on "Compensatory 'AND' " (1992). It should be recalled that when meta-linguistic expressions are represented in terms of FDCF and FCCF, Fuzzy Disjunctive Canonical Form and Fuzzy Conjunctive Canonical Forms, Table 3, they are no longer equivalent, i.e., $FDCF_i(.)TFCCF_i(.)$, i=1,…,16, for the sixteen basic expressions of any set and logic theory (Türkşen, 1986, 2002) which are shown in Table 4. This is true in particular for 3rd and 6th expressions shown in Table 4, i.e., "A OR B", and "A AND B", respectively. These two expressions and their FDCF and FCCF expressions are essential in reinterpreting the Meta-Linguistic axioms stated in Table 2.

As it can be observed, it should be recalled that $FDCF_i(.)=DNF_i(.)$ and $FCCF_i(.)=CNF_i$, i=1,…,16 **in form only but not in content.**

Table 3. Classical Disjunctive Normal and Fuzzy Disjunctive Canonical Forms, DNF and FDCF and Classic Conjunctive Normal and Fuzzy Conjunctive Canonical Forms, CNF and FCCF, where ∩ is a conjunction, ∪ is a disjunction and c is a complementation operator in the set domain.

No.	Fuzzy Disjunctive Canonical Forms/Disjunctive Normal Forms
1	(A∩B) ∪ (A∩c(B)) ∪ (c(A) ∩B) ∪ (c(A) ∩c(B))
2	∅
3	(A∩B) ∪ (A∩c(B)) ∪ (c(A) ∩B)
4	(c(A) ∩c(B))
5	(A∩c(B)) ∪ (c(A) ∩B) ∪ (c(A) ∩c(B))
6	(A∩B)
7	(A∩B) ∪ (c(A) ∩B) ∪ (c(A) ∩c(B))
8	(A∩c(B))
9	(A∩B) ∪ (A∩c(B)) ∪ (c(A) ∩c(B))
10	(c(A) ∩B)
11	(A∩B) ∪ (c(A) ∩c(B))
12	(A∩c(B)) ∪ (c(A) ∩B)
13	(A∩B) ∪ (A∩c(B))
14	(c(A) ∩B) ∪ (c(A) ∩c(B))
15	(A∩B) ∪ (c(A) ∩B)
16	(A∩c(B)) ∪ (c(A) ∩c(B))

No.	Fuzzy Conjunctive Canonical Forms/Conjunctive Normal Forms
1	I
2	(A∪B) ∩ (A∪c(B)) ∩ (c(A) ∪B) ∩ c(A) ∪c(B))
3	(A∪B)
4	(A∪c(B)) ∩ (c(A) ∪B) ∩ c(A) ∪c(B))
5	(c(A) ∪c(B))
6	(A∪B) ∩ (A∪c(B)) ∩ (c(A) ∪B)
7	(c(A) ∪B)
8	(A∪B) ∩ (A∪c(B)) ∩ (c(A) ∪c(B))
9	(A∪c(B))
10	(A∪B) ∩ (c(A) ∪B) ∩ (c(A) ∪c(B))
11	(A∪c(B)) ∩ (c(A) ∪B)
12	(A∪B) ∩ (c(A) ∪c(B))
13	(A∪B) ∩ (A∪c(B))
14	(c(A) ∪B) ∩ (c(A) ∪c(B))
15	(A∪B) ∩ (c(A) ∪B)
16	(A∪c(B)) ∩ (c(A) ∪c(B))

Table 4. Sixteen Possible Combinations of any two sets, A and B.

Number	Meta-Linguistic Expressions
1	UNIVERSE
2	EMPTY SET
3	A OR B
4	NOT A AND NOT B
5	NOT A OR NOT B
6	A AND B
7	A IMPLIES B
8	A AND NOT B
9	A OR NOT B
10	NOT A AND B
11	A IF AND ONLY IF B
12	A EXCLUSIVE OR B
13	A
14	NOT A
15	B
16	NOT B

3. Consequences of the Proposed Meta-Linguistic Axioms

In order to appreciate the consequences of the proposed Meta-Linguistic Axioms for CWW, we first very briefly review the classical axioms and the habit of usual use of them in our investigation. After this, we state the consequences of the proposed Meta-Linguistic Axioms for CWW.

3.1. Classical Axioms

It is well known that $DNF_i(.) \equiv CNF_i(.)$, $i=1,\ldots,16$, in classical theory. Thus, in classical applications, one always use the shortest of these two forms.

For example:

(1) For "A AND B", we get:

$$\begin{cases} \text{DNF(A AND B)} = A \cap B, \\ \text{CNF(A AND B)} = (A \cup B) \cap (c(A) \cup B) \cap (A \cup c(B)), \end{cases}$$

together with the equivalence of DNF and CNF, i.e., DNF(A AND B) ≡ CNF(A AND B).

But in all our calculations, we use only "A∩B" for a representation of "A AND B" in the classical set domain.

(2) For "A OR B", we get:

$$\begin{cases} \text{DNF(A OR B)} = (A \cap B) \cup (c(A) \cap B) \cup (A \cap c(B)), \\ \text{CNF(A OR B)} = A \cup B, \end{cases}$$

together with equivalence of DNF(A OR B) ≡ CNF(A OR B).

But again in all our calculations, we use only "A∪B" for a representation of "A OR B" in the classical set domain.

This short hand form use is applied to all the remaining linguistic combination. Furthermore, this habit of using the short hand form of these combinations was carried out by most fuzzy researchers in their applied as well as theoretical investigations. However, the investigations carried out by Türkşen (1986-2002) and Türkşen, et.al.(1998, 1999) Resconi and Türkşen (2001) indicate that we ought to use both the $FDCF_i(.)$ and $FCCF_i(.)$, i=1,…,16, because the equivalence no longer holds in fuzzy theory, i.e., $FDCF_i(.)$ T $FCCF_i(.)$. We next investigate the consequences of this non-equivalence for each of the proposed meta-linguistic axioms for CWW stated in Table 2.

3.2. Meta-Linguistic Axioms

In fuzzy theory and its applications in CWW, most researcher continue the usual habit of using the shortest form of classical axioms by directly fuzzifying all the classical axioms. That is as in the classical theory, "A AND B" is directly taken to be "A∩B" but fuzzified, "A OR B" is directly taken to be "A∪B" but fuzzified. But the other longer form is ignored or not considered either because of habit or because most of us are generally short sighted.

In turn, we state that certain axioms hold and others do not hold. But such a stance is not in the spirit of fuzzy theory. As all of us believe, in fuzzy theory all are a matter of degree. This usual habit of use continues to persist in most of the current research and applications despite the fact that the equivalences break down in fuzzy theory, i.e., $FDCF_i(.)$ T $FCCF_i(.)$, i=1,…,16, that have been published in various paper over about the last twenty years or so (Türkşen, 1986-2001). If we take into the account the fact that these non-equivalences of the Fuzzy Disjunctive and Conjunctive Canonical forms, then we have to realize that the interpretation of the proposed Meta-Linguistic Axioms must be expressed in two distinct forms in set symbolic notation and must give two distinct results in computational, numeric, domain with the application of t-norms and t-conorms. We next express this realization for each of the proposed Meta-Linguistic Axioms.

Fuzzy Involution: Since for the purposes of this paper we have taken NOT(.) = c(.) then involution holds as specified. That is there is no new interpretation of this axiom at this writing. In the future, when we investigate other linguistic negation operators, this will probably produce some new results as it should.

Fuzzy Commutativity: In Table 2, there are two Meta-Linguistic Commutativity axioms:

"A AND B = B AND A", and
"A OR B" = B OR A"

Now, we know that
FDCF(A AND B) T FCCF(A AND B), and

FDCF(A OR B) T FCCF(A OR B).

Therefore, we obtain two set theoretic axioms of the commutativity in fuzzy theory for CWW for these two Meta-Linguistic Axioms and similarity for all the other axioms shown in Table 2 as follows.

3.3. Fuzzy Set Theoretic Axioms for CWW

Here we state both the FDCF and FCCF versions of the whole set of Meta-Linguistic Axioms for CWW.

Fuzzy Commutativity with "AND":

(a) FDCF(A AND B) = FDCF(B AND A), by a substitution of their fuzzy set symbols, we get:
i.e., $A \cap B = B \cap A$.

(b) FCCF(A AND B) = FCCF(B AND A), again by a substitution, we get:
i.e., $(A \cup B) \cap (c(A) \cup B) \cap (c(B) \cup A) = (B \cup A) \cap (c(B) \cup A) \cap (B \cup c(A))$

Fuzzy Commutativity with "OR":

(a) FDCF(A OR B) = FDCF(B OR A),
i.e., $(A\cap B)\cup(c(A)\cap B)\cup(A\cap c(B)) = (B\cap A)\cup(c(B)\cap A)\cup(B\cap c(A))$

(b) FCCF(A OR B) = FCCF(B OR A), i.e., $A\cup B = B\cup A$

Therefore, fuzzy commutativity holds fuzzily as a matter of degree in two separate axioms. This in turn exposes an uncertainty region for the fuzzy commutativity axioms.

Fuzzy Associativity with "AND"

(a) Let us first investigate the fuzzy associativity with FDCF's:

FDCF[(A AND B) AND C] = FDCF[A AND (B AND C)]

which is to be derived from:

FDCF[FDCF(A AND B) AND C)] = FDCF[A AND FDCF(B AND C)]

i.e., we get $(A\cap B)\cap C = A\cap(B\cap C)$

This version holds to a fuzzy degree in analogy to the classical theory.

(b) Let us next investigate the fuzzy associativity with FCCF's:

FCCF[(A AND B) AND C] = FCCF[A AND (B AND C)]

which is to be derived from:

FCCF[FCCF(A AND B) AND C)] = FCCF[A AND FCCF(B AND C)] (3.1)

Recall that, we have:

FCCF(A AND B) = $(A\cup B)\cap(c(A)\cup B)\cap (A\cup c(B))$
FCCF(B AND C) = $(B\cup C)\cap(c(B)\cup C)\cap(B\cup c(C))$

Therefore, on the left hand side of (3.1) we get:

FCCF[FCCF(A AND B) AND C]
= [FCCF(A AND B)$\cup$C]$\cap$[c[FCCF(A AND B)]$\cup$C]$\cap$[FCCF(A AND B)$\cup$c(C)]

$=\{[(A\cup B)\cap(c(A)\cup B)\cap(A\cup c(B))]\cup C\}\cap\{c[(A\cup B)\cap(c(A)\cup B)\cap(A\cup c(B)]\cup C\}$

$\cap\{[(A\cup B)\cap(c(A)\cup B)\cap(A\cup c(B))]\cup c(C)\}$
(3.2)

On the right hand side of (3.1), we get:

FCCF[A AND FCCF(B AND C)]
=[A∪FCCF(B AND C)]∩[c(A)∪FCCF(B AND C)]∩{A∪c[FCCF(B AND C)]}
$=\{A\cup[(B\cup C)\cap(c(B)\cup C)\cap(B\cup c(C))]\}\cap\{c(A)\cup[(B\cup C)\cap(c(B\cup C)\cap(B\cup c(C)]\}$
$\cap\{A\cup c[(B\cup C)\cap(c(B\cup C)\cap(B\cup c(C)]\}$
(3.3)

It should be noted and it is clear and straight forward to drive and observe that the first, i.e., (a), interpretation of the commutativity equality holds. However, it is also clear that in general the second, i.e., (b), interpretation of assoativity does not hold even to a fuzzy degree.

Fuzzy Associativity with "OR"

(a) First let us investigate the fuzzy associativity for ((A OR B) OR C)=(A OR (B OR C)) with FDCF's:

FDCF[(A OR B) OR C] = FDCF[A OR (B OR C)]

which is to be derived from:

FDCF[FDCF(A OR B) OR C)] = FDCF[A OR FDCF(B OR C)]
(3.4)

Recall that, we have:

FDCF(A OR B) = (A∩B)∪(c(A)∩B)∪ (A∩c(B))
FDCF(B OR C) = (B∩C)∪(c(B)∩C)∪(B∩c(C))

Therefore, on the left hand side of (3.4) we get:

FDCF[FDCF(A OR B) OR C]
= [FDCF(A OR B)∩C]∪[c[FDCF(A OR B)]∩C]∪[FDCF(A OR B)∩c(C)]
(3.5)

$=\{[(A\cap B)\cup(c(A)\cap B)\cup(A\cap c(B))]\cap C\}\cup\{c[(A\cap B)\cup(c(A)\cap B)\cup(A\cap c(B)]\cap C\}$
$\cup\{[(A\cap B)\cup(c(A)\cap B)\cup(A\cap c(B))]\cap c(C)\}$

As well on the right hand side (3.4), we get:

FDCF[A OR FDCF(B OR C)]

$$= \{A\cap[(B\cap C)\cup(c(B)\cap C)\cup(B\cap c(C))]\}\cup\{c(A)\cap[(B\cap C)\cup(c(B)\cap C)\cup(B\cap c(C))]\} \quad (3.6)$$
$$\cup\{A\cap c[(B\cup C)\cup(c(B)\cap C)\cup(B\cap c(C))]\}.$$

Again, it is clear that in general, in this case the first form, i.e., (a), interpretation of associativity does not hold even to a fuzzy degree.

(b) Next let us investigate the fuzzy associativity for ((A OR B) OR C)=(A OR (B OR C)) with FCCF's:

FCCF[(A OR B) OR C] = FCCF[A OR (B OR C)]

which is to be derived from:
FCCF[FCCF(A OR B) OR C)] = FCCF[A OR FCCF(B OR C)]

Since FCCF(A OR B) = $A\cup B$, and FCCF(B OR C) = $B\cup C$,

Thus, we get:

$(A\cup B)\cup C = A\cup(B\cup C)$,

which holds in a straight forward manner but naturally to a fuzzy degree!

Fuzzy Distributivity for "A AND (B OR C)=(A AND B) OR (A AND C)"

Works for a particular FDCF and FCCF combination as follows:

FDCF[(A AND FCCF(B OR C)] = FCCF[FDCF(A AND B) OR FDCF(A AND C)]

By substituting known values of FDCF(.) and FCCF(.), we get

$A\cap(B\cup C) = (A\cap B)\cup(A\cap C)$

Hence

FDCF[A AND FCCF(B OR C)] = FCCF[FDCF(A AND B) OR FDCF(A AND C)]

which holds to a fuzzy degree.

Fuzzy Distributivity for "A OR (B AND C)=(A OR B) AND (A OR C))"

Again this works but this time for a particular FCCF and FDCF combination:

FCCF[(A OR FDCF(B AND C)] = $A\cup(B\cap C)$

FDCF[FCCF(A OR B) AND FCCF(A OR C)] = $(A\cup B)\cap(A\cup C)$

Therefore, we get:

FCCF[A OR FDCF(B AND C)] = FDCF[FCCF(A OR B) AND FCCF(A OR C)]

By substituting known values FCCF(.) and FDCF(.), we get:

$A\cup(B\cap C) = (A\cup B) \cap (A\cup C)$

which holds to a fuzzy degree.

Generally, other FDCF and FCCF combinations do not hold

Fuzzy Idempotency "A AND A=A"

FDCF(A AND A) = $A\cap A$
FCCF(A AND A) = $(A\cup A)\cap(c(A)\cup A)\cap(A\cup c(A))$

These were investigated in Türkşen, et.al.(1999) in detail and in general expose a region of uncertainty between these two expressions of fuzzy idempotency. However, in that paper there is a duplication error which should be noted, and it was connected in later papers Türkşen (2001, 2002).

Fuzzy Idempotency "A OR A=A"

FDCF(A OR A) = $(A\cap A)\cup (c(A)\cap A)\cup(A\cap c(A))$
FCCF(A OR A) = $A\cup A$

Again these were investigated in Türkşen, et.al.(1999) in detail and again expose a region of uncertainty.

These Fuzzy idempotency laws were discussed under the heading of "Reaffirmation ..." in Türkşen, et.al.(1999) again there is a duplication error which is to be noted. These duplication errors were corrected in Türkşen (2001, 2002).

Fuzzy Absorption by "A OR X=X"

FDCF(A OR X) = $(A\cap X)\cup(c(A)\cap X)\cup(A\cap c(X))$
FCCF(A OR X) = $A\cup X = X$

Fuzzy Absorption by "A AND ∅=∅"

FDCF(A AND ∅) = A∩∅ = ∅
FCCF(A AND ∅) = (A∪∅)∩(c(A)∪∅)∩(A∪c(∅))

Fuzzy Absorption by "A OR ∅=A"

FDCF(A OR ∅) = (A∩∅)∪(c(A)∩ ∅)∪(A∩c(∅))
FCCF(A OR ∅) = A∪∅ = A

As it can be observed fuzzy absorption exposes a region of uncertainty.

Fuzzy Identity "A AND X=A"

FDCF(A AND X) = A∩X=A
FCCF(A AND X) = (A∪X)∩(c(A)∪X)∩(A∪c(X))

Law of Fuzzy Contradiction "A AND NOT(A) ⊇ ∅"

FDCF(A AND NOT(A)) = A∩c(A) ⊇ ∅
FCCF(A AND NOT(A)) = (A∪c(A))∩(c(A)∪c(A))∩(A∪A) ⊆ X

which was named "The Law of Fuzzy Contradiction". Again this was discussed in detail in Türkşen et.al.(1999).

Law of Fuzzy Middle "A OR NOT(A)⊆X"

FDCF(A OR NOT(A)) ⊆ (A∩c(A))∪(c(A)∩c(A))∪(A∩A) ⊆ X
FCCF(A OR NOT(A)) = A∪c(A) ⊆ X

which was named "The Law of Fuzzy Middle" and it was discussed in detail in Türkşen et.al.(1999). Once again, it is noted that in all these topics that were discussed in Türkşen (1999) there are duplication errors which were corrected in later papers in Türkşen (2001, 2002).

Fuzzy De Morgan Laws
"NOT(A AND B)=NOT(A) OR NOT(B)"

holds where NOT(.) is taken as the standard complementation in set domain and as the standard negation, in membership domain, i.e., n(a)=1-a such that

NOT(FDCF(A AND B)) = FCCF(NOT(A) OR NOT(B))

$= c(A \cap B) = c(A \cup B)$

"NOT(A OR B) = NOT(A) AND NOT(B)"

holds again in the same manner such that

NOT(FCCF(A OR B)) = FDCF[NOT(A) AND NOT(B)]

$= c(A \cup B) = c(A)$ AND $c(B)$

Once again these hold as a matter of degree.

We also check to see if (i):

NOT(FDCF(A OR B)) = FCCF[NOT(A) AND NOT(B)]

and if (ii)

NOT[FCCF(A AND B)] = FDCF[NOT(A) OR NOT(B)]
holds.

By substituting their set symbolic equivalents we get for

$c[(A \cap B) \cup (c(A) \cap B) \cup (A \cap c(B))] = [(c(A) \cup (c(B)) \cap (A \cup c(B)) \cap (c(A) \cup B))]$,

with the application of c(.) to the left hand side we get:

$(c(A) \cup (c(B)) \cap (A \cup c(B)) \cap (c(A) \cup B)) = (c(A) \cup (c(B)) \cap (A \cup c(B)) \cap (c(A) \cup B))$

Also, for (ii) by again with substitution we obtain

$c[(A \cup B) \cap (c(A) \cup B) \cap (A \cup c(B))]$

Again with the application of c(.) to the left hand side, we get $(c(A) \cap c(B)) \cup (A \cap c(B)) \cup (c(A) \cap B)$
$= (c(A) \cap c(B)) \cup (A \cap c(B)) \cup (c(A) \cap B)$

Thus we observe that there are two De Morgan Laws in fuzzy theory for each of the De Morgan Laws of classical theory.

4. Conclusion

We have proposed a Meta-Linguistic Set of Axioms as a foundation for Computing With Words. In these meta-linguistic axioms, linguistic terms of linguistic variable are symbolized by A, B, C,...,etc, representing fuzzy sets that capture the meaning of the linguistic terms in a precisiated natural language expression where the meaning representation is precisiated via fuzzy set membership functions. In addition linguistic connectives "AND", "OR" are not mapped in a one-to-one correspondence to the set symbols "$\cap$", "$\cup$", respectively.
Rather, linguistic connectives are represented by interval-valued Type 2 fuzzy sets generated by FDCF and FCCF, Fuzzy Disjunctive and Fuzzy Conjunctive Canonical Forms.

As a result all the meta-linguistic axioms generate two distinct expressions to represent meaning specification of these axioms. This reveals an uncertainty interval in the interpretation of the meta-linguistic axioms. But provides a more expressive power that gets closer to the rich meaning representation associated with natural language expressions.

It is expected that the proposed set of meta-linguistic axioms and their applications will provide a sound grounding for Computing With Words at this early stages of development in furthering fuzzy set theory to provide a good bases for future intellectual developments.

Finally, in the light of the analysis developed in this paper, it should be noted that it is not enough to start with the classical axioms and make statements that suggest certain axioms do not hold.

In turn, its myopic to state that "there are only four axioms for t-norms and t-conorms, i.e., boundary, commutativity and associativity and monotoincity".

References

[1] Bilgic, T. (1995), Measurement-Theoretic Frameworks for Fuzzy Set Theory with Applications to Preference Modelling, PhD thesis, University of Toronto, Dept. of Industrial Engineering Toronto, Ontario M5S 1A4 Canada.(supervisor, I.B. Türkşen)

[2] Dempster A.P (1967), "Upper and Lower Probabilities Induced by a Multivalued Mapping", In: Annals of Mathematical Statistics, 38, 325-339.

[3] Resconi, G., I.B. Türkşen (2001), "Canonical Forms of Fuzzy Truthoods by Meta-Theory Based Upon Modal Logic", Information Sciences, 131, 157-194.

[4] Türkşen I.B. (2002), "Interval-valued Type 2 Fuzzy Sets, Multi-valued Maps and Rough Sets", (in) A Grmela and N.E.Mastorakis (eds.), Advances in Intelligent Systems, Fuzzy Sets, Evolutionary Computation, WSEAS, 142-146.

[5] Türkşen I.B. (2002), "Upper and Lower Set Formulas: Restriction and Modification of Dempster-Pawlak Formalism", Special Issue of the International Journal of Applied Mathematics and Computer Science, V.12, No.3, 101-111.

[6] Türkşen I.B. (2001), "Computing with Descriptive and Veristic Words: Knowledge Representation and Reasoning", in: Computing With Words, P.P. Wang(ed.), Chapter 10, Wiley, New York, 297-328.

[7] Türkşen, I.B. (1999), "Theories of Set and Logic with Crisp or Fuzzy Information Granules", J.of Advanced Computational Intelligence, 3,4, 264-273.

[8] Türkşen, I.B., A. Kandel, Y-Q. Zhang (1999), "Normal Forms of Fuzzy Middle and Fuzzy Contradiction", IEEE-SMC, 29-2, Part B, Cybernetics, 237-253.

[9] Türkşen, I.B., A. Kandel, Y-Q. Zhang (1998), "Universal Truth Tables and Normal Forms", IEEE-Fuzzy Systems, 6-2, 295-303.

[10] Türkşen, I.B. (1992), "Interval Valued Fuzzy Sets and 'Compensatory AND' ", Fuzzy Sets and Systems, 51, 295-307.

[11] Türkşen I.B. (1986), "Interval-Valued Fuzzy Sets based on Normal Forms", Fuzzy Sets and Systems, 191-210.

[12] Zadeh, L.A. (2001), "From Computing with Numbers to Computing with Words - From Manipulation of Measurements to Manipulation of Perceptions", in: P.P. Wang(ed.) Computing With Words, Wiley Series on Intelligent Systems, Wiley and Sons, New York, 35-68.

[13] Zadeh, L.A. (1999), "From Computing with Numbers to Computing with Words—From Manipulation of Measurements to Manipulation of Perceptions", IEEE-Trans on Curciuts and Systems, 45, 105-119.

[14] Zadeh, L.A. (1979), "A Theory of Approximate Reasoning", in J. Hayes, D. Michie, and L.I. Mikulich (eds) Machine Intelligence, Halstead Press, New York, Vol. 9, 149-194.

[15] Zadeh, L.A. (1978), "Fuzzy Sets as a Basis for a Theory of Possibility", Fuzzy Sets and Systems, 3-28.

[16] Zadeh, L.A. (1973), "Outline of a New Approach to the Analysis of Complex Systems and Decision Processes", IEEE Trans. Systems Man Cybernet, 3, 28-44.

[17] Zadeh, L.A. (1968), "Probability Measures of Fuzzy Events", J.Math. Analysis and Appl., 10, 421-427

[18] Zadeh, L.A. (1965), "Fuzzy Sets", Information and Control Systems, Vol.8, 338-353.

[19] Zimmermann, H.J., P. Zysno (1980), "Latent Connectives in Human Decision Making", Fuzzy Sets and Systems, 4, 37-51.

Augmented Fuzzy Cognitive Maps Supplemented with Case Based Reasoning for Advanced Medical Decision Support

Voula Georgopoulos [(1)] and Chrysostomos Stylios [(2)]
[(1)] Department of Speech and Language Therapy, Technological Educational Institute of Patras, Patras, Greece
Email: voula@teipat.gr, Tel: +302610322812, Fax: +302610369167
[(2)] Computer Science Department, University of Ioannina, Email: stylios@cs.uoi.gr Tel: +302651098818

Abstract: Fuzzy Cognitive Maps (FCMs) have been used to design Decision Support Systems and particularly for medical informatics to develop Intelligent Diagnosis Systems. Even though they have been successfully used in many different areas, there are situations where incomplete and vague input information may present difficulty in reaching a decision. In this chapter the idea of using the Case Based Reasoning technique to augment FCMs is presented leading to the development of an Advanced Medical Decision Support System. This system is applied in the speech pathology area to diagnose language impairments..

1. Introduction

This chapter presents how the Soft Computing technique of Fuzzy Cognitive Maps (FCMs) can be combined with Case Based Reasoning methods in order to develop an Advanced Medical Decision Support System. FCM is a knowledge-based methodology suitable to describe and model complex systems and handle information from an abstract point of view (Kosko 1986). Soft computing techniques such as FCMs have been successfully used to model complex systems that involve discipline factors, states, variables, input, output, events and trends. FCM modeling can integrate and include in the decision-making process the partial influence of controversial factors, can take under consideration causal effect among factors and evaluate the influence from different sources, factors and other characteristics using fuzzy logic reasoning. Each one of the involved factors has a different degree of importance in determining (or influencing) the decision, which increases the complexity of the problem. Thus, soft computing methods are ideal for developing Decision Support systems in Medical Informatics where humans use mainly differential diagnosis based on fuzzy factors some of which are complementary, others similar and others conflicting, and all are taken into consideration when a decision is reached (Kasabov 1996, 2002; Zeleznikow and Nolan 2001).

Fuzzy Cognitive Maps develop a behavioral model of the system exploiting the experience and knowledge of experts. Fuzzy Cognitive Maps applicability in modeling complex systems has been successfully used in many different application areas (Stylios et al. 1999). An FCM is a signed fuzzy graph with feedback, consisting of concepts-nodes and weighted interconnections. Nodes of the graph stand for concepts that are used to describe main behavioral characteristics of the modeled system. Nodes are connected by signed and fuzzy weighted arcs representing the cause and effect relationship existing among concepts. Thus, an FCM is a fuzzy-graph structure, which allows systematic causal propagation, in particular forward and backward chaining (Stylios and Groumpos 2000). Fuzzy Cognitive Maps have been successfully used to develop a Decision Support System (FCM-DSS) for differential diagnosis (Georgopoulos et al. 2003), to determine the success of radiation therapy process estimating the final dose delivered to the target volume (Papageorgiou et al. 2003) and many other application areas. But there are cases where the input information is not adequate and FCM-DSS cannot discriminate and reach a decision; this surfaces the need of a mechanism to supplement the FCM-DSS.

In this research work we combine FCMs with methods and approaches that have been used for Case-Based Reasoning (CBR) (Noh et al. 2000; Kolodner et al. 1993). This is a successful methodology for managing implicit knowledge (Watson 1999; Lopez de Mantaras 2001), which has also been used in medical informatics (Schmidt et al. 1999, 2001). CBRs embed a considerable amount of previous solved instances of problems (cases). The problem solving experience is explicitly taken into account by storing past cases in a database (case base), and by suitably retrieving them when a new problem has to be tackled (Noh et al. 2000). It simply makes decisions on new cases by their similarity to old cases stored in its case-base rather than using some derivative representation, as is done for example in adaptive-type methodologies. But, if the new case has no match with the stored cases, the CBR has no solution. Similarly to FCMs, CBRs have been applied to medical diagnosis and patient treatment outcomes. Despite the limitations of CBRs, they are usually assumed to have a certain degree of richness of stored knowledge, and a certain degree of complexity due to the way they are organized.

This chapter is divided into 7 sections. Section 2 describes Fuzzy Cognitive Maps, how they model systems and how they are developed. Section 3 presents why Case Based Reasoning (CBS) is important in Medical Decision Systems and how CBR could be combined with FCMs. Section 4 proposes an algorithm to develop an Advanced Medical Decision System, implementing Case Base Reasoning to augment Competitive Fuzzy Cognitive Maps (CFCM); the CFVM developing algorithm is also presented. Section 5 presents an application of the proposed model to speech and language pathology and in section 6 the results of the example are presented. Finally section 7 concludes the chapter.

2. Fuzzy Cognitive Maps

Fuzzy Cognitive Maps (FCM) are a soft computing tool that is a result of the synergy of Fuzzy Logic and Neural Network methodologies and is based on the exploitation of the integrated experience of expert-scientists (Stylios et al. 1999). The graphical illustration of a FCM is a signed, weighted graph with feedback that consists of nodes and weighted arcs. Nodes of the graph are the concepts that correspond to variables, states, factors and other characteristics incorporated in the model, which describe the behavior of the system. Directed, signed and weighted arcs, which represent the causal relationships that exist between the concepts, interconnect the FCM concepts. Each concept represents a qualitative characteristic, state or variable of the system; concepts stand for events, actions, goals, values, and/or trends of the system being modeled as an FCM. Each concept is characterized by a numeric value that represents a quantitative measure of the concept's presence in the model. A high numeric value indicates the strong presence of a concept. The numeric value results from the transformation of the real value of the system's variable, for which this concept stands, to the interval [0,1]. All the values in the graph are fuzzy, so weights of the arcs are described with linguistic values that can be defuzzified and transformed to the interval [-1,1].

Between concepts, there are three possible types of causal relationships that express the type of influence of one concept on the others. The weight of an interconnection, denoted by W_{ij}, for the arc from concept C_i to concept C_j, can be positive, ($W_{ij}>0$), which means that an increase in the value of concept C_i leads to the increase of the value of concept C_j, and a decrease in the value of concept C_i leads to the decrease of the value of concept C_j. Or there is negative causality ($W_{ij}<0$), which means that an increase in the value of concept C_i leads to the decrease of the value of concept C_j and vice versa. When, there is no relationship from concept C_i to concept C_j, then $W_{ij}=0$ (Kosko 1991).

When the Fuzzy Cognitive Map starts to model the system, concepts take their initial values and then the system is simulated. At each step, the value of each concept is determined by the influence of the interconnected concepts on the corresponding weights:

$$A_i^t = f\left(\sum_{\substack{j=1 \\ j\neq i}}^{n} A_j^{t-1} W_{ji} + A_i^{t-1}\right) \qquad (1)$$

where A_i^t is the value of concept C_i at step t, A_{j-1}^t is the value of the interconnected concept C_j at step t-1, and W_{ji} is the weighted arc from C_j to C_i and f is a threshold function.

Fuzzy Cognitive Maps represent the human knowledge on the operation of the system, so in order to develop an FCM one expert is asked to do so; thus, FCMs rely on the exploitation of experts' experience on system's model and behavior. Experts determine the number and kind of concepts of FCM and the interrelation among concepts. Experts know the main factors that determine the behavior of the system, each one of these factors is represented by a concept. So an expert draws an FCM according to his experience, he determines the concepts, which for example stand for events, actions, goals, values, and trends of the system. The expert knows which elements of the system influence other elements; for the corresponding concepts he determines the negative or positive effect of one concept on the others, with a fuzzy degree of causation. The determination of the degree of casual relationship among concepts can be improved by the application of learning rules for choosing appropriate weights for the FCM (Kosko 1986). In this way, an expert decodes his own knowledge on the behavioral model of the system and transforms this knowledge in a weighted graph. After the construction of the map, the FCM starts to simulate the operation of the system and each concept interacts with other concepts.

The major advantage of fuzzy cognitive maps is that they can handle even incomplete or conflicting information. This is very important in the decision-making and diagnosis in the area of medical informatics. Especially, in the case of language/communication disorders it is very difficult to reach a conclusion and frequently important information may (Georgopoulos et al. 2003): i) be missing, ii) be unreliable, iii) be vague or conflicting, and/or iv) be difficult to integrate with other information.

3. Case Based Reasoning

Even though successful medical Decision Support FCMs have been developed (Georgopoulos et al. 2003; Papageorgiou et al. 2003), there are situations where the patient data to be input into the system presents a very rare configuration of symptoms where most of the nodes of the FCM would not be active. In other words, for example, although the FCM-Model of a Medical Decision Support System has been designed to include all possible symptoms and causative factors (nodes-concepts) and the relationship between them (weights) for some medical condition, there are particular situations where very few symptoms are available and are taken into consideration. Thus, in such a diagnosis or prognosis model Decision Support FCM, the decision would be made only using a very small subset of the concepts of the entire system. Such a system could lead to either an erroneous decision or difficulty in reaching stability since the weighting of the active nodes reflects only a small amount of the experts' stored knowledge.

Using a CBR-augmented FCM Decision support system, as shown in Figure 1, in such situations, the decisions support system would draw upon cases that are maximally similar according to distance measures and would use the CBR subsystem to generate a sub-FCM emphasizing the nodes activated by the patient data and thus redistributing the causal weightings between the concept-nodes.

The advantage of CBR-augmented FCMs lies in the ability to represent rare occurrences of medical conditions/symptoms, which may not be adequately represented in an FCM due to its design methodology, which is dependent on human experts and learning algorithms (Georgopoulos and Stylios 2003).

There are a variety of approaches that determine the similarity between an input case and the stored cases. Some similarity measures rely on only the shared features between input and stored cases (Rosch and Mervis 1975) whereas others determine similarity by adding up the features that are shared and subtracting the features that are not shared between the input case and each stored case (Tversky 1977). The most common techniques used in CBR diagnostic systems are based on nearest-neighbor retrieval since it is a simple approach that computes the similarity between stored cases and an input case based on weight features. The similarity of the problem (input case) to a case in the case-library for each case attribute is determined. This measure may be multiplied by a weighting factor. The weighted sum of the similarity of all attributes provides a measure of the similarity of each case in the library to the input case, as given by (Noh et al. 2000; Kolodner et al. 1993):

$$Similarity(I,R) = \frac{\sum_{i=1}^{m} f(I_i, R_i) \times w_i}{\sum_{i=1}^{m} w_i} \quad (2)$$

where I is the input case; R the retrieved case; m the number of attributes in each case; i an individual attribute from 1 to m; f a similarity function for attribute i in cases I and R; and w the importance weighting of attribute i. This calculation is repeated for every case in the case library to rank cases by similarity to the input. The normalization is used so that similarity values fall within a range of zero to one, where zero is totally dissimilar and one is an exact match (Watson 1999).

Since the CBRs are used to augment FCMs, linguistic variables are used to represent the attributes of each case in the CBR and the similarity measures are calculated based on fuzzy combination rules, according to well-defined operators called triangular norms (Watson 1999; Lopez de Mantaras 2001).

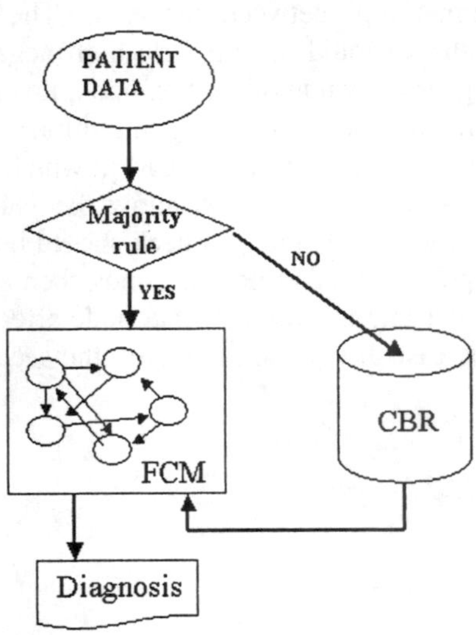

Figure 1. Schematic illustration of CBR augmented FCM.

4. Algorithm to augment CFCM combined with CBR

In this research a special type of Fuzzy Cognitive Maps is used in conjunction with CBR methods to develop a Medical Decision Support System (MDSS). This type is called a Competitive Fuzzy Cognitive Map (CFCM) and it consists of two main kinds of concepts:

- the n decision-concepts
- the m factor-concepts

Each one of the decision concepts stands for one decision/diagnosis, which means that these concepts must be mutually exclusive if our intention is to infer always only one diagnosis. This is the case of most medical applications, where, according to symptoms, medical professionals conclude to only one diagnosis and then decide accordingly concerning the treatment.

The factor-concepts can be considered as inputs of the DSS from patient data, observed symptoms, patient records, experimental and laboratory tests etc, which can be dynamically updated based on the system interaction, whereas the decision-concepts are considered as outputs where their estimated values outline the possible diagnosis for the patient.

However, the real strength of FCMs is their ability to describe systems and handle situations where there are feedback relationships and relationships between the factor concepts. Thus, interrelations between factor-concepts can be included in the proposed medical decision-support model.

In addition to this, another important quality of the proposed FCM for medical decision support system is that it includes connections (arcs) between the decision-concepts (outputs) themselves. These are not cause-effect connections, but inhibitory connections. These decision concepts must "compete" against each other in order for only one of them to dominate and be considered the correct decision (e.g. diagnosis with the highest probability). Here a new idea is proposed for achieving this "competition" between concepts. The interaction of each of these nodes with the others should have a very high negative weight (even -1). This implies that the higher the value of a given node, this should lead to a lowering of the value of competing nodes, i.e. strong inhibition.

Another novel consideration is that in the FCM in which there are nodes that do not accept feedback, it is important not to allow the values of those nodes to change. In order for this to be achieved, a check should be made of each node to examine if it accepts inputs from other nodes. If not, then a self-feedback value of the node should be set at 1 and the value of that node after each repetition should remain the same. In this case at equation (1), only the second term inside the parenthesis is non-zero.

4.1 The CFCM algorithm

Therefore, the following algorithm is proposed, which describes how to develop a Competitive Fuzzy Cognitive Map (CFCM), which is suitable for decision support systems:

- Set values A_i of nodes according to the input factors involved in the decision process. These values are described using fuzzy linguistic degrees similar to: i.e. none, very-very low, very low, low, medium, high, very high, and very-very high. These linguistic degrees are around to the numerical weights 0, 10%, 20%, 35%, 50%, 65%, 80%, and 90%, respectively, as shown by the membership functions of Figure 2. The decision-concepts are given the initial value of 0 because there is no initial diagnosis.
- The connection weights between the factor-concepts and the decision-concepts are taking their initial values. These connection linguistic weights have been proposed by experts who inferred them using IF THEN rules (Stylios et al. 1999). For the current research, the linguistic weights are defuzzified and transformed in the range are between 0 and 1. Then these numerical weights are then placed in a matrix *W* of size (*n*+*m*)x(*n*+*m*). The values in the first *n* columns correspond to the weighted connections from all the concepts towards the n decision-concepts. The values in the remaining m columns correspond to the weighted connections from all the concepts towards the factor-concepts. Also included in this matrix are the -1 weight values for competition between output decision concepts, as described earlier.
- Use equation (1) to calculate the updated value of each concept, where the sigmoid nonlinearity ensures that values of concepts are between 0 and 1 by the implementation of the unipolar sigmoid:

$$f(x) = \frac{1}{1+e^{-\lambda x}} \quad (3)$$

  where $\lambda \succ 0$ determines the steepness of the sigmoid.
- Repeat steps until equilibrium has been reached and the values of concepts no longer change
- The procedure stops and the final values of the decision-concepts are found, the maximum of which is the chosen decision.

4.2 Algorithm to combine CFCM along with CBR

In the Competitive Fuzzy Cognitive Map (CFCM) model, which is used for medical decision support there are some factors that are considered most important and are the main factors determining a particular decision-diagnosis. These factors are called Critical Factor Concepts and they are dependent on the specific application. When experts develop the CFCM for an application, they determine factor-concepts and the decision-concepts; they are also asked to select among the factor-concepts, the Critical Factor Concepts that are more prevalent in the assigning of the diagnoses. Critical Factor Concepts play important role in reaching any decision but the most important is that the lack of information on a number of them may forbid any decision.

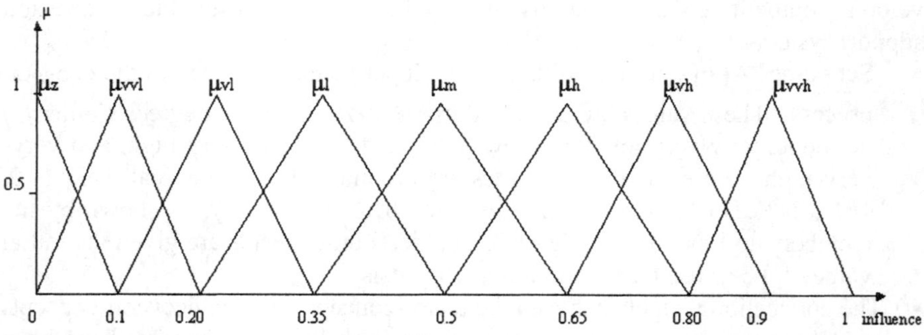

Figure 2. Membership functions

Therefore, when the patient input is entered into the system a logical majority operation rule is applied. The logical majority rule operation is applied to the total of all Critical Factors involved in the decision. This means that if the majority of the Critical Factors for all the possible decision outcomes are activated then the inputs are provided to the Competitive FCM (CFCM) Decision System, otherwise the CBR is called upon. This is shown in Figure 1 where a decision box is included.

When, the CBR is called, the input values describing the problem under examination are compared to the cases stored in the CBR and the case with the highest similarity is selected. Then, the CFCM is updated according to the case with the highest similarity, i.e. in the CFCM only the concepts corresponding to the information of the similar case in CBR are included. Then the updated CFCM is used to suggest a decision/diagnosis, which combines expert's knowledge (CFCM) and previous tested cases (CBR), thus, leading to a more reliable decision. It should be noted that this step is actually performed before the update rule of equation (1) is applied for the first time.

Figure 3 illustrates the implementation of the combination algorithm and the effect that it has in the structure of FCM. Part 3.a of the figure presents the CFCM that was initially developed to suggest one of the three different diagnoses, which are represented by the three striped concepts in the center of CFCM (Georgopoulos et al. 2003). Figure 3.b illustrates an intermediate stage, when the majority rule does not apply, so the CBR is called and a similar case is found in the case base. Thus, updating of the connections between CFCM concepts occurs; that actually means that some weights become zero and the corresponding concepts do not play any role in the decision. Therefore, for this specific case, the concepts with zero influence are removed as is depicted in figure 3.c and only the remaining concepts are used to provide the decision. It should be mentioned that for the next forthcoming problem the CFCM is restored to its initial structure.

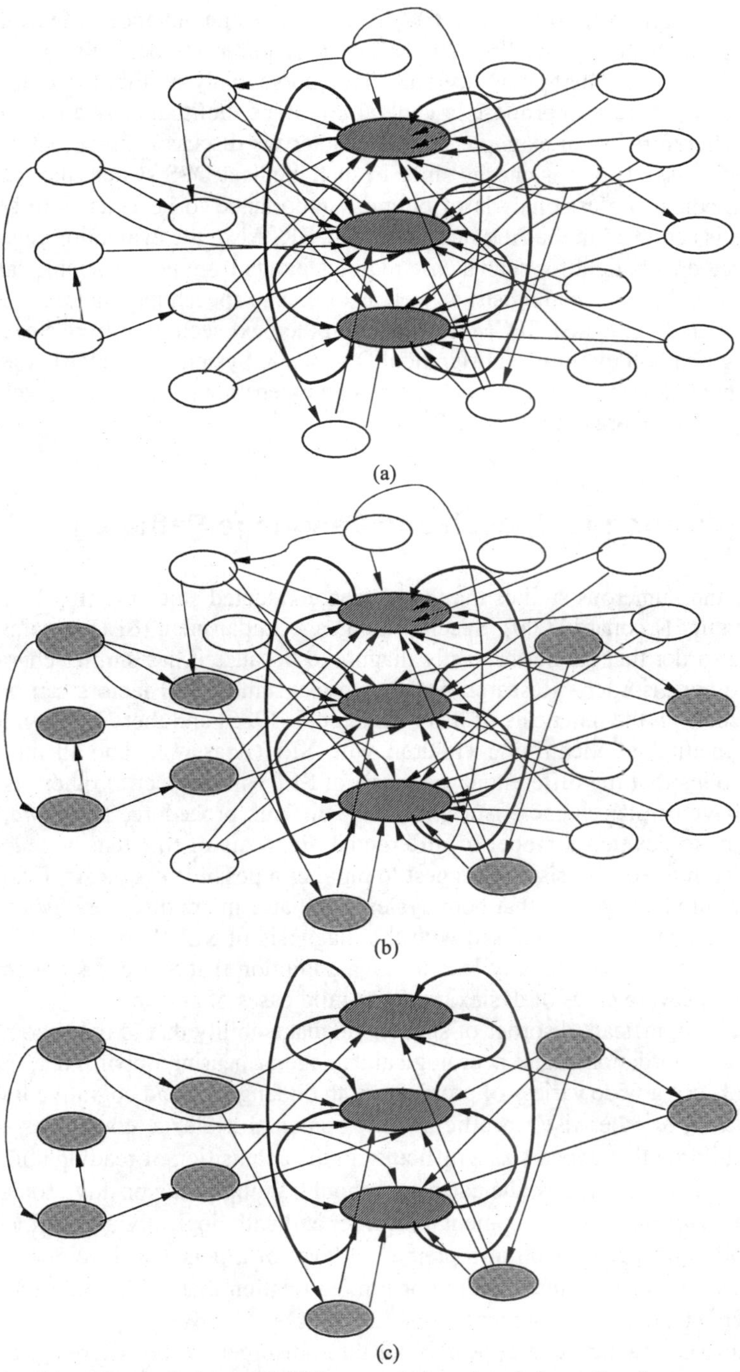

Figure 3 a, b, c. The evolution of CFCM structure based on information from case-base for a specific problem

Even though CBR Augmented CFCM and creates an advanced Medical Decision Support Systems (MDSS); this MDSS is required to perform such distinct tasks as diagnosis, therapy advice and time course analysis, that it would be too ambitious to attempt to propose a general prototype tool that can handle all these tasks. Therefore, as an example, in this chapter we discuss a single but complex diagnostic task. This was chosen since in such medical DSS system the reasoning of the medical professionals is of outmost importance to be taken into account, and this is achieved in the Augmented FCM-CBR. Also, the evaluation required to be carried on a patient for such a case requires inputs from pediatricians, ear-nose-throat specialists, psychologists, as well as of course speech pathologists. The example of an Augmented CFCM developed in the next section is from Speech and Language Pathology. It is a Differential Diagnostic System for Specific Language Impairment, Autism and Dyslexia. This is an extension of the CFCM, which was developed in (Georgopoulos et al. 2003).

5. Application to Speech and Language Pathology

Despite the numerous studies that have been conducted since the first half of the 19th century (Leonard 2000), Specific Language Impairment (SLI) remains a language disorder that cannot be easily diagnosed because it has similar characteristics to other disorders. Research has shown that almost 160 factors can be taken into account in the diagnosis of SLI (Tallal et al. 1985) and there is no widely accepted method of identifying children with SLI (Krasswski and Plante 1997). This implies that the differential diagnosis of SLI with respect to other disorders, which have similar characteristics, is a very difficult procedure. Therefore, it was necessary to develop a model of differential diagnosis of SLI that would aid the specialist in the diagnosis and suggest to him/her a possible diagnosis. Findings in the literature have shown that both dyslexia and autism are disorders, whose diagnoses often have been confused with the diagnosis of SLI (Leonard 2000). Particularly, the data has initially lead to the assumption that SLI cases are confused either with severe cases of dyslexia or with mild cases of autism.

SLI is a significant disorder of spoken language ability that is not accompanied by mental retardation, frank neurological damage or hearing impairment. Children with SLI face a wide variety of problems both on language and cognitive levels.

Dyslexia, or otherwise, specific or developmental dyslexia, constitutes a disorder of children that appears as a difficulty in the acquisition of reading ability, despite their mental abilities, the adequate school training or the positive social environment. *Autism* is a developmental disorder and pathologically it is defined as an interruption or a regression at a premature level of a person's development. The main idea in autism is the impaired or limited relation that exists between the autistic person and its environment

Some basic factors that appear in all three disorders with different frequency and severity in most cases were included in this study. The considered factors are either causative factors or symptoms of the disorders. The factors within each disorder were taken into consideration in a comparative way in the development of the model. The significance of each factor as a diagnostic criterion is defined with the following fuzzy variables: a) Very-very important, b) very important, c) important, d) medium, e) not very important, and f) minimally important. These criteria are represented in the Competitive Fuzzy Cognitive Map Differential Diagnosis Model as the fuzzy weight with which each factor influences every one of the three diagnoses. The advanced MDSS consisted of CFCM and CBR is shown in Figure 4.

Table I shows the information for four case examples stored in the Fuzzy CBR Database used to augment the CFCM Differential Diagnosis system. The first case in the Table is a case with SLI as the final diagnosis, the second and third cases with Dyslexia as a diagnosis, and the fourth case has a diagnosis of autism. The names of the attributes of the CBR are the same as the Factor-Concepts of the CFCM. The critical factors for each disorder have been defined in our previous work (Georgopoulos et al. 2003), as having weightings of very-very high. These are the attributes of Table I, 1, 2, 3, and 9 for SLI, attributes 4, 6, and 9 for Dyslexia and 1, 2, 3, 5, 7, 8, 11, 12, 13, and 15 for Autism. Thus, non-critical factors for all 3 disorders are only 10 and 14 and are not included in the majority test performed in the beginning of the CBR-Augmented CFCM algorithm (i.e. the majority rule imposed here would require *majority=(critical factors)/2 + 1* which in our case is *8* factor-concepts).

It should be noted that the CBR that includes cases of Table I is a general one concerning differential diagnosis of SLI/Dyslexia/Autism and does not only include cases for which the majority rule does not apply.

6. Example

As an example we consider an input case, which is described with the initial values for the factors as shown in Table II. These values are based on the patient's history and test results.

We can try to obtain a diagnosis for this input case using the CFCM model that was developed in (Georgopoulos et al. 2003) and the CBR augmented CFCM model proposed here. If we use the input information of Table II in the CFCM model, after simulation equilibrium is reached where decision concepts have the values:

$$SLI = 0.8700 \quad Dyslexia = 0.6550 \quad Autism = 0.8989$$

It is apparent, that two of the three possible diagnoses have values very close each other and so it is difficult to suggest a diagnosis.

Then, we test the same input case for the CBR Augmented CFCM. This input case does not meet the majority rule, so the CBR component in the MDSS is activated. Then a comparison of this input case to the stored cases in the case-base of CBR is performed. When a good match is found, the attributes of the case found in the CBR are used to reconstruct the CFCM. Then this reconstructed CFCM is run and it reaches the following equilibrium:

SLI= 0.8763 *Dyslexia*=0. 6878 *Autism*=0. 9526

It is obvious that the concept of 'Autism' dominates over values of 'SLI' and 'Dyslexia' concepts and thus the diagnosis of Autism is proposed for this case. With this simple example, is suggested that a sufficient MDSS model was developed which, under constraints, processes the information about a case in such a way that out of three possible diagnoses we are lead to the diagnosis of the most probable disorder.

Table I. Sample Clinical Cases Stored in Fuzzy CBR used to Augment FCM

Attributes	Case 1	Case 2	Case 3	Case 4
1. Reduced Lexical Abilities	VERY - VERY HIGH	HIGH	MEDIUM	VERY HIGH
2. Problems in Syntax	VERY-VERY HIGH	HIGH	HIGH	VERY-VERY HIGH
3. Problems in Grammatical Morphology	VERY HIGH	HIGH	HIGH	VERY-VERY HIGH
4. Impaired or Limited Phonological Development	HIGH	MEDIUM	HIGH	VERY HIGH
5. Impaired Use of Pragmatics	MEDIUM	0	0	VERY-VERY HIGH
6. Reading Difficulties	0	MEDIUM	VERY-VERY HIGH	-HIGH
7. Echolalia	0	0	0	VERY HIGH
8. Reduced Ability of Verbal Language Comprehension	0	0	0	VERY-VERY HIGH
9. Difference between Verbal - Nonverbal IQ	HIGH	HIGH	HIGH	0
10. Heredity	0	0	0	0
11. Impaired Sociability	MEDIUM	0	VERY LOW	VERY-VERY HIGH
12. Impaired Mobility	MEDIUM	0	LOW	VERY HIGH
13. Attention Distraction	0	0	LOW	VERY HIGH
14. Reduced Arithmetic Ability	MEDIUM	0	MEDIUM	-HIGH
15. Limited Use of Symbolic Play	0	0	0	VERY-VERY HIGH

Table II. Values for example

Attributes	Example
1. Reduced Lexical Abilities	HIGH
2. Problems in Syntax	HIGH
3. Problems in Grammatical Morphology	VERY HIGH
4. Impaired or Limited Phonological Development	-
5. Impaired Use of Pragmatics	-
6. Reading Difficulties	-
7. Echolalia	-
8. Reduced Ability of Verbal Language Comprehension	MEDIUM
9. Difference between Verbal - Nonverbal IQ	MEDIUM
10. Heredity	-
11. Impaired Sociability	MEDIUM
12. Impaired Mobility	-
13. Attention Distraction	-
14. Reduced Arithmetic Ability	-
15. Limited Use of Symbolic Play	LOW

7. Conclusions

In this chapter, we described an advanced Medical Decision Support System (MDSS) which is based on the augmentation of Competitive Fuzzy Cognitive Map (CFCM) with Case Based Reasoning (CBR) methods. The proposed Decision System of CBR-Augmented Competitive FCM is applied and tested as a Medical Decision System for Speech and Language Disorders. For one problem case the CBR-Augmented Competitive FCM is compared with the simple CFCM and the results show the advantages of the new proposed system. In essence, this CBR-Augmented Competitive Fuzzy Cognitive Map is capable on its own to perform a comparison and lead to a decision based on expert knowledge and experience (structure of CFCM) and well known tested previous cases (CBR).

Acknowledgements

Funding for this research was provided by the program EPEAEK II: Archimedes - Research Support in TEI of the Ministry of National Education and Religious Affairs of Greece.

References

Georgopoulos VC and Stylios CD (2003) Augmented fuzzy cognitive maps based on case based reasoning for decisions in medical informatics. In: Proceedings BISC FLINT-CIBI 2003 International joint workshop on Soft Computing for Internet and Bioinformatics, 15-19 December 2003, University of California, Berkeley, California, USA

Georgopoulos VC, Malandraki GA, and Stylios CD (2003) A fuzzy cognitive map approach to differential diagnosis of specific language impairment. Journal of Artificial Intelligence in Medicine. 29:261–278

Kasabov N (1996) Foundations of Neural Networks, Fuzzy Systems and Knowledge Engineering. MIT Press, Cambridge.

Kasabov N (2002) Decision support systems and expert systems. In: M. Arbib (ed) Handbook of brain study and neural networks. MIT Press, Cambridge.

Kolodner J (1993) Case-based reasoning. Morgan Kaufmann Publishers, San Mateo.

Kosko B (1986) Fuzzy cognitive maps. International J. of Man-Machine Studies. 24:65-75.

Kosko B (1991) Neural networks and fuzzy systems. Prentice-Hall, Englewood Cliffs,.

Krasswski E and Plante E (1997) IQ variability in children with SLI: implications for use of cognitive referencing in determining SLI. J. of Communication Disorders, 30:1-9.

Leonard LB (2000) Children with specific language impairment. MIT Press, Cambridge.

Noh JB, Lee KC, Kim JK, Lee J.K, and Kim SH (2000) A case-based reasoning approach to cognitive map-driven tacit knowledge management. Experts Systems with Applications. 19:249-259.

Lopez de Mantaras R (2001) Case-based reasoning. In: Paliouras G, Karkaletsis V, and Spyropoulos CD (eds.): LNAI 2049, Springer-Verlag Berlin Heidelberg, pp 127-145.

Papageorgiou E, Stylios C, and Groumpos P (2003) An integrated two-level hierarchical system for decision making in radiation therapy using fuzzy cognitive maps. IEEE Transactions on Biomedical Engineering. 50:1326-1339.

Rosch E and Mervis CB (1975) Family resemblance: studies in the internal structures of categories. Cognitive Psychology 7:573-605.

Schmidt R, Pollwein B, and Gier L (1999) Experiences with case-based reasoning methods and prototypes for medical knowledge-based systems In: Horn W et al. (eds.) LNAI 1620, pp 124-132.

Schmidt R, Montani S, Bellazzi R, Portinale L, and Gierl L (2001) Cased-based reasoning for medical knowledge-based systems. International Journal of Medical Informatics. 64:355–367.

Stylios CD, Groumpos PP, and Georgopoulos VC (1999) A fuzzy cognitive maps approach to process control systems. J. Advanced Computational Intelligence. 3:409-417.

Stylios CD and Groumpos PP (2000) Fuzzy cognitive maps in modeling supervisory control systems. Journal of Intelligent & Fuzzy Systems. 8:83-98.

Tallal P, Stark R, and Mellitis E (1985) Identification of language-impaired children on the basis of rapid perception and production skills, Brain and Language 25:351-357.

Tversky A (1977) Features of similarity. Psychological Review 84:327-352.

Watson I (1999) Case-based reasoning is a methodology not a technology. Knowledge-Based Systems. 12:303–308.

Zeleznikow J and Nolan J (2001) Using soft computing to build real world intelligent decision support systems in uncertain domains. Decision Support Systems, 31: 263-285.

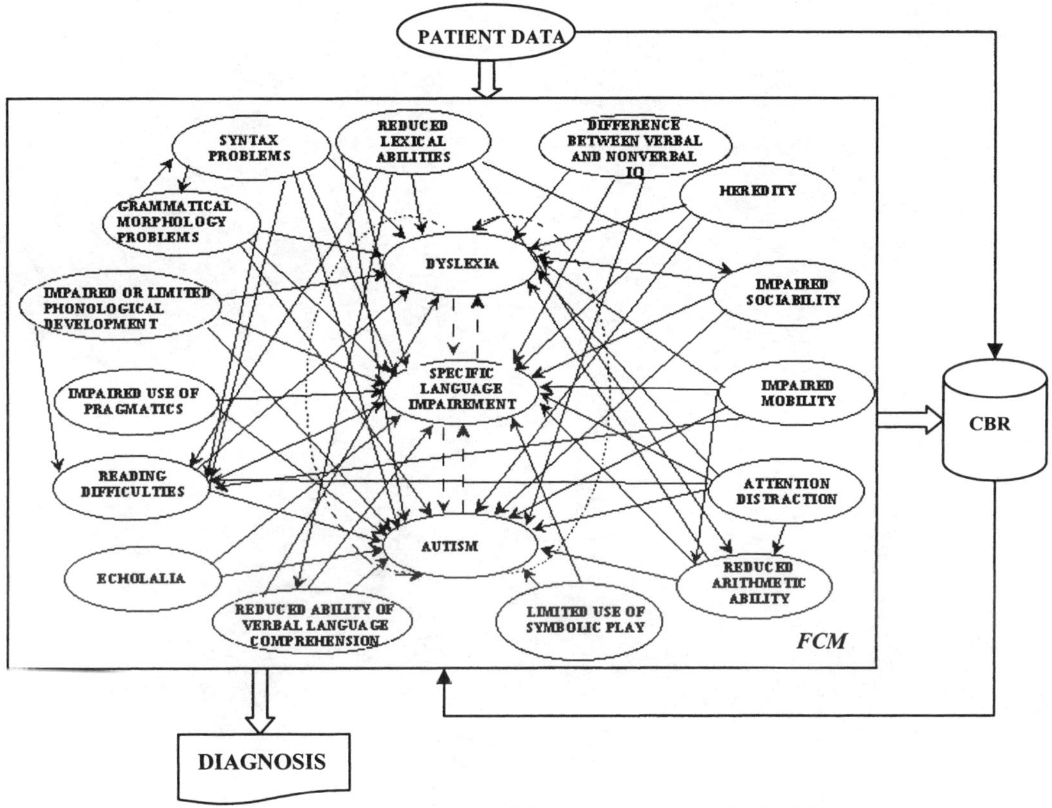

Figure 4. An abstract illustration with a view of the CBR augmented CFCM.

Pruning, Selective Binding and Emergence of Internal Models: Applications to ICA and Analogical Reasoning

Syozo Yasui, Kyushu Institute of Technology, Graduate School of Life Science and Systems Engineering, Kitakyushu, Japan, yasui@brain.kyutech.ac.jp

Abstract: Pruning of multi input/output neural networks is discussed and a pruning algorithm called CSDF is described. CSDF acts to induce internal models as a result of redundancy elimination and selective bindings. CSDF is used in a new ICA method based on an auto-encoder performing sensor-signal identity mapping. An internal model of the external signal-mixing situation emerges due to the CSDF pruning, and the hidden units that survive the CSDF pruning reconstruct the blind source signals. This ICA method which requires no pre-processing such as whitening is characterized by its high adaptability and robustness, as is demonstrated by trouble cases such as sudden increase of the source signals, sudden failure of sensors and so on. As another example, CSDF is applied in a neural network for analogical learning/inference. Internal abstraction models together with abstraction/de-abstraction bindings are generated as a result of the CSDF structural learning coupled with the backpropagation training. The internal abstraction model acts as an attractor for new relevant dataset, a process corresponding to analogical memory retrieval.

1. Introduction

The synaptic density of the kitten visual cortex increases rapidly following the birth, and then decreases as the visual experience proceeds. This decrease does not, however, occur in the visually deprived kittens. There are a number of similar findings in the developing brain [1]. These observations indicate that a pruning mechanism is at work during early stages of visual learning. And such pruning is thought to manifest the Principle of Redundancy Reduction (PRR) advocated by many neuroscientists as one of the fundamental strategies underlying the brain mechanisms.

Pruning is an important strategy in the field of artificial neural networks as well. In neural network engineering, one wishes to find the minimum (necessary and sufficient) size/complexity for the neural network structure, for reasons relevant to (1) economy, (2) understanding / insight, (3) generalization ability and (4) avoidance of local minima. Generally speaking, however, the minimum structure for the given task is not a priori known. A solution of this problem is pruning whereby a sufficiently large and complex structure is prepared initially and unnecessary connections are eliminated during the training phase. As such, neural network pruning can be viewed as a connectionist's reflection of PRR. Pruning algorithms such as *Weight Decay* [2] and *Optimum Brain Damage* [3] are widely known.

These algorithms have been applied mostly for single-output neural nets, in which case elimination of a hidden-output connection would imply removal of a hidden unit. This is not necessarily the case if the network has two or more output units. When it comes to the problem of pruning a multi-output neural network, one would have to specify what kind of pruning is desired. Under this consideration, the present author proposed a pruning algorithm called *Convergence Suppression and Divergence Facilitation* (CSDF) whose objective is not only to minimize the number of active hidden units but also to make each of the surviving hidden units be utilized jointly by as many as output units possible [4].

Four keywords (1)-(4) have been mentioned in the first paragraph as merits of pruning. The CSDF algorithm gives a few more merits, namely in keyword, (5) optimum modularization,(6) dynamical binding/association , and (7) abstraction ability.

This paper describes applications of the CSDF pruning to analogical reasoning and independent component analysis (ICA).

2. CSDF Pruning

The CSDF pruning algorithm is briefly reviewed below. If w_{ij} denotes the synaptic weight parameter of the path from j th hidden unit to i th output unit (Figure 1), then the iterative correction Δw_{ij} is given as follows.

$$\Delta w_{ij}.(t) = -\eta\, \partial E^2/\partial w_{ij} + \alpha\, \Delta w_{ij}.(t\text{-}1) + S + F \tag{1}$$

$$S = -\varepsilon\, sgn(w_{ij}) (|w_{i1}| + \ldots + |w_{im}| - |w_{ij}|) \tag{2}$$

$$F = -\gamma\, \partial E^2/\partial w_{ij} (|w_{1j}| + \ldots + |w_{nj}| - |w_{ij}|) \tag{3}$$

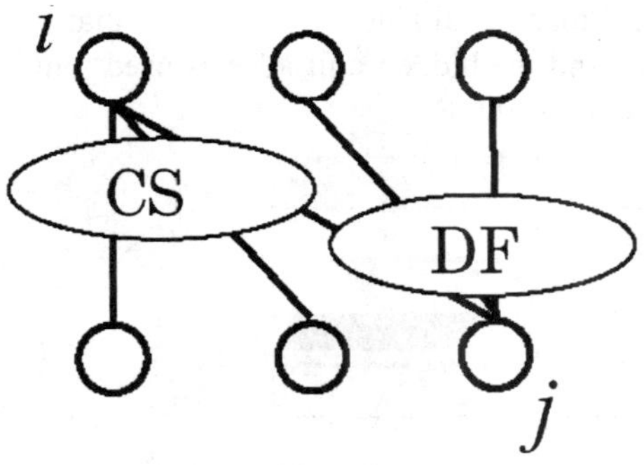

Figure 1: CSDF pruning.

Here, it is assumed that m hidden units and n output units are made available for use. E denotes the Euclidean distance between the output activity vector and teacher signal vector. η , α , ε and γ are positive constant parameters. The first two terms on RHS of eq.1 constitute the standard backpropagation (BP) algorithm. The term S of eq.2 represents the convergence suppression (CS) which is applied across the receptive field of each output unit (Fig.1). Thus, the pathways converging to the same output unit try to prevent each other from growing. The term F of eq.3, on the other hand, represents the divergence facilitation (DF) which is applied across the projective field of each hidden unit (Fig.1). Thus, the pathways diverging from the same hidden unit attempt to promote each other for growing by reinforcing the BP learning. Every hidden-output pathway receives both CS and DF actions. A pathway will eventually disappear if the net effect of CS and DF is suppressive enough against the error-reducing BP action. Such elimination would happen to all synaptic paths originating in every insignificant or unnecessary hidden unit.

In the CSDF method, therefore, the fate of a weight parameter is affected by the magnitudes of other weights. Such property is absent in the *Weight Decay* [2] and *Optimal Brain Damage* [3] pruning algorithms. Actually, the CSDF algorithm not only eliminates unnecessary connections but also can create new connections if useful, as will be shown later in the application examples. Thus, the CSDF algorithm is both destructive and constructive (*cf.* [5], and the term "CSDF *pruning* algorithm" which is frequently used in this paper is actually somewhat misleading

As some minor modifications of the CSDF algorithm, $sgn\ (w_{ij})$ in eq.2 may be replaced by w_{ij}, and $|\ w_{ij}\ |$ in eqs. 2 and 3 by w_{ij}^2 . These changes do not affect the pruning performance significantly.

A test example is shown in Figure 2. The neural network was trained to learn AND, OR and XOR binary logic operations by using the BP-CSDF algorithm. Initially, a total of six hidden units were made available. Fig.2 shows that four of them were eliminated due to CSDF. And the hidden unit #2 was used jointly by AND and XOR, and #4 was used by OR and XOR.

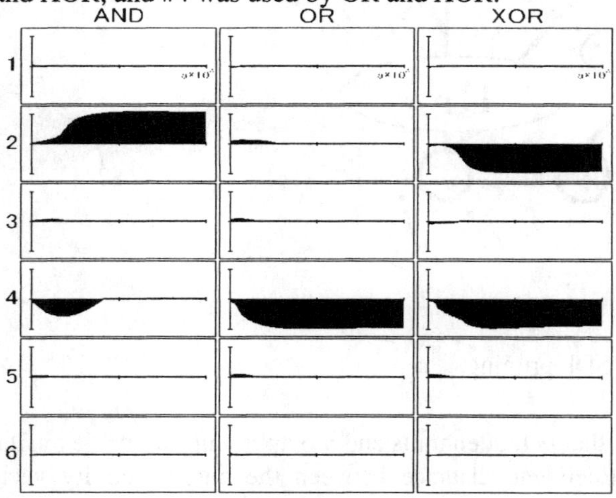

Figure 2: Evolution of synaptic weights ***W***.

3. Application to Analogical Reasoning

3.1 Background

Analogy allows intelligent being to find correspondences between aspects of two or more situations. The corresponding aspects are recognized as such when they play the same role in relationships with other aspects within the respective situations. Let the ordered triple (α, π, β) denote that item α is related to item β through relationship π. The arguments α and β are the role fillers.

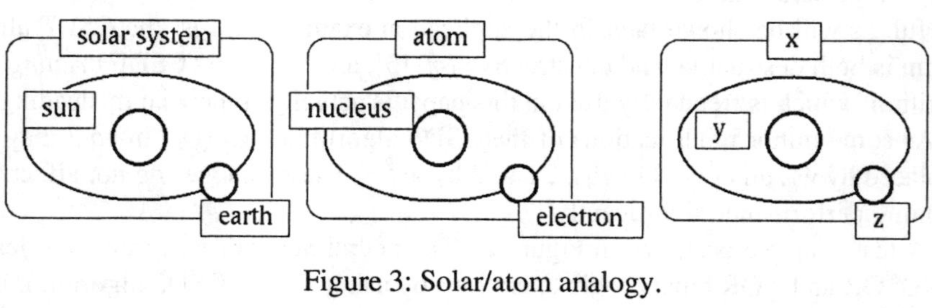

Figure 3: Solar/atom analogy.

For example, the solar/atom analogy (Rutherford model) shown in Figure 3 is described below.

Solar System
(sun, **attract,** earth)
(earth, **revolve around**, sun)
(sun, **more massive than**, earth)
(solar system, **more massive than**, (sun, earth))
(sun, **included in**, solar system)
(earth, **included in**, solar system)

Atomic System
(nucleus, **attract,** electron)
(electron, **revolve around**, nucleus)
(nucleus, **more massive than**, electron)
(atomic system, **more massive than**, (nucleus, electron))
(nucleus, **included in**, atomic system)
(electron, **included in**, atomic system)

Clearly, there is a relational isomorphism between the solar and atomic systems.
There have been a great deal of computational research attempts in AI for the study of analogy. SME (Structure Mapping Engine) is one such example [6] related to the structure mapping theory of Gentner [7]. SME operates on the AI tradition involving rule-based reasoning and symbolic processing. The connectionist approach has drawn much attention in resent years. ACME (Analogical Constraint Mapping Engine)due to Holyork and Thagard [8] and Hinton's network[9] are such attempts. The present approach called AB-CAP (Abstraction Based Connectionist Analogy Processor) is a more recent attempt due to the present author's group [10-15]. In AB-CAP, the CSDF pruning plays a crucial role in selective dynamical bindings. The basic idea underlying AB-CAP is similar to Kantian notion of

schema. This is described as follows for the solar-atom analogy.

Abstraction Model
(*X*, **attract,** *Y*)
(*Y*, **revolve around**, *X*)
(*X*, **more massive than**, *Y*)
(*Z*, **more massive than**, (*X*,*Y*))
(*X*, **included in**, *Z*)
(*Y*, **included in**, *Z*)

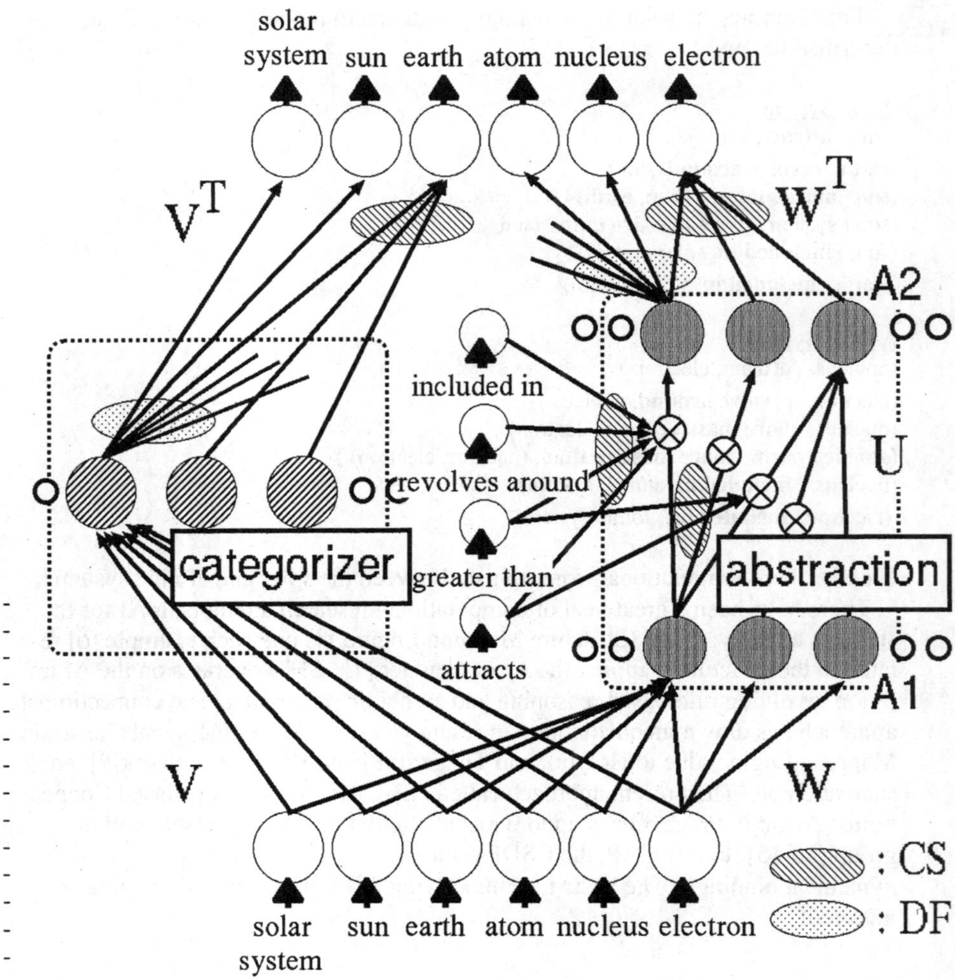

Figure 4: Architecture of AB-CAP.

Abstraction Mapping
sun, nucleus $\longrightarrow$ X; earth, electron $\longrightarrow$ Y; solar system, atomic system $\longrightarrow Z$

De-Abstraction Mapping
$X \longrightarrow$ sun, nucleus; $Y \longrightarrow$ earth, electron; $Z \longrightarrow$ solar system, atomic system

3.2 Architecture and Operation of AB-CAP

The architecture of AB-CAP is shown in Figure 4. The training data are a set of specific relational predicates, *e.g*., (sun, **attract**, earth). For (α, π, β), the item-input slot for α and relation-input slot for β are set to the value of "*1*" and all other input slots "*0*". CS (Convergence Suppression) and DF (Divergence Facilitation) are applied as shown in

Fig.4 across appropriate receptive and projective fields, respectively. As a result of the CSDF-coupled BP training, the weight matrices $\boldsymbol{W}$ and $\boldsymbol{W}^T$ are expected to evolve into the abstraction and de-abstraction mappings, respectively (*cf*., Fig.3). The A1-layer units that survive the CSDF pruning will represent the abstraction items such as *X, Y* and *Z* of Fig. 3. The same is true for the role of the A2-laye units. The connections from A1 to A2 are gated by signals from the relation-input layer. This gating is mediated by multipliers. Thus, the block formed by the A1, A2 and relation-input layers will become the internal abstraction model (*cf*., Fig.3).

Another group of signal flows from the item-input layer to the output layer form the auto-categorizer. In the case of Rutherford analogy, for example, two hidden units of this module are expected to survive the CSDF pruning, to represent the solar and atomic systems. Thus, the weight matrices, $\boldsymbol{V}$ and $\boldsymbol{V}^T$, will evolve into the classification and de-classification mapping, respectively. All hidden units have a sigmoidal activation function. Further details about the AB-CAP can be found elsewhere [10,13].

3.3 Evolution during Training

Figure 5 shows a set of records obtained during the BP-CSDF training of AB-CAP. Five auto-categorizer hidden units and six units for each of the A1 and A2 layers were made available for use. The initial values of the connection weights were random. Two categorizer hidden units survived the CSDF pruning, to represent the solar and atomic systems. Three units in each of the A1 and A2 layers survived to represent the three abstracted items corresponding to *X, Y* and *Z* of Fig.3. The pruned $\boldsymbol{W}$ pattern from the item-input layer to the A1 layer agrees with the abstraction mapping expressed in Fig.3. The record of $\boldsymbol{U}$ needs explanation. Thus, $u_{ijk} \in \boldsymbol{U}$ denotes the synaptic weight from the *k* th predicate input to the multiplier that controls gating of the signal from the *i* th A1 unit to the *j* th A2 unit. In Fig. 5, the 3-D

record of u_{ijk} describing the potential relationships between any two abstracted items is coded in alphabet, *i.e*., attract → A,a , included in → I,i, more massive than → M,m , revolve around → R,r . The capital or lower-case letter is used if the corresponding u_{ijk} value exceeds the higher(th_2) or lower

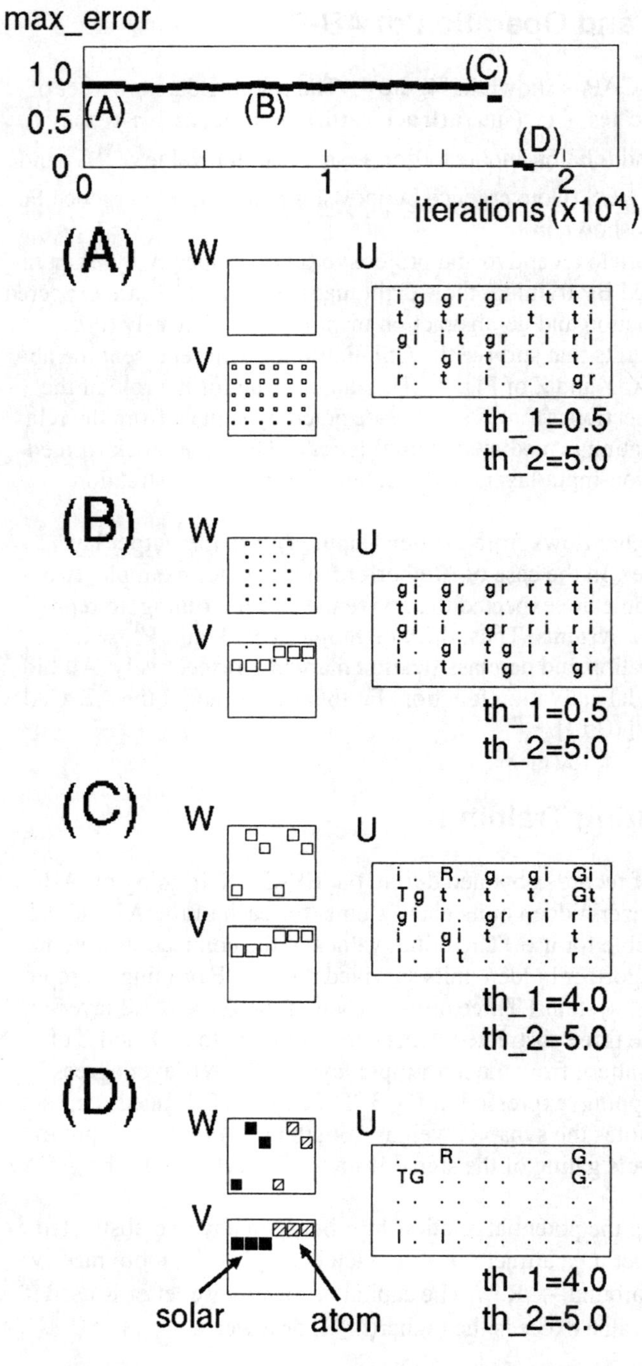

Figure 5: Structure evolution of AB-CAP learning the solar/atom analogy.

(th_1) preset threshold, respectively. The ***U*** pattern of Fig.5 conforms to the abstraction model of Fig.3.

3.4 Analogical Inference

Actually, the information of (sun,**attract**,earth) was not included in the training dataset for the training run of Fig. 5. Nevertheless, everything for the analogy was produced in the final structure of AB-CAP, showing proper formation of the abstraction and de-abstraction mappings as well as the abstraction model. In fact, a test asking (sun, **attract**, ?) made after the successful training produced the correct answer of ?=earth. This means that the connections needed for (sun, **attract**, earth) were induced from the training dataset. This is because the bindings relevant to (*X*, **attract**,*Y*) were induced from (nucleus,**attract**, electron). And the abstraction of sun $\rightarrow X$ and de-abstraction of $Y \rightarrow$ earth were generated from the data involving sun and earth such as (earth, **revolve around**, sun). This way, the untaught proposition (sun, **attract**, earth) is captured in the internal structure finally acquired by AB-CAP. Further study on analogical inference by AB-CAP is made elsewhere [12-14].

3.5 Incremental Analogical Learning

Good performance quality of AB-CAP can be demonstrated by incremental analogical learning as well. It is shown in this and earlier studies [11,12] that the internal abstraction model acquired from prior analogical learning is a potent attracter that binds dynamically new datasets of the same isomorphic structure. In order to show all this, a new analogy example is introduced, is illustrated in Figure

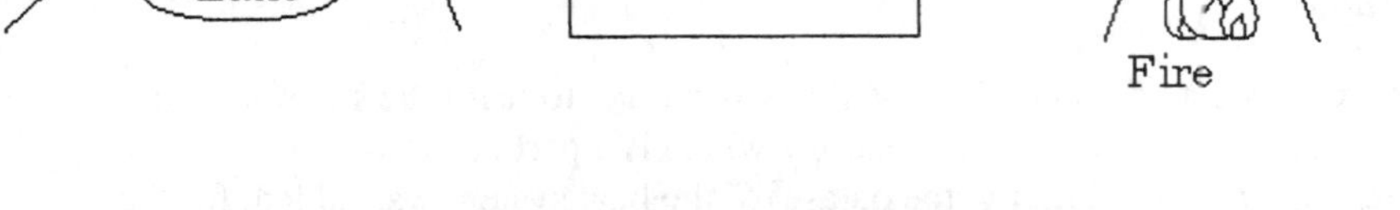

Figure 6: Water/electric/heat analogy.

Water System
(spring, **is higher than**, lake)
(spring, **supply**, water)
(river, **conduct**, water)
(lake, **accumulate,** water)
(water, **flow through**, river)
(water, **flow into**, lake)

Electric System
(battery, **is higher than** (in voltage), capacitor)
(battery, **supply**, charge)
(register, **conduct**, charge)
(capacitor, **accumulate,** charge)
(charge, **flow through**, register)
(charge, **flow into**, capacitor)

Heat System
(fire, **is higher than** (in temperature), coffee)
(fire, **supply**, *heat*)
(metal, **conduct**, heat)
(coffee, **accumulate,** *heat*)
(heat, **flow through**, *metal*)
(heat, **flow into**, coffee)

Abstraction Model
(*A*, **is higher than**, C)
(*A*, **supply**, *D*)
(*B*, **conduct**, *D*)
(*C*, **accumulate,** *D*)
(*D*, **flow through**, *B*)
(*D*, **flow into**, *C*)

Abstraction Mapping
spring, battery, fire → *A*; river, register, metal → *B*; lake, capacitor, coffee → *C*; water, charge, heat → *D*

De-Abstraction Mapping
A → spring, battery, fire ; *B* → river, register, metal; *C* → lake, capacitor, coffee; *D* → water, charge, heat

In the experiment of Figure 7, AB-CAP was trained to learn the solar/atomic analogy first and then water/electric analogy which is a part of the water/electric/heat analogy. Finally, the dataset of the heat system was added for the training. At stage **b**, learning of the solar/atomic analogy was nearly complete. At stage **c**, two additional categorizer units emerged, to represent the water and electric systems (see *V*). Also, new binding patterns were induced for ***W*** and ***U***. The part of the internal structure that had been obtained from the solar/atomic analogical learning was preserved. Thus, two abstraction models became present (see ***U***).

At stage **d**, the final round of learning ended successfully. The new training dataset for this final round was supplied from the heat system. The internal structure including the two abstraction models did not change except for the emergence of another categorizer unit and a new repeated pattern in ***W*** which bound the heat system to the abstraction model that was captured by learning of the water/electric analogy, not the solar/atom analogy. In other words, the abstraction model acts as a selective attracter for the new relevant data.

3.6 Toward Analogical Data Mining

In the reality, analogical relationships are entities that may exist in the presence of numerous irrelevant
data. Before analogical reasoning, therefore, one needs to find and extract the sub-dataset that are relevant. This is important for analogical data mining. A preliminary attempt using a simple example with AB-CAP toward this direction is made elsewhere[15].

4. Application to Blind Source Separation

4.1 Background

Blind Source Separation (BSS), which is also called as Independent Component Analysis (ICA), refers to the problem of recovering unknown signals

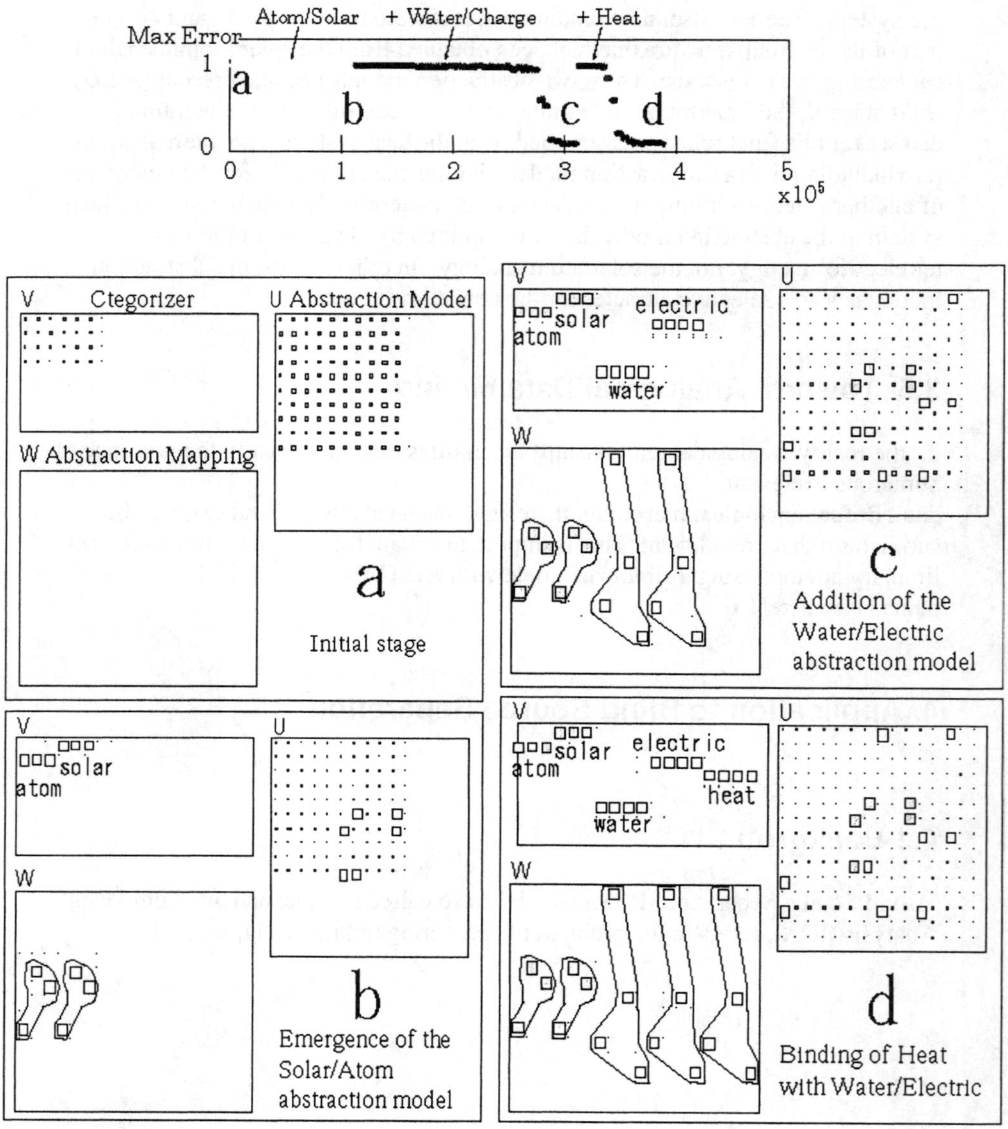

Figure 7: Structure evolution of AB-CAP engaged in incremental analogical learning.

from their mixtures that can be observed as sensor signals. In BSS, the properties as well as the number of sources are not known. Also, the source-sensor mixing matrix is unknown. The only clue is the assumption that the sources are statistically independent. The major approach include informax, maximum likelihood estimation, negaentropy maximization, nonlinear PCA ([16] for review) and fast ICA[17]. The first three are information theoretic and have been shown to be equivalent. The nonlinear PCA can also be viewed from information theoretic principles. The fast ICA is probability-theoretic and operates on the basis of the de-Gaussianization principle due to the central limit theory.

4.2 Architecture and Algorithms

The present approach is not information/probability theoretic and is fundamentally different. As shown in Figure 8, it is based on an auto-encoder which incorporates the CSDF pruning mechanism [18-21].Previous applications of the auto-encoder include data compression, dimensionality reduction and PCA. In these applications, it is well known that the non-linearity is not helpful [22]. In the present application, by contrast, the non-linearity (*e.g.*, tanh(cz), tanh(cz^3), z^3)of hidden units coupled with

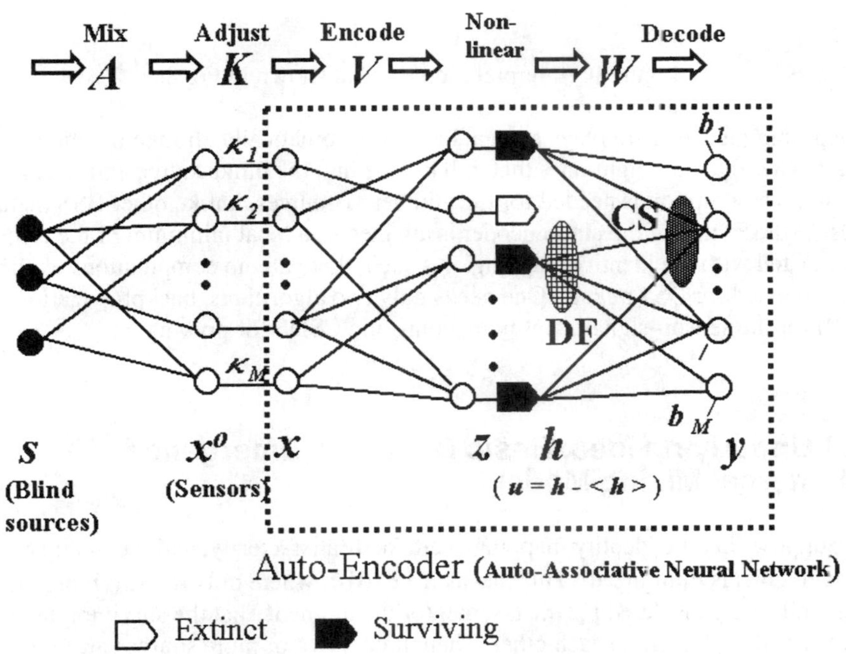

Figure 8: Architecture for BSS.

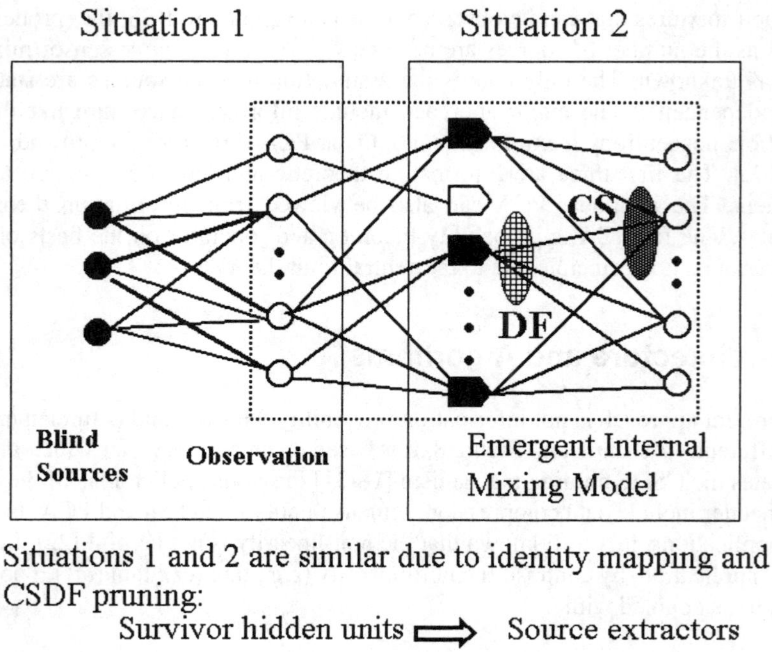

Figure 9: Interpretation of what shown in Fig.8.

the pruning mechanism plays a crucial role in automatically eliminating hidden units other than the right ones that will evolve into the blind source extractors. As such, no assumption is needed for the number of sources, unlike other BSS methods. A BSS state of the auto-encoder is attained as a local minimum of the error associated with the identity mapping. As such, there are no computations explicitly intended for BSS *per se*. One needs only two algorithms, backpropagation (BP) for the sensor–signal identity mapping and CSDF for pruning.

4.3 Underlying Idea, Basic Behavior, Emergence of Internal Mixing Model

Suppose that the identity mapping is attained satisfactorily, and the number of hidden units is minimized. This means $\boldsymbol{KAx} \approx \boldsymbol{Wu}$, where only the surviving hidden units are considered for $\boldsymbol{u}$. Assume, for the moment, that the surviving hidden units are independent to each other, Then, unless two or more sources are Gaussian, the surviving hidden units would represent the blind sources and the decoder matrix $\boldsymbol{W}$ would correspond to the adjusted mixing matrix $\boldsymbol{KA}$. In other words, since $\boldsymbol{KA}$ and $\boldsymbol{W}$ are constant whereas $\boldsymbol{x}$ and $\boldsymbol{u}$ vary with time, $\boldsymbol{KAx} \approx \boldsymbol{Wu}$ implies that $\boldsymbol{W}$ and $\boldsymbol{u}$ are proportional to $\boldsymbol{KA}$ and $\boldsymbol{x}$, respectively. This is illustrated in Figure 9. Thus, Situation 2 would be "similar" to Situation 1 and the former would be an internal mixing model to recontruct the whole external circumstance, *i.e.,* not only the blind sources but also the mixing relations.

The use of linear hidden units does not work for BSS although the identity mapping can be made exactly ($\boldsymbol{WV=I}$ in Fig.8). This is because $\boldsymbol{WV=WPP^{-1}V=I}$ where $\boldsymbol{P}$ is an arbitrary non-singular matrix. The question then is why the use of non-linearity works for BSS. This point is discussed as follows. For simplicity, there are two sources so that $N=L=2$. Assume that the hidden units are not independent, so that at least one hidden unit operates on a linear combination of s_1 and s_2, *i.e.*, non-BSS state Then, the nonlinear hidden unit generates the terms of $s_1^p s_2^q$ where p and q are non-negative integers and $p+q$ is odd. The coefficients of these are statistically determined, for instance, by the Wiener-like stochastic orthogonal expansion [23]. The nonlinear terms (*e.g.*, s_1^3, s_1^5, s_2^3, $s_1^2 s_2$, $s_1^3 s_2^2$) go to the output layer, but are absent in the sensor signals. This means that these nonlinear components are not desired for the accuracy of the identity mapping.

There are two possibilities for minimizing the nonlinear products. One is to keep the input to the surviving hidden units staying small in amplitude, so that the hidden units behave nearly linearly. The other is BSS which eliminates the all cross terms such as $s_1 s_2^2$, $s_1^2 s_2^3$ and so on. Actually, both effects can be made to occur by appropriate choice of the non-linearity and parameters, as shown in the following example.

Figure 10 shows what happens during a successful simulation run for a case in which there are two sources, four sensors and five nonlinear hidden units. Initially, the decoder matrix is random. At some point, the CSDF pruning becomes effective enough to eliminate three hidden units completely. About at the same time, the signals from the remaining two hidden units become quite small in comparison with those from the extinct hidden units. However, if the former ones are enlarged, then one finds reconstructed source signals as shown in Fig.10. Thus, the two survivor hidden units are the source extractors. Also, one can notice that the decoder matrix linearly corresponds to the mixing matrix $\boldsymbol{KA}$ (up to sign and scaling). Thus, an internal mixing model has emerged, as is expected. The surviving hidden units behave almost linearly because of their small activities in amplitude (*e.g.*, h_2 and h_4 of Fig.10).

The extinct nonlinear hidden units continue to respond with large amplitude (*e.g.*, h_1, h_3, h_5 of Fig.10). They are eliminated (*i.e.*, no decoder connection), because nonlinear distortions such as saturation make only a negative contribution to the identity mapping. One interpretation from all this is that the hidden–layer acts as a nonlinear gating array to select the proper hidden units for the final source extractors.

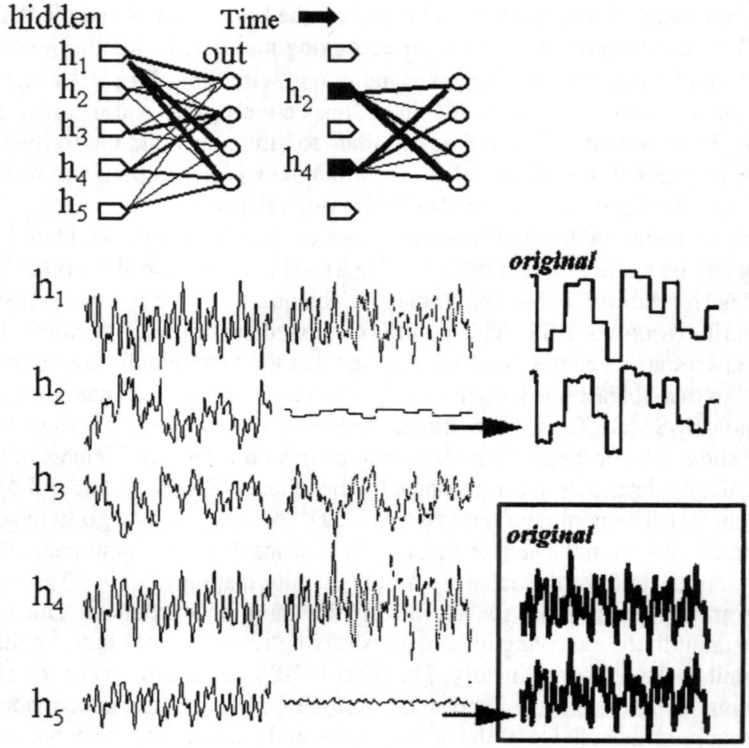

Figure 10: Typical behavior.

4.4 BSS Performance: Two Blind Sources

Figure 11 shows the BSS performance results obtained for the cases of two white-noise blind signals having uniform or sub-Gaussian or super-Gaussian probability density functions. The BSS success criterion was set as follows. If BSS were perfect, then the normalized cross -correlation matrix between the source signals and the inputs to the surviving hidden units would be the identity matrix or one of its permutation matrices. The BSS was judged as success, when the elements that should be *1* were greater than *0.98* and those that should be *0* were less than *0.1*.

For the hidden-unit non-linear activation function, $0.5z+\tanh(0.1z^3)$ was used this time. Also, the de-mixing matrix *V* was updated such that the mean square of the input to each hidden unit was minimized, in addition to the iterative computation to minimize the identity mapping error. These aspects improved the BSS performance significantly in comparison to the previous tests of [18-20].

4.5 Adaptability Tests

The auto-encoder approach to BSS is expected to have high adaptability. The previous papers [18,19] have demonstrated this by testing a time-varying mixing matrix, as well as a case in which a new source suddenly appears to join the mixing before it disappears again. The main purpose here is to examine what will happen if one of the sensors abruptly fails. A sample record is shown in Figure 12.

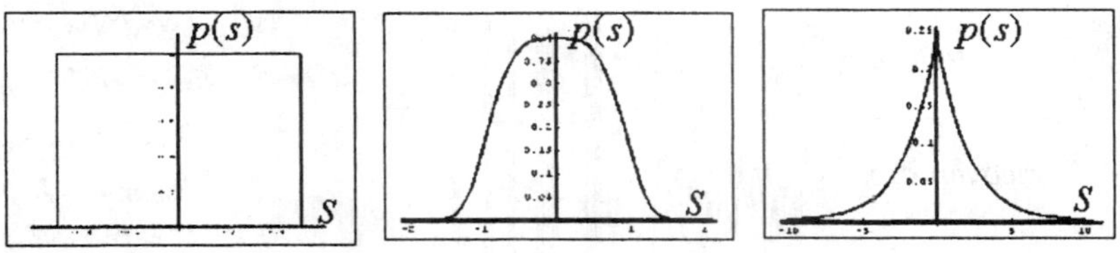

A. *Uniform* B. *Sub-Gaussian* C. *Super-Gaussian*

	A	B	C
A	100.0		
B	100.0	100.0	
C	99.6	98.6	100.0

Figure 11: BSS success rate (%). 1,000 trials for each test.

Initially, there are two sources, P (sum of sinusoids) and Q (Poisson-Like process), five sensors and five hidden units. The decoder connections were still random at (a). By time (b), two hidden units had been selected to reconstruct P and Q. Shortly after (b), a third source (Gaussian) R was suddenly added, and correspondingly the identity-mapping error increased somewhat. At (c), the network settled and re-activated a different hidden unit from which R was recovered. After that, the sensor #5 was suddenly destroyed. There was a transient increase of the identity mapping error. However, the same three hidden units continued to represent the three sources, even though the connection pattern changed somewhat. Other examples including applications to separations of blind pictures also show adaptability/robustness can be found elsewhere [19,21].

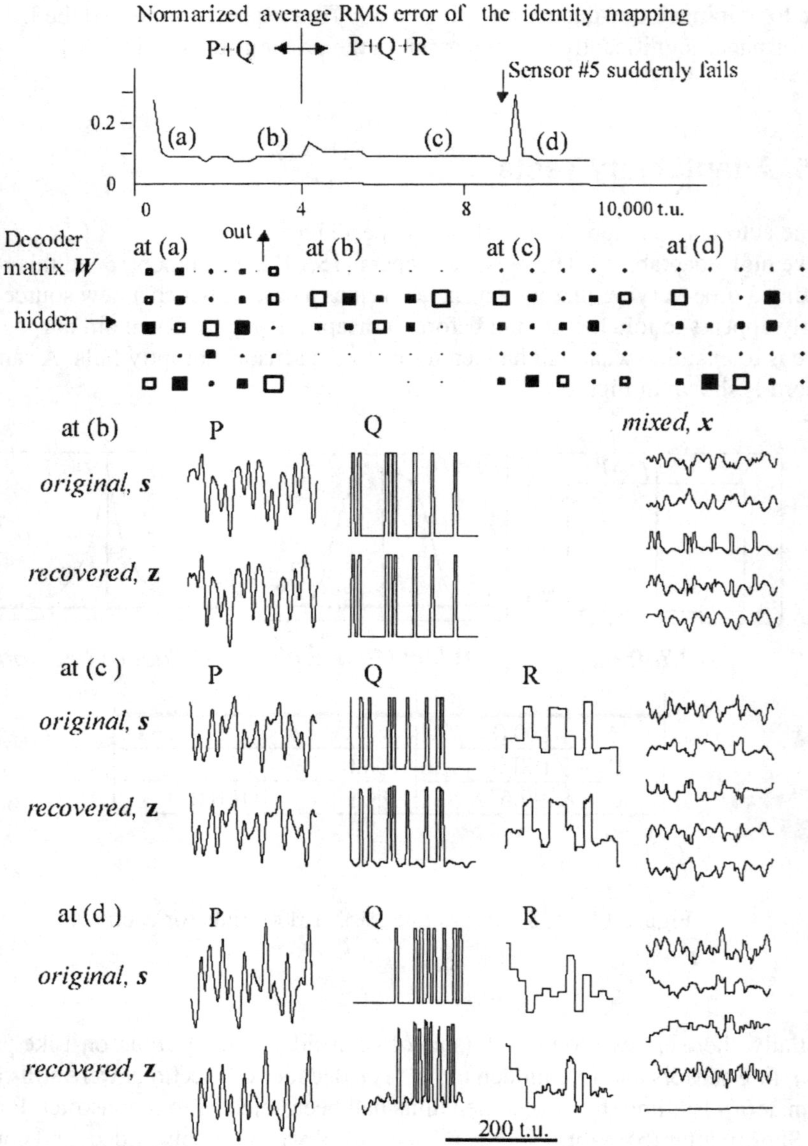

Figure 12: Sudden appearance of a third source followed by sudden failure of one sensor. Open and filled squares for positive and negative weights, respectively. The polarity and scale of each trace is adjusted.

5. Closing Remarks

A multi-output newral network architecture with fewest hidden units would mean that some hidden units are used jointly by two or more outputs units. The CSDF pruning algorithm is devised to find such architectures by introducing two types of internal structural interactions, namely CS and DF that exert opposite effects upon the development of synaptic weights. The balance between CS and DF together with the BP learning determines the fate of each hidden unit. As such, the rise and fall of a synaptic connection is controlled dynamically in relationship to other synaptic connections. Such interactive aspect is absent in WD[2],OBD[3] as well as other pruning methods reviewed in [5].

Selection mechanisms similar to the CSDF algorithm have been described in the literature of developmental neurobiology. That is, the facilitation mechanism (DF) operating in the projective fields seem to have a neuro-embryological counterpart of the axonal protrusion involving the growth cones and nerve growth factors. On the other hand, the mutual suppression mechanism (CS) acting across the receptive fields appears to be similar in notion to the "anatomical lateral inhibition" that has been found in the developing compound eye [24].

Finally, it should be noted that "CSDF *pruning* " is actually a misleading naming, since new connections and new hidden units may emerge in the middle of synaptic development; for example, the case of incremental analogical learning in the section 3,as well as the case of sudden appearance of a third blind source signal in the Section 4.

Neural network pruning is a connectionist's PRR (Principle of Redundancy Reduction). One would agree that PRR is indispensable for good generalization ability. What is needed beyond generalization ability would be "abstraction ability" which would require the acquisition of internal model(s) and related association bindings. This is what this study has attempted to pursue by applying the CSDF pruning algorithm.

References

[1] P.R.Huttenlocher, *Neural Plasticity*, Harvard Univ. Press, 2002.

[2] M. Ishikawa, "A Structural Learning Algorithm with Forgetting of Link Weights", *Proc. Int. Joint Conf. on Neural Networks*, 1989, Vol.2 ,p.626.

[3] Y. LeCann, J.S. Denker and S.A. Solla, "Optimal Brain Damage", In D.S. Touretzky (Ed.), *Advances in Neural Information Processing Systems*, Morgan Kaufman, Vol. 2, 1990, pp.598-605.

[4] S. Yasui, "Convergence Suppression and Divegence Facilitation: Minimum and Joint Use of Hidden Units by Multiple Outputs," *Neural Networks*, Vol.10, No.2, 1997, pp.353-367.

[5] R.Reed,"Pruning Algorithms – A Survey", *IEEE Trans. Neural Networks*, Vol. 4, 1993,pp.740-747.

[6] B. Falkenhainer, K.D. Forbus, and D. Gentner, "The Structural Mapping Engine", *Artificial Intelligence*, Vol.41, 1989, pp.1-63.

[7] D. Gentner, " Structure Mapping: A Theoretical Framework", *Cognitive Science*, Vol. 7, 1983, pp.155-170.

[8] K.J. Holyoak and P. Thagard, "Analogical Mapping by Constraint Satisfaction", *Cognitive Science*, Vol.13, 1989a, pp.295-355.

[9] G.E. Hinton, "Learning Distributed Representations of Concepts", *8th Annual Conference on Cognitive Science,* 1986, pp.1-12.

[10] S. Yasui, "Connectionist Analogical Inference in Relational Isomorphism Paradigm", Proc. *ICONIP '97 Conference,* Vol.1, 1997, pp.526-530.

[11] S. Yausi, "Connectionist Abstraction for Machine Learning by Analogy", *Proc. ICONIP '98 Conference* Vol. 3, 1998, pp.1453-1458.

[12] T. Watanabe, H. Fujimura and S. Yasui, "Connectionist Incremental Learning by Analogy", *Proc. ICONIP'99*, Vol.3, 1999, pp.940-945.

[13]S.Yasui, "Abstraction-Based Connectionist Analogy Processor", *Int. J. applied Math. And Computer Sciences*, Vol.10, No4, 2000, pp.791-812.

[14] H. Fujimura and S. Yasui, "Connectionist Analogical Inference on Predicates and Their Arguments", *Proc. ICONIP 2000* , Vol.1, 2000, pp.482-487.

[15] S.Yasui and T.Watanabe,"Connectionist Analogical Learning of Relational Isomorphism Obscured by Irrelevant Relationships", *Proc.6th Int'l Conf. On Soft Computing (Iizuka 2000)*, 2000, CD ROM.

[16] T-W. Lee, M. Girolami, A.J. Bell, and T.J. Sejnowski,. "A Unifying Information - Theoretic Framework for Independent Component Analysis," *Computers and Mathematics with Applications*, Vol. 39, 2000, pp.1-21.

[17] A. Hyvarinen and E. Oja, " A Fast Fixed-Point Algorithm for Independent Component Analysis", *Neural Computation,* Vol.9, No.7, 1997, pp.1483-1492.

[18] S. Yasui, " A Conventional Auto-Associative Neural Network Separates Blind Sources without Adding Intentional Algorithms Other Than Pruning", *Neural Networks for Signal Processing XI*, IEEE, 2001, pp.313-322.

[19] S. Yasui, "Adaptive Blind Source Separation by Auto-Associative Neural Network with Pruning", *Proc. ICONIP 2001*, Vol.2, 2001, pp.807-812.

[20] S. Yausi, "Blind Source Separation by Sensor-Signal Identity Mapping with Hidden-Layer Pruning", *IJCNN'02*, Vol.2, 2002, pp.1305-1309.

[21] S. Yasui, S. Takahashi, and T. Furukawa, "Picture Blind Source Separation by Auto-Encoder Identity Mapping with Structural Pruning", *ICONIP'02*, Vol.3, 2002, pp.1393-1397.

[22] K.I.Diamantaras and S.Y.King, *Principal Component Neural Neyworks*, John Wiley & Sons,1996, Analysis.

[23] S. Yasui, "Stochastic Functional Fourier Series, Volterra Series and Nonlinear Systems Analysis", *IEEE Trans. on Automatic Control*, Vol. AC-24, No.2, 1979, pp.230-242.

[24] E.B. Baker, M.Mlodzik and G.M.Rubin, "Spacing Differentiation in the Developing *Drophila* Eye: a Fibrinogen-Related Lateral Inhibitor Encoded by *Scabrous*", *Science*,Vol.250,1990,pp.1370-1376.

EVOLUTION OF THE LAWS THAT DEAL WITH THE UTILIZATION OF INFORMATION NETWORKS

Babak Hodjat[1]
Adam Cheyer[2]
1 Dejima Inc., babak@dejima.com
1 SRI International, adam.cheyer@sri.com

Abstract. Three Laws are used to explain how the potential value of a network increases as the network expands: Sarnoff's Law, Metcalf's Law, and Reed's Law. How accurately do these laws predict the actual value of information networks? We will take a closer look at the application of the laws to information networks and derive corollaries based upon which we shall propose certain attributes that will increase the value of an information network much more profoundly than the number of nodes, which is the primary concern of the laws mentioned above.

1. Introduction

Value or *Utility* is a measure of the satisfaction gained from the consumption of a "package" of goods and services. Today, three Laws are used to explain how the potential value of a network increases: Sarnoff's law, Metcalf's law [1], and Reed's law [2]. As Reed puts it: "There are at least three categories of value that networks can provide: the linear value of services that are aimed at individual users, the "square" value from facilitating transactions, and exponential value from facilitating group affiliations. The dominant value in a typical network tends to shift from one category to another as the scale of the network increases." (Figure 1).

[1] Dejima Inc., babak@dejima.com
[2] SRI International, adam.cheyer@sri.com

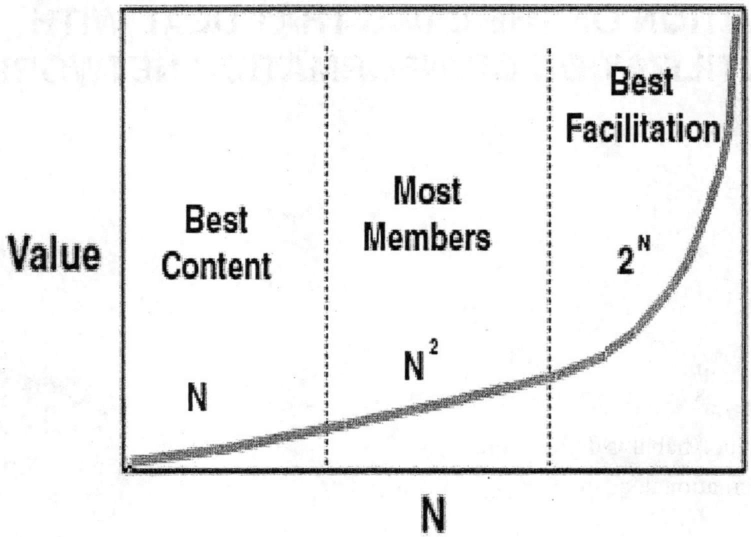

Figure 1. Three famous laws concerning value of networks.

The advent of such generalized laws has had profound effects on the perceived value of information networks and the strategies undertaken to increase the value of such networks [3].

How accurately do these laws predict the actual value of information networks? In this paper we will take a closer look at the application of the laws to information networks and derive corollaries based upon which we shall propose certain attributes that will increase the value of an information network much more profoundly than the number of nodes, which is the primary concern of the laws mentioned above.

We will first review the laws and discuss some observation with regards to them and the assumptions they make in order to have a better perspective over how the laws can be interpreted in real world networks. We will then proceed to define "information networks" and consider the application of the three laws to this class of networks.

2. Sarnoff's law

The value of the network grows with the number of nodes:

$V(n) \sim n$

In the real world n is limited by the following:
Cost of access: In the cellular phone networks, for example, the cost of the handset and the monthly subscription fee are barriers to adoption.
Perceived value of access: In the example above, many people buy cell phones for safety reasons (e.g., being able to call for emergency).
Perceived ease of access: Many people do not enable WAP services on their cell phones because it is assumed to be hard to use.

3. Metcalf's law

The total value of a network where each node can reach every other node grows with the square of the number of nodes:

$V(n) \sim n^2$

In many cases, for each user on such a network, a maximum a nodes are accessible at any given time. This may be a limitation on the user's part, or as a consequence of the network layout and cost of navigation[3], both of which are not proportional to n when n is sufficiently large. In these cases, the total value of the network is computed as:

$V(n) \sim na \sim n$ (Sarnoff's law)

4. Reed's Law

The value of the Group Forming (GF) network grows exponentially to the number of users:

$V(n) \sim 2^n$

This law is based on the fact that certain configurations (i.e., groupings) of node connections in a network yield a higher value than others. A Group Forming network resembles a network with smart nodes that, on-demand, form into such con-

[3] Of course this logic does not hold in the case of mass broadcasts such as spam, but it is debatable as to how spam affects the value of a network.

figurations. Reed mentions social networks as the catalyst. If we take the Internet as an example, if we replace a passive web page with an active human representative that forms and utilizes links with other human represented nodes depending on information demand at hand, Reed's law predicts exponential growth in the potential value of the network by achieving relevant network groupings. E-bay could be considered as an example of this phenomenon.

This, of course, is quite a controversial law, predicting that the addition of a single user to a GF network, can potentially double the value of the whole network. Observations such as the following show that the actual value of such networks may not always yield such promising value[4]:

Reed's law counts the number of possible unique groups that can be formed in a GF network of n nodes. Will a new group always increase the value of the network? In most networks, forming new groups of value is difficult for large n. We have no reason to assume that the number of groups that are valuable is a function of n. It is quite possible that in many situations the maximum possible number of valuable groups is much less than n for large n.
Finding existing groups of value may be difficult, making it difficult for a new member of the network to join groups of value, thus increasing the value of the network as a whole[5]. The maximum number of valuable groups a user can join is not necessarily a function of n.

5. Information Networks

Using the World Wide Web as our guiding example, we shall define an information network as a network with nodes that have one or more of the following content or behavior:

Raw information: It is assumed that there is at least one node on the network, for which the access of this raw information has potential value.
Transactional (e.g., e-commerce, banking): Information content on the network is manipulated using transactional nodes. Such manipulations are deemed valuable to some nodes on the network.

---

[4] Reed does discuss the effect of supply and demand on the three laws, assuming that Money and attention resources scale linearly with n. He does not, however, consider the number of possible valuable groups as discussed here.

[5] Reed's law also assumes that a user joining a group results in two groups, one without the user, and one now with the new user. In reality, this is not how we calculate the number of valuable groups and usually joining a group does not create a new group in addition to the one before the user joined.

Computational (e.g., calculator): Processing that does not necessarily effect the information content of the network, but is valuable to some nodes on the network.
Navigational information (e.g., classifications): Navigational information help nodes on the network access information content, transactional or computational nodes on the network.
User (e.g., human, bots): Derive value from a network by consuming information content, creating or transacting on existing information content, or processing information.

In the example of the World Wide Web, currently access to nodes is quite primitive and access is facilitated at the location of the service node. An analogy here is driving an early model of a car: In the early days access was facilitated at the physical location of a device. To honk the horn you actually squeezed the horn itself. To start the car you would get out, go to the front of the car, and use a handle to rotate the pistons in the cylinders. In the case of the Web, a user needs to navigate to where the information or service resides in order to utilize it. There seems to be a need for reversing this paradigm, and bring the service to the user.

A user node is said to have acquired *utility knowledge* of a node when it learns to locate and utilize the node.

6. The Role of Knowledge

Due to the cost of acquisition of utility knowledge, the full benefits of the ever-expanding network of content, services, and applications available to a user remains dormant, and does not conform to the value curves described in the Three Laws.

We propose the following perspective to complement the laws mentioned in the last section:

The value of a network grows as a function of the number of nodes for which access and/or utility knowledge has been acquired, not the number of nodes.

In other words the number of nodes that exist in a network, for which no knowledge is acquired, has no relevance to the actual value of the network. This, of course, is another way of stating that raw information is always a cost unless it is transformed into knowledge. The potential value of a network can only be achieved once knowledge is acquired for all nodes in the network, but there is a cost associated with this knowledge acquisition process. This cost can be measured in processing power and speed, cost of access, and usability.

If k is used to denote the number of nodes for which utility knowledge has been acquired[6], the three laws can now be rewritten more accurately for information networks:

$V(n) \sim k$ (Sarnoff)
$V(n) \sim k^2$ (Metcalf)
$V(n) \sim 2^k$ (Reed)

Currently, for large n, k seems to be much less than n.

7. The Usability Angle

The cost of accessing information on a network is proportional to the cost of mapping the user's model (human or machine) to the actual network.

The above statements attempt to explain the cost of complexity of a network and the concept of "value" referred to in the three laws, as a function of a certain mapping between a user's model and the reality of a network. Knowledge is acquired through a mapping function between an internal model of intent on the part of the user, and the network nodes and topology. Value is therefore measured relative to the user's knowledge, and not as an abstract existence. This perspective states that in the absence of a perfect mapping, there is a cost to be paid to access nodes in a network.

The value of a network is in its effective use. The cost of using a network is proportional to the size and complexity of the network, but this cost is measured against the user's knowledge of the network: the more unfamiliar the user, the more costly the use. Knowledge of a network can therefore be defined as the cost of mapping the user's model of the network to the actual network. An example of this is the mental model that a human user has for a certain classification hierarchy, which may not necessarily map with the actual model as implemented in a content hierarchy.

Technology can and should be used to facilitate the transformation of information to knowledge. A faster database is worthless if the information content cannot efficiently be transformed into knowledge. If each node in a network actively works

[6] k is an oversimplification as it is denoting the acquisition of knowledge as a binary notion when in reality such knowledge acquisition per node is more of a fuzzy membership function.

to conform to the mapping the user expects, the mapping cost can be dramatically decreased, increasing the value.

8. Knowledge Networks

If each node in a network reacts to usage with the goal to conform to the mapping the user expects, the mapping cost can be reduced, increasing the value of the network. We shall refer to such networks as *Knowledge Networks*. The ultimate incarnation of such systems would allow minimization of the cost of conforming to predefined, rigid, and complex network configurations, by taking the burden of the mapping off of the user's shoulders and distributing it over the network nodes.

Each node in such a network is represented by an agent, responsible for mapping user requests, formulated based on the user's model, to the ontology encapsulated in that node. If the mapping cannot be performed, in other words, if the agent determines that it alone is not capable of offering value to the user, it should collaborate with neighboring nodes that may.

In an information network, we shall call nodes represented by such agents as *active ontologies.*

Knowledge networks can be built by striking a usable balance between:
facilitating the input of the user intent and translating it to the network configuration, and,
using context to predict what the user intent will be and to present it to the user in a usable manner.

Another aspect of a knowledge network is the navigational cost. Users of conventional networks always pay a cost of navigation in order to access a node, even if the user has perfect knowledge of the network configuration. This cost can be reduced in a knowledge network if the network nodes propagate the user intent throughout the network, giving all nodes a chance to contribute to the delivery of value to the user. This is assuming that all network nodes are capable of aggregating value contributed by other nodes, and of delivering it to the user.

As an example, let's say a human user would like to see a picture she knows is sold by an e-commerce site, and she is at the home page of the site. Even if she knows where exactly to find the picture, she would have to navigate to it by clicking through the pages and getting to the web page (i.e., network node) containing the picture. If we replace the company web site with a knowledge network of active ontologies representing the information and functionality of each web page, and if we give the user a means to express her intent, say a text box where she can type in what she needs, then, upon entering her request, all active ontologies in this network would collaborate to understand, and facilitate this request. Also, the result of her request, i.e., the picture, should be presentable at the node she is at. In this example, a search box feature would simulate this behavior by indexing web site content and facilitating navigation to pages containing the desired content. In many cases, however, utilizing a search box requires specialized knowledge as to how to formulate the search and how to navigate through the resulting hits, therefore the user is still taking steps to map her internal model (intent) to the network. Search engines also do little in facilitating transactions.

9. An Agent-Oriented Approach

A number of agent-based approaches are being proposed recently that show promise in creating the basis for true knowledge networks [4] [5] [6]. The Adaptive Agent oriented Software Architecture (AAOSA), is an agent-based infrastructure, which uses an adaptable, user-centric approach to rapidly construct an accurate representation of the user's task model, as well as the mapping from this to a specific application's functionality and interfaces [7]. An AAOSA agent network looks like a static network of nodes, but each node can be activated regardless of the distance from the entry node, and based on utility.

AAOSA builds upon the generative nature of knowledge, utilizing a user's existing knowledge (i.e., intent) to enable acquisition of knowledge, as opposed to information, over computerized networks.

An agent may be able to break a problem into sub-problems, and ask other agents to help solve them. Therefore, agents have communication capabilities and an inter-agent communication language (ACL). In order to ensure localization, reusability, dynamic addition and removal of agents to networks, and distributability, the registration of agents is localized to the agents themselves, or within limited domains (i.e., agent sub-networks).

Figure 2 shows the internals of a cross section of an agent-oriented system for a home entertainment and broadcasting system. Users will be able to enter the network and query it from any node, establishing the context of their request.

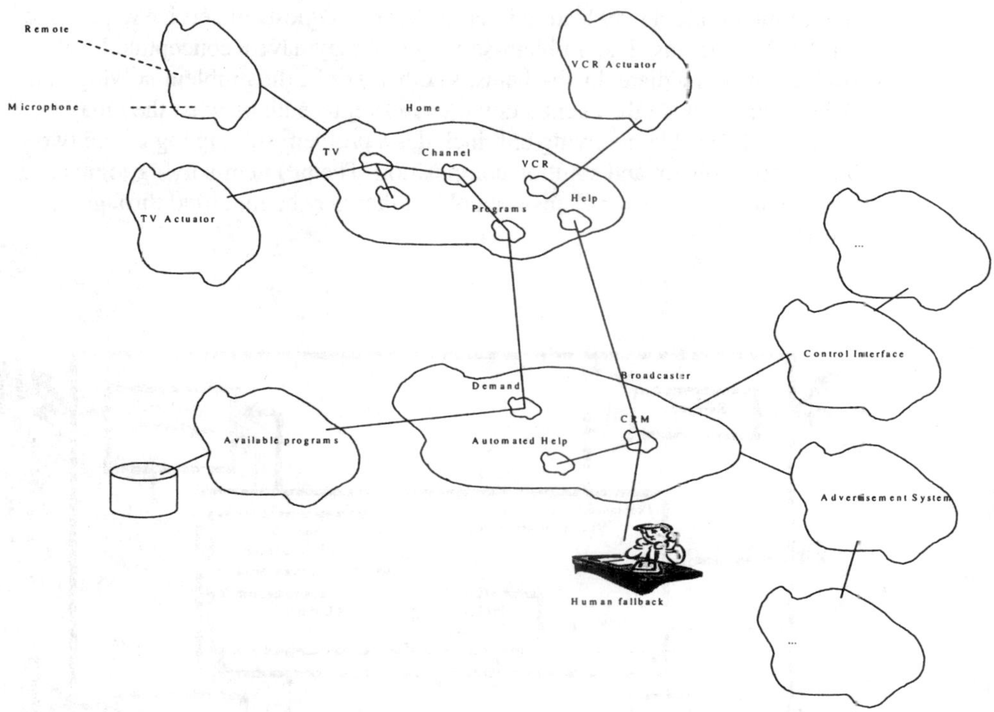

Figure 2. Example of an actual agent-oriented system for home entertainment.

A sub-network is a subset of a network of existing agents in a system. An outsider module, which may itself be an AAOSA agent, decides when to start a session. It goes on to pose problems to a sub-network. Agents providing a solution or parts of a solution also assert their relevance to deal with follow-up problems. If an agent determines irrelevance, it reroutes the request to its immediate up-chain within the path established by the session. This mechanism guarantees the traversal of the agent network to locate agents responsible for solving a problem, even though the entry point for posing the problem can be any agent in the network.
A time-out or depth of propagation is used to ensure a response by the network within a reasonable time frame. For information networks of the scale of the Web, search bots should be paired with the active ontology agents to help identify relevant user entry points by indexing the network under a generalized classification hierarchy.

Figure 3 shows a schematic of an AAOSA agent. The most basic capability of an AAOSA agent is to provide services to outside service requests. The service request-processing unit processes service requests, which may be local objects internal to the agent. The problem-solving unit is responsible for solving problems posed to it from outside the agent. This unit is more customized to the specifics of the problem domain than other modules in an AAOSA-agent and the actual processing of the problem may lead to internal service requests or service requests to other AAOSA-agents. The problem-solving unit may have a conceptual level knowledge of immediate down-chains. In other words, the problem-solving unit may be aware of what the agent's down-chains represent and what they may be able to do. The problem-solving unit includes a problem solving logic, and two sub-units for problem and solution composition. The problem-solving logic used by the agent to process problems is an object that may be modified through service requests.

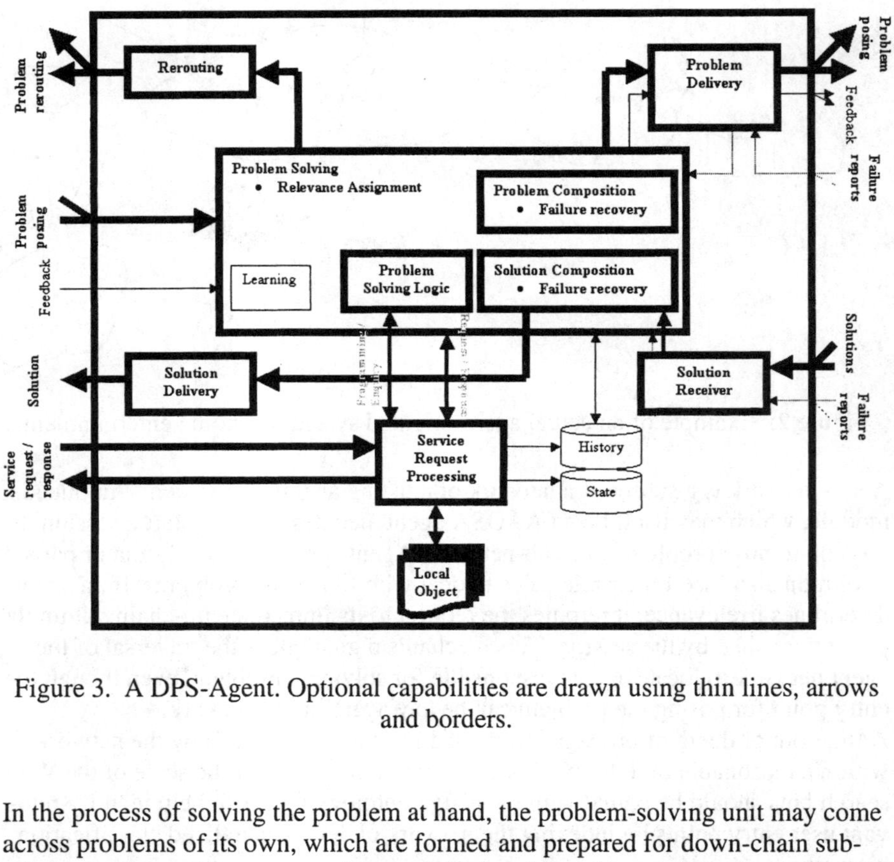

Figure 3. A DPS-Agent. Optional capabilities are drawn using thin lines, arrows and borders.

In the process of solving the problem at hand, the problem-solving unit may come across problems of its own, which are formed and prepared for down-chain submission in the problem-composition unit. Using the problem-solving logic, the agent may choose to decompose a problem into sub-problems, some of which may also be handed down-chain. In the solution-composition unit solutions received from down-chain agents are considered, filtered, and composed into the solution to be provided by the agent. The problem-delivery unit is responsible for posing a problem to down-chain agents and is triggered by the problem-solving unit. This unit includes a mechanism to identify down-chains. The solution-receiver unit communicates with down-chains in order to receive their solutions to problems posed to them by the agent, and hands these solutions over to the solution-composition unit. The solution-delivery unit hands a solution or set of solutions up to the initial problem poser. Each solution is paired with a relevance object, which includes a confidence in the solution, along with the scope, or subset of the problem addressed by this solution. Relevance assignment is also handled in the problem-solving unit. If an agent determines irrelevance to solve a problem, it may re-route it to its session up-chain.

An agent, being somewhat higher level and more dependable than an object, requires elaborate failure recovery mechanisms built into it. An agent should be able to solve problems in spite of changes to the agent network, unpredictability of the problems, or when there are no responses or slow responses to problems posed down-chain.

This architecture lends itself well to capabilities such as distributability and learning and users should have the option of incorporating and utilizing them in their implementations.

10. Actual Deployments

AAOSA has been deployed at Salesforce.com as their Wireless Edition (www.wireless.salesforce.com). Regardless of what modality or device the Salesforce.com user prefers, AAOSA takes their query, makes transactions against Salesforce.com on behalf of the user, and presents the user with the results of their query, in an intuitive, useful manner. Since deployment, the AAOSA-NLI for Sales Force Automation has enjoyed a strong uptake. In the first 120 days since the announcement of the service more than 300 companies signed up for use of the system. The success-rate of the system has been consistently above 90%, and more than 95% of the queries have been within the functional scope of the system capabilities. The average use per user of the system in the first few weeks of deployment was around 5.2 hits per week. Trends show a steady growth in usage and subscriptions since deployment.

11. Conclusion

The Three Laws mentioned in this paper all imply optimal communication and navigational efficiency for the described network effects to work. For example, a Group Forming network made up of people who all speak different languages will collectively produce little value. In order to achieve the potential value promised by the Three Laws, nodes in the value network -- be they human, applications, or

information content -- must possess a shared communication/information model, and there is a cost to acquiring this knowledge.
Today, certain agent-oriented technologies provide a glimpse into how systems can minimize the mapping cost mentioned above.

References

[1] G. Gilder, "George Gilder's Telecosm: Metcalfe's Law and Legacy," *Forbes ASAP*, September 13,
1993. (http://www.seas.upenn.edu/~gaj1/metgg.html)

[2] D. P. Reed, That Sneaky Exponential—Beyond Metcalfe's Law to the Power of Community Building, 1999 (http://www.reed.com/Papers/GFN/reedslaw.html)

[7] H. Rheingold, Smart Mobs: The Next Social Revolution, Perseus Publishing; October 15, 2002.

[4 R. J. Bayardo, Jr., W. Bohrer, R. Brice, A. Cichocki, J. Fowler, A. Helal, V. Kashyap T. Ksiezyk, G. Martin, M. Nodine, M. Rashid, M. Rusinkiewicz, R. Shea, InfoSleuth: Agent-Based Semantic Integration of Information in Open and Dynamic Environments, Proceedings of the ACM SIGMOD International Conference on Management of Data, 1997.

[5 Snehal Thakkar, Craig A. Knoblock, Jose-Luis Ambite, and Cyrus Shahabi, Dynamically Composing Web Services from On-line Sources, In Proceeding of 2002 AAAI Workshop on Intelligent Service Integration, Edmonton, Alberta, Canada, pp. 1-7, July 2002.

[6 Genesereth, M. R., Keller, A. M. and Duschka, O. M. 1997. Infomaster: An Information Integration System. In Proceedings of ACM SIGMOD-97

[7] Introducing the Adaptive Agent Oriented Software Architecture and its Application in Natural Language User Interfaces, Hodjat B., Amamiya M., IJCAI Work-shop on Distributed Constraint Reasoning held at IJCAI'2001, pp.109-122, Seattle, 2001.

Intelligent Type-2 Fuzzy Inference for Web Information Search Task

Yuchun Tang and Yan-Qing Zhang

Department of Computer Science, Georgia State University, Atlanta, GA 30302-4110, USA

Abstract. This chapter focuses on using interval TSK type-2 fuzzy inference to execute a Web Information Search Task (WIST). Type-2 fuzzy inference is helpful to address the "rule uncertainty problem" to improve the performance of a WIST because less prediction error can be achieved. On the other hand, type-2 fuzzy inference is generally computational intensive; this chapter proposes a simple idea to simplify the computation for interval TSK type-2 fuzzy inference.

1. Introduction

Finding the desired information on the World Wide Web is not an easy task because the information available on the WWW is inherently unordered, distributed, and heterogeneous. As a result, the ability to search and retrieve information from the Web efficiently and effectively is a key technology for realizing the Web's full potential [1][2].

Expressing a search request is the first thing we need to solve. At almost all search scenarios, the desired information is on some Web pages (especially HTML/XML formatted). For some similar search requests, the desired Web pages share some similar "content characteristics" and/or "structure characteristics". Traditional search methods let users to submit "keywords" that may be displayed on the desired Web pages to express their search requests. So we call them "keyword-based search" or "content-based search". There are 2 problems: 1 in many cases, a search request is inherently fuzzy and thus difficult or even impossible to be expressed by "crisp" keywords. Furthermore, many "partial related" Web pages cannot be retrieved in a crisp way. And 2 some keywords provide some "structure characteristics" while others only provide "content characteristics". Unfortunately, traditional search methods do not differentiate the two kinds of keywords. In the following, "keywords" will be referred as those words that only provide content characteristics.

Retrieving information effectively is another issue we need to think about. Current search engines are known for poor accuracy: they have both low recall (fraction of desired Web pages that are retrieved) and low precision (fraction of retrieved Web pages that are desired). Furthermore, the query result is usually not listed in a desired order.

The appeal of Fuzzy Logic, Neural Networks, and Genetic Algorithms, as efficient tools featuring computational intelligence, which is already acknowledged in many areas of information technology, plays an important role on addressing these issues [3].

Tang and Zhang proposed an intelligent Web information search and retrieval model called Web Information Search Task (WIST) [4]. WIST model has two goals: one is to make the interface of a search service more expressive and another is to make information retrieval more effective. Many search requests have different content characteristics but share similar structure characteristics. These structure characteristics are expressed by simple "structure rules" in WIST model. Basically, a structure rule is an IF-THEN formula defined on Web pages' URL, Title, Text, Input Links, Output Links, or other related sections. If the IF part is satisfied by a Web page, it may be desired, otherwise it may be not desired. The THEN part will give a function to compute the "desirability", the possibility whether a Web page is desired and how much is the possibility, of a Web page. The function in the THEN part can be type-0 fuzzy function (crisp), type-1 fuzzy function or type-2 fuzzy function. These structure rules will function as an input fuzzifier so we can make fuzzy inference to derive the desirability based on the input attributes. In this way, all search requests with similar (and usually fuzzy) structure characteristics can be "grouped" into a WIST, which is implemented as an intelligent software agent using FL, NN, GA, and other technologies to automatically find all relevant Web pages based on the relevance inferred from structure rules and user-submitted keywords, and automatically rank them in a desired way. Essentially, a WIST agent uses a TSK-based Fuzzy Neural Network (FNN) to infer the desirability. The agent is "intelligent" because it can 1 learn to get better parameters of FNN, 2 learn to get better structure of FNN, and 3 learn to define structure characteristics by adding/modifying structure rules.

2. Type-2 Fuzzy Inference for Homepage-Finder

Homepage-Finder is a special WIST to find and retrieve researchers' homepages intelligently. In this chapter, we focus on applying type-2 fuzzy inference to implement Homepage-Finder and compare the performance with type-1 fuzzy inference based Homepage-Finder [4]. Type-2 fuzzy sets are fuzzy sets whose membership values themselves are ordinary type-1 fuzzy sets. The characteristic of type-2 fuzzy sets is especially useful to execute a WIST because in a WIST, it is usually difficult (and not reasonable) to determine an exact membership function for a fuzzy set.

Homepage-Finder defines 3 structure rules for URL, 2 structure rules for Title, 4 structure rules for Text. In type-1 fuzzy inference, each structure rule's consequent part (THEN part) gives an unknown parameter in the field of [0,1] to denote the desirability. So there are 9 unknown parameters called "premise parameters" and denoted by url1, url2, url3, title1, title2, text1, text2, text3, text4, respectively. Three examples of structure rules are

If a Web page's URL string includes "/~" and the URL string's last character is "/", then its URLscore is url1.

If a Web page's Title string includes "homepage" or "home page", then its Titlescore is title1.

If a Web page's Text string includes "homepage" and "welcome", then its Textscore is text1.

If title1=0.7, the second structure rules means "estimated from the Title, the desirability the Web page is a personal homepage is 70%". It also means "the possibility the Web page's Titlescore is HIGH is 70%" and "the possibility the Web page's Titlescore is LOW is 30%". The linguistic variables "HIGH" and "LOW" are defined on the 3 input scores:

$$\mu_{HIGH}(URLscore) = URLscore, \quad \mu_{LOW}(URLscore) = 1 - URLscore,$$
$$\mu_{HIGH}(Titlescore) = Titlescore, \quad \mu_{LOW}(Titlescore) = 1 - Titlescore,$$
$$\mu_{HIGH}(Textscore) = Textscore, \quad \mu_{LOW}(Textscore) = 1 - Textscore,$$

"HIGH" means "how much possibility a Web page is a personal homepage", "LOW" means "how much possibility a Web page is NOT a personal homepage". Essentially, the fuzzification is discrete because Homepage-Finder scores a Web page according to the discrete structure rules.

In above type-1 system, we arbitrarily set a premise parameter to be just a crisp value, while in type-2 fuzzy system, each premise parameter will be a type-1 fuzzy set itself to address the "rule uncertainty problem": according to a structure rule, we can only approximately derive the desirability of a Web page to be in a fuzzy interval but not a precise value. In this way, we expect a WIST can make better prediction thus improve the performance of search. Here we adopt interval type-2 fuzzy set, that is, each premise parameter is just a (crisp) interval in the field of [0,1].

Following are fuzzy rules defined in Homepage-Finder system:

FR(1) IF URLscore is LOW and Titlescore is LOW and Textscore is LOW,

THEN $\text{Totalscore} = 0$

FR(2) IF URLscore is LOW and Titlescore is LOW and Textscore is HIGH,
THEN $\text{Totalscore} = p_{21} * \text{URLscore} + p_{22} * \text{Titlescore} + p_{23} * \text{Textscore}$

FR(3) IF URLscore is LOW and Titlescore is HIGH and Textscore is LOW,
THEN $\text{Totalscore} = p_{31} * \text{URLscore} + p_{32} * \text{Titlescore} + p_{33} * \text{Textscore}$

FR(4) IF URLscore is LOW and Titlescore is HIGH and Textscore is HIGH,
THEN $\text{Totalscore} = p_{41} * \text{URLscore} + p_{42} * \text{Titlescore} + p_{43} * \text{Textscore}$

FR(5) IF URLscore is HIGH and Titlescore is LOW and Textscore is LOW,
THEN $\text{Totalscore} = p_{51} * \text{URLscore} + p_{52} * \text{Titlescore} + p_{53} * \text{Textscore}$

FR(6) IF URLscore is HIGH and Titlescore is LOW and Textscore is HIGH,
THEN $\text{Totalscore} = p_{61} * \text{URLscore} + p_{62} * \text{Titlescore} + p_{63} * \text{Textscore}$

FR(7) IF URLscore is HIGH and Titlescore is HIGH and Textscore is LOW,
THEN $\text{Totalscore} = p_{71} * \text{URLscore} + p_{72} * \text{Titlescore} + p_{73} * \text{Textscore}$

FR(8) IF URLscore is HIGH and Titlescore is HIGH and Textscore is HIGH,
THEN $\text{Totalscore} = 1$

$p_{ij} \in [0,1]$ are called "consequence parameters", $\sum_{j=1}^{3} p_{ij} = 1$, $i \in \{2,3,4,5,6,7\}$. Each consequence parameter will be a crisp value in the field of [0,1].

Type-2 FLSs are computational intensive [5]. In Homepage-Finder, we use a simple idea to simplify the computation of interval TSK type-2 fuzzy inference. The premise parameter intervals for one input attribute, say URL, are supposed to have the same size, so we could keep the original type-1 premise parameters (url1, url2, url3) as center values and define span value q1 for URL. Correspondingly, we define span values q2, q3 for Title and Text. q1, q2, q3 are limited in the field of [0,0.2]. Now the three structure rules aforementioned are modified as

If a Web page's URL string includes "/~" and the URL string's last character is "/", then its URLscore is [url1-q1, url1+q1].

If a Web page's Title string includes "homepage" or "home page", then its Titlescore is [title1-q2, title1+q2].

If a Web page's Text string includes "homepage" and "welcome", then its Textscore is [text1-q3, text1+q3].

To make fuzzy inference, for a page whose URLscore is in [urlvalue1, urlvalue2], Titlescore is in [titlevalue1, titlevalue2], Textscore is in [textvalue1, textvalue2], we can get 8 3-tuples

(urlvalue1, titlevalue1, textvalue1),
(urlvalue1, titlevalue1, textvalue2),
(urlvalue1, titlevalue2, textvalue1),
(urlvalue1, titlevalue2, textvalue2),
(urlvalue2, titlevalue1, textvalue1),
(urlvalue2, titlevalue1, textvalue2),
(urlvalue2, titlevalue2, textvalue1),
(urlvalue2, titlevalue2, textvalue2),

then we adopt original type-1 fuzzy inference to compute the Totalscore for each 3-tuples to get 8 Totalscores. Suppose the minimum one is mintotalscore, the maximum one is maxtotalscore, and the average value of the 8 Totalscores is avgtotalscore.

If avgtotalscore>0.65,
Then final output of Totalscore is maxtotalscore;
If avgtotalscore<=0.45,
Then final output of Totalscore is mintotalscore;
If avgtotalscore is in (0.45,0.65],
Then final output of Totalscore is avgtotalscore.

Homepage-Finder uses a Genetic Algorithm that is one of the most popular derivative-free optimization methods [6] to optimize these unknown parameters because

- GA is parallel. As a result, it can be implemented by using multi-thread technology to massively speed up the learning procedure;
- The performance of GA does not depend on the initial values of the unknown parameters.

Homepage-Finder is a multi-thread Java Application so it can efficiently utilize the bandwidth to quickly retrieve web pages. Homepage-Finder also uses multiple threads to speed up the GA. Fig. 1 shows the GUI of Homepage-Finder.

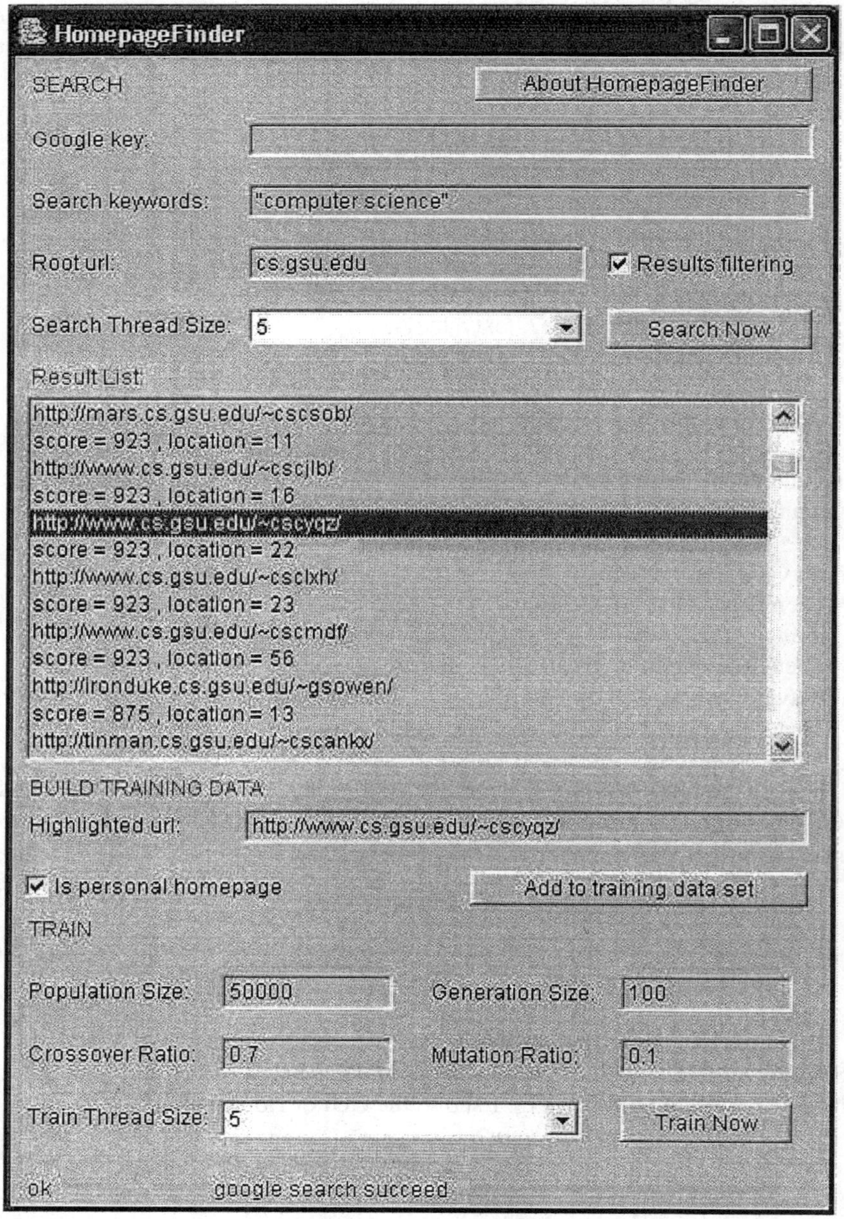

Fig. 1 User Interface of Homepage-Finder

3. Performance Evaluations

3 web information search systems are evaluated for performance comparison:

- Google without Homepage-Finder;
- Type-1 FNN: Homepage-Finder based on type-1 fuzzy inference;
- Type-2 FNN: Homepage-Finder based on type-2 fuzzy inference.

Type-1 FNN and Type-2 FNN use the same training data set that consists of 87 Web pages selected from the Department of Computer Science of University of Georgia (cs.uga.edu). The parameters optimized by GA are given as follows:

```
- <parameter fitness="1771022">
    <url1>900</url1>
    <url2>700</url2>
    <url3>200</url3>
    <title1>900</title1>
    <title2>700</title2>
    <text1>900</text1>
    <text2>700</text2>
    <text3>800</text3>
    <text4>600</text4>
    <f2p1>200</f2p1>
    <f2p2>0</f2p2>
    <f2p3>800</f2p3>
    <f3p1>500</f3p1>
    <f3p2>100</f3p2>
    <f3p3>400</f3p3>
    <f4p1>0</f4p1>
    <f4p2>100</f4p2>
    <f4p3>900</f4p3>
    <f5p1>700</f5p1>
    <f5p2>0</f5p2>
    <f5p3>300</f5p3>
    <f6p1>800</f6p1>
    <f6p2>0</f6p2>
    <f6p3>200</f6p3>
    <f7p1>900</f7p1>
    <f7p2>0</f7p2>
    <f7p3>100</f7p3>
    <q1>0</q1>
    <q2>0</q2>
    <q3>0</q3>
  </parameter>
```

```
- <parameter fitness="629098">
    <url1>900</url1>
    <url2>700</url2>
    <url3>300</url3>
    <title1>900</title1>
    <title2>700</title2>
    <text1>900</text1>
    <text2>700</text2>
    <text3>800</text3>
    <text4>500</text4>
    <f2p1>300</f2p1>
    <f2p2>0</f2p2>
    <f2p3>700</f2p3>
    <f3p1>300</f3p1>
    <f3p2>100</f3p2>
    <f3p3>600</f3p3>
    <f4p1>100</f4p1>
    <f4p2>800</f4p2>
    <f4p3>100</f4p3>
    <f5p1>600</f5p1>
    <f5p2>300</f5p2>
    <f5p3>100</f5p3>
    <f6p1>900</f6p1>
    <f6p2>0</f6p2>
    <f6p3>100</f6p3>
    <f7p1>600</f7p1>
    <f7p2>300</f7p2>
    <f7p3>100</f7p3>
    <q1>170</q1>
    <q2>130</q2>
    <q3>170</q3>
  </parameter>
```

Fig. 2 Parameters Values; the left side lists parameter values of type-1 FNN, the right side lists parameter values of type-2 FNN

The desired Web pages in precision are decided manually, that is, a person decides whether each retrieved Web page is a personal homepage or at least it is mainly used to provide the information of a researcher or a group of researchers. The desired Web pages in recall are defined as all personal homepages of which the URLs are in the sites referred by Root URLs, include the search keywords, and are displayed on the "list pages" as shown in TABLE 1-4. The titles of the 4 tables are formatted as "search KEYWORDS in ROOT URLS". The first columns of the tables list the numbers of retrieved Web pages; the second columns give precision, that is, how many retrieved Web pages are desired; the third columns give recall, that is, how many desired Web pages are retrieved; the fourth columns give the average prediction error of the desired Web pages in recall which are retrieved. For a desired Web page, the higher its Totalscore is, the lower its prediction error is, and thus the higher it is ordered in the result list. For example, in TABLE 1, Type-1 FNN retrieves 35 Web pages, in which 31 are desired, thus the precision is 88.6%; in the 18 desired Web pages defined in recall, 16 are retrieved, thus the recall is 88.9%; the average prediction error of the 16 Web pages is 0.077, which means that Totalscore is averagely 0.923 for each of the 16 Web pages.

	length	precision	recall	avg(Ei)
Google	32	22/32=68.8%	12/18=66.7%	N/A
FNN1	35	31/35=88.6%	16/18=88.9%	0.077
FNN2	33	30/33=90.9%	16/18=88.9%	0.007

Table 1. Search "computer science" in cs.gsu.edu
list page: http://www.cs.gsu.edu/people/faculty.html

	length	precision	recall	avg(Ei)
Google	25	17/25=68%	0/16=0%	N/A
FNN1	49	47/49=95.9%	16/16=100%	0.304
FNN2	48	47/48=97.9%	16/16=100%	0.116

Table 2. Search "computer science" in cs.caltech.edu
list page: http://www.cs.caltech.edu/people.html

	length	precision	recall	avg(Ei)
Google	101	45/101=44.6%	1/20=5%	N/A
FNN1	78	76/78=97.4%	17/20=85%	0.122
FNN2	79	78/79=98.7%	17/20=85%	0.009

Table 3. Search "computer" in csee.usf.edu
list page: http://www.csee.usf.edu/faculty_pages/faculty_list.html

	length	precision	recall	avg(Ei)
Google	247	161/247=65.2%	41/44=93.2%	N/A
FNN1	287	277/287=96.5%	41/44=93.2%	0.095
FNN2	284	279/284=98.2%	41/44=93.2%	0.003

Table 4. Search "computer science" in cs.berkeley.edu
list page: http://www.eecs.berkeley.edu/Faculty/Lists/CS/

	length	precision	recall	avg(Ei)
Google	405	245/405=60.5%	54/98=55.1%	N/A
FNN1	449	431/449=96.0%	90/98=91.8%	0.134
FNN2	444	434/444=97.7%	90/98=91.8%	0.025

Table 5. Simulation results including the above 4 departments

The simulation results show that

Type-1 FNN has much higher precision/recall than Google.

Type-2 FNN can make much better estimation than type-1 FNN, while keep the number of retrieved and desired documents and precision/recall at the same level.

4. Conclusions

Finding desired information from the Web efficiently and effectively is a key technology to realize its full potential and is not an easy task [1][2]. The chapter focuses on applying interval TSK type-2 fuzzy inference to improve the performance of Homepage-Finder, an intelligent software agent which uses CI technologies including FL, NN, and GA to define and implement a specific WIST to automatically find all relevant researchers' homepages based on the possibility whether a Web page is a personal homepage and the relevance with keywords, given a root URL, some keywords and some structure rules. The simulation results show that Homepage-Finder with type-2 fuzzy inference can find more personal homepages (from 245 to 434) with much higher precision (from 60.5% to 97.7%), much higher recall (from 55.1% to 91.8%) than Google and list them in a desired order (average error is 0.025). In the future, we want to try some more complicated type-2 fuzzy inference while keep the computation complexity from heightened and thus keep the search speed from lowered.

References

[1] Sankar K. Pal, “Web Mining in Soft Computing Framework: Relevance, State of the Art and Future Directions,” IEEE Transactions on Neural Networks, vol. 13, no. 5, 2002, pp 1163-1177.

[2] Ivan Ricarte, “A Reference Model for Intelligent Information Search,” In Proceedings of the 2001 BISC International Workshop on Fuzzy Logic and the Internet, August, 2001, pp 80-85.

[3] Soumen Chakrabarti, "Data Mining for Hypertext: A Tutorial Survey," SIGKDD: SIGKDD Exploration: Newsletter of the Special Interest Group (SIG) on Knowledge Discovery & Data Mining, ACM1(2), 2000, pp 1-11.

[4] Y. Tang, Y. -Q. Zhang, “Smart Homepage-Finder – A TSK-Based Genetic Fuzzy Information Filtering Agent for Searching Homepages Intelligently,” Enhancing the Power of the Internet – Studies in Fuzziness and Soft Computing (M. Nikravesh, L. A. Zadeh, B. Azvine, R. R. Yager), Springer, 2003, pp 379-389.

[5] Liang, Q. and J. M. Mendel, "Interval Type-2 Fuzzy Logic Systems: Theory and Design," IEEE Trans. On Fuzzy Systems, Vol. 8, 2000, pp 535-550.

[6] J.-S. R. Jang, C.-T. Sun, E. Mizutani, Neuro-Fuzzy and Soft Computing, A Computational Approach to Learning and Machine Intelligence, Prentice Hall, Upper Saddle River, NJ, 1st edition, 1996, pp 81-84.

Causality In An Inherently Ill Defined World

Lawrence J. Mazlack
Applied Artificial Intelligence Laboratory
University of Cincinnati
Cincinnati, OH 45221-0030
mazlack@uc.edu

Abstract. Commonsense causal reasoning occupies a central position in human reasoning. It plays an essential role in both informal and formal human decision-making. Causality itself as well as human understanding of causality is imprecise, sometimes necessarily so. Our common sense understanding of the world tells us that we have to deal with imprecision, uncertainty and imperfect knowledge. A difficulty is striking a good balance between precise formalism and commonsense imprecise reality. Clearly, an algorithmic method of handling imprecision is needed. Today, data mining holds the promise of extracting unsuspected information from very large databases. In many ways, the interest is the promise (or illusion) of causal, or at least, predictive relationships. However, the most common data mining rule forms only calculate a joint occurrence frequency; they do not express a causal relationship. Without understanding the underlying causality, a naïve use of data mining rules can lead to undesirable actions.

1 Introduction

Commonsense causal reasoning occupies a central position in human reasoning. It plays an essential role in human decision-making. Considerable effort has been spent examining causation. Philosophers, mathematicians, computer scientists, cognitive scientists, psychologists, and others have formally explored questions of causation beginning at least three thousand years ago with the Greeks.

Whether causality can be recognized at all has long been a theoretical speculation of scientists and philosophers. At the same time, in our daily lives, we operate on the commonsense belief that causality exists.

Causal relationships exist in the commonsense world. If an automobile fails to stop at a red light and there is an accident, it can be said that the failure to stop was the accident's cause. However, conversely, failing to stop at a red light is not a certain cause of a fatal accident; sometimes no accident of any kind occurs. So, it can be said that knowledge of some causal effects is imprecise. Perhaps, complete knowledge of all possible factors might lead to a crisp description of whether a causal effect will occur. However, in our commonsense world, it is unlikely that all possible factors can be known. What is needed is a method to model imprecise causal models.

Another way to think of causal relationships is counterfactually. For example, if a driver dies in an accident, it might be said that had the accident *not* occurred; they would still be alive.

Our common sense understanding of the world tells us that we have to deal with imprecision, uncertainty and imperfect knowledge. This is also the case of our scientific knowledge of the world. Clearly, we need an algorithmic way of handling imprecision if we are to computationally handle causality. Models are needed to algorithmically consider causes. These models may be symbolic or graphic. A difficulty is striking a good balance between precise formalism and commonsense imprecise reality.

1.1 Data Mining, Introduction

Data mining is an advanced tool for managing large masses of data. It analyzes data previously collected. It is *secondary* analysis. Secondary analysis precludes the possibility of experimentally varying the data to identify causal relationships.

There are several different data mining products. The most common are *conditional rules* or *association rules*. Conditional rules are most often drawn from induced trees while association rules are most often learned from tabular data. Of these, the most common data mining product is association rules; for example:

- ***Conditional rule:***
  IF Age < 20
  THEN Income < $10,000
  with {belief = 0.8}
- ***Association rule:***
  Customers who
  buy beer and sausage
  also tend to buy mustard
  with {confidence = 0.8}
  in {support = 0.15}

At first glance, these association rules seem to imply a causal or cause-effect relationship. That is: *A customer's purchase of both sausage and beer <u>causes</u> the customer to also buy mustard.* In fact, all that is discovered is the *existence* of a statistical relationship between the items. The *nature* of the relationship is not specified. We do not know whether the presence of an item or sets of items causes the presence of another item or set of items; or the converse, or some other phenomenon causes them to occur together.

When typically developed, association rules do not *necessarily* describe causality. Also, the strength of causal dependency may be very different from a respective association value. All that can be said is that associations describe the strength of joint co-occurrences. Sometimes, the relationship might be causal; for example, if someone eats salty peanuts and then drinks beer, there is probably a causal relationship. On the other hand, it is unlikely that a crowing rooster causes the sun to rise.

1.2 Naive Association Rules Can Lead To Bad Decisions

One of the reasons why association rules are used is to aid in making retail decisions. However, simple association rules may lead to errors. It is common for a food store to put one item on sale and then to raise the price of another item whose purchase is assumed to be associated. This may work if the items are truly associated; but it is problematic if association rules are blindly followed [Silverstein, 1998].

Example: At a particular store, a customer buys:

- *hamburger 33%* of the time
- *hot dogs* 33% of the time
- both *hamburger* and *hot dogs* 33% of the time
- *sauerkraut** only if *hot dogs* are also purchased

This would produce the transaction matrix:

	hamburger	hot dog	sauerkraut
t_1	1	1	1
t_2	1	0	0
t_3	0	1	1

This would lead to the associations:

- (hamburger, hot dog) = 0.5
- (hamburger, sauerkraut) = 0.5
- (hot dog, sauerkraut) = 1.0

If the merchant:

- Reduced price of hamburger (as a sale item)
- Raised price of sauerkraut to compensate (as the rule *hamburger* ⟹ *sauerkraut* has a high confidence.
- The offset pricing compensation would not work as the sales of sauer-kraut would not increase with the sales of hamburger. Most likely, the sales of hot dogs (and consequently, sauer-kraut) would likely decrease as buyers would substitute hamburger for hot dogs.

1.3 False Causality

Complicating causal recognition are the many cases of false causal recognition. For example, a coach may win a game when wearing a particular pair of socks, then always wear the same socks to games. More interesting, is the occasional false causality

* Sauerkraut is a form of pickled cabbage. Some people greatly enjoy using sauerkraut as a garnish with sausages. However, it is rarely consumed as a garnish with hamburger. For more about sauerkraut, see: *http://www.sauerkraut.com/*

between music and motion. For example, Lillian Schwartz developed a series of computer generated images, sequenced them, and attached a sound track (usually Mozart). While there were some connections between one image and the next, the music was not scored to the images; however, a person viewing the assemblage viewing them, the music appeared to be connected. All of the connections were observer supplied.

An example of non-computer illusionary causality is the choreography of Merce Cunningham. To him, his work is non-representational and without intellectual meaning. He often worked with John Cage, a randomist composer. Cunningham would rehearse his dancers, Cage would create the music; only at the time of the performance would music and motion come together. However, the audience usually conceived of a causal connection between music and motion and saw structure in both.

1.4 Recognizing Causality Basics

A common approach to recognizing causal relationships is by manipulating variables by experimentation. How to accomplish causal discouvery in purely observational data is not solved. (Observational data is the most likely to be available for data mining analysis.) Algorithms for discouvery in observational data often use correlation and probabilistic independence. If two variables are statistically independent, it can be asserted that they are not causally related. The reverse is not necessarily true.

Real world events are often affected by a large number of potential factors. For example, with plant growth, many factors such as temperature, chemicals in the soil, types of creatures present, etc., can all affect plant growth. What is unknown is what causal factors will or will not be present in the data; and, how many of the underlying causal relationships can be discouvered among observational data.

Some define cause-effect relationships as: When α occurs, β <u>always</u> occurs. This is inconsistent with our commonsense understanding of causality. A simple environment example: When a hammer hits a bottle, the bottle *usually* breaks. A more complex environment example: When a plant receives water, it *usually* grows.

An important part of data mining is understanding whether there is a relationship between data items. Sometimes, data items may occur in pairs but may not have a deterministic relationship; for example, a grocery store shopper may buy both bread and milk at the same time. Most of the time, the milk purchase is not caused by the bread purchase; nor is the bread purchase caused by the milk purchase.

Alternatively, if someone buys strawberries, this may causally affect the purchase of whipped cream. *Some* people who buy strawberries want whipped cream with them; of these, the desire for the whipped cream varies. So, we have a conditional primary effect (whipped cream purchase) modified by a secondary effect (desire). How to represent all of this is open.

A largely unexplored aspect of mined rules is how to determine when one event causes another. Given that α and β are variables and there appears to be a statistical covariability between α and β, is this covariability a causal relation? More generally, when is any pair relationship causal? Differentiation between covariability and causality is difficult.

Some problems with discouvering causality include:

- Adequately defining a causal relation
- Representing possible causal relations
- Computing causal strengths
- Missing attributes that have a causal effect
- Distinguishing between association and causal values
- Inferring causes and effects from the representation.

Beyond data mining, causality is a fundamentally interesting area for workers in intelligent machine based systems. It is an area where interest waxes and wanes; in part because of definitional and complexity difficulties. The decline in computational interest in cognitive science also plays a part. Activities in both philosophy and psychology [Glymour, 2001] overlap and illuminate computationally focused work. Often, the work in psychology is more interested in how people *perceive* causality as opposed to whether causality actually exists. Work in psychology and linguistics [Lakoff, 1990] [Mazlack, 1987] show that categories are often linked to causal descriptions. For the most part, work in intelligent computer systems has been relatively uninterested in grounding based on human perceptions of categories and causality. This paper is concerned with developing commonsense representations that are compatible in several domains.

2 Causality

Centuries ago, in their quest to unravel the future, mystics aspired to decipher the cries of birds, the patterns of the stars and the garbled utterances of oracles. Kings and generals would offer precious rewards for the information soothsayers furnished. Today, though predictive methods are different from those of the ancient world, the knowledge that dependency recognition attempts to provide is highly valued. From weather reports to stock market prediction, and from medical prognoses to social forecasting, superior insights about the shape of things to come are prized [Halpern, 2000].

Democritus, the Greek philosopher, once said: "Everything existing in the universe is the fruit of chance and necessity." This seems self-evident. Both randomness and causation are in the world. Democritus used a poppy example. Whether the poppy seed lands on fertile soil or on a barren rock is chance. If it takes root, however, it will grow into a poppy, not a geranium or a Siberian Husky [Lederman, 1993].

Beyond computational complexity and holistic knowledge issues, there appear to be inherent limits on whether causality can be determined. Among them are:

- *Quantum Physics:* In particular, Heisenberg's uncertainty principle
- Knowledge of the world might never be complete because we, as observers, are integral parts of what we observe

- *Gödel's Theorem:* Which showed in any logical formulation of arithmetic that there would always be statements whose validity was indeterminate. This strongly suggests that there will always be inherently unpredictable aspects of the future.
- *Turing Halting Problem:* Turning (as well as Church) showed that any problem solvable by a step-by-step procedure could be solved using a Turing machine. However, there are many routines where you cannot ascertain if the program will take a finite, or an infinite number of steps. Thus, there is a curtain between what can and cannot be known mathematically.
- *Chaos Theory:* Chaotic systems appear to be deterministic; but are computationally irreducible. If nature is chaotic at its core, it might be fully deterministic, yet wholly unpredictable [Halpern, 2000, 139].
- *Space-Time:* The malleability of Einstein's space time that has the effect that what is "now" and "later" is local to a particular observer; another observer may have contradictory views.
- *Arithmetic Indeterminism:* Arithmetic itself has random aspects that introduce uncertainty as to whether equations may be solvable. Chatin [1987, 1990] discovered that Diophantine equations may or may not have solutions, depending on the parameters chosen to form them. Whether a parameter leads to a solvable equation appears to be random. (Diophantine equations represent well-defined problems, emblematic of simple arithmetic procedures.)

Given determinism's potential uncertainty and imprecision, we might throw up out hands in despair. It may well be that a precise and complete knowledge of causal events is uncertain. On the other hand, we have a commonsense belief that causal effects exist in the real world. If we can develop models tolerant of imprecision, it would be useful. Perhaps, the tools found in soft computing may be useful.

3 Problems With Using Probability

There has been significant work in using various forms of Bayesian networks for causal discovery. A *Bayesian network* is a combination of a probability distribution and a structural model that is a directed acyclic graph in which the nodes represent the variables (attributes) and the edges (arcs) represent probabilistic dependence. A *causal Bayesian network* is a Bayesian network where the predecessors of a node are interpreted as directly causing the variable associated with a node. However, Bayesian networks can be computationally expensive. Inferring *complete* causal Bayesian networks is essentially impossible in large-scale data mining with thousands of variables.

Restricted algorithms [Cooper, 1997] have been suggested that might be useful for causal discovery in market basket data. However, the restrictions on the data and the assumptions made about the relationships are overly limiting. The restrictions are:

- Discrete or continuous data must be reduced to Boolean values
- There is no missing data

• Causal relationships are not cyclic, either directly or indirectly (through another attribute)

4 Epilogue

Causality occupies a central position in human commonsense reasoning. In particular, it plays an essential role in common sense human decision-making by providing a basis for choosing an action that is likely to lead to a desired result. In our daily lives, we make the commonsense observation that causality exists. Carrying this commonsense observation further, the concern is how to computationally recognize a causal relationship.

Data mining holds the promise of extracting unsuspected information from very large databases. Methods have been developed to build rules. In many ways, the interest in rules is that they offer the promise (or illusion) of causal, or at least, predictive relationships. However, the most common form of data mining rules (association) only calculates a joint occurrence frequency, not a causal strength. A fundamental question is determining whether or not recognizing an association can lead to recognizing a causal relationship.

An interesting question is how to determine when causality can be said to be stronger or weaker. Either in the case where the causal strength may be different in two independent relationships; or, where in the case where two items each have a causal relationship on the other.

Causality is a central concept in many branches of science and philosophy. In a way, the term "causality" is like "truth" -- a word with many meanings and facets. Some of the definitions are extremely precise. Some of them involve a style of reasoning best be supported by fuzzy logic.

Defining and representing causal and potentially causal relationships is necessary to applying algorithmic methods. A graph consisting of a collection of simple directed edges will most likely not offer a sufficiently rich representation. Representations that embrace some aspects of imprecision are necessary.

A deep question is when anything can be said to cause anything else. And if it does, what is the nature of the causality? There is a strong motivation to attempt causality discouvery in association rules. The research concern is how to best approach the recognition of causality or non-causality in association rules. Or, if there is to recognize causality as long as association rules are the result of secondary analysis?

References

G. Chatin [1987] **Algorithmic Information Theory**, Cambridge University Press, Cambridge, United Kingdom

G. Chatin [1990] "A Random Walk In Arithmetic," New Scientist 125, n 1709 (March, 1990), 44-66

G. Cooper [1997] "A Simple Constraint-Based Algorithm For Efficiently Mining Observational Databases For Causal Relationships," Data Mining and Knowledge Discovery, v 1, n 2, 203-224

C. Glymour [2001] **The Mind's Arrows, Bayes Nets And Graphical Causal Models In Psychology**, MIT Press, Cambridge, Massachusetts

P. Halpern [2000] **The Pursuit Of Destiny**, Perseus, Cambridge, Massachusetts

G. Lakoff [1990] **Women, Fire, And Dangerous Things: What Categories Reveal About The Mind**, University of Chicago Press

L. Lederman, D. Teresi [1993] **The God Particle: If the Universe Is the Answer, What Is the Question?** Delta, New York

L. Mazlack [1987] "Machine Conceptualization Categories," *Proceedings 1987 IEEE Conference on Systems, Man, and Cybernetics*

C. Silverstein, S. Brin, R. Motwani, J. Ullman [1998] "Scalable Techniques for Mining Causal Structures," Proceedings. 1998 International Conference Very Large Data Bases, New York, NY, August 1998, 594--605,

L. Zadeh [2000] "Abstract Of A Lecture Presented At The Rolf Nevanilinna Colloquium, University of Helsinki," reported to: *Fuzzy Distribution List*, fuzzy-mail@dbai.tuwien.ac.at, August 24, 2000

Printing: Strauss GmbH, Mörlenbach
Binding: Schäffer, Grünstadt